M000072903

Open Questions in
Relativistic Physics

Edited by Franco Selleri
Università di Bari
Bari, Italy

Apeiron
Montreal

Published by Apeiron
4405, rue St-Dominique
Montreal, Quebec H2W 2B2 Canada
http://redshift.vif.com

First Published 1998

Canadian Cataloguing in Publication Data

Main entry under title:

Open questions in relativistic physics

Includes index.
Proceedings of an international conference on Relativistic Physics and Some of its Applications, June 25-28, 1997, Athens, Greece
ISBN 0-9683689-1-3

1. Special relativity (Physics). I. Selleri, Franco

QC173.5.O64 1998   530.11   C98-900498-8

Contents

Cosmology and Astrophysics

Quantum Theory and Relativity

Preface

The true conceptual background of the two relativistic theories (special and general) has been re-emerging in recent times, after more than half a century of domination of the neopositivist conception of science. Einstein himself was strongly influenced by positivism in his youth, and admitted that the special theory of relativity was based on a direction of thought conforming with Mach's ideas [1]. The hegemony of logical empirism had, as a first consequence, that Einstein's role was somewhat inflated, while the contributions of other authors (Lorentz, Larmor, Poincaré, *etc.*) were generally underestimated. More than experimental evidence, this was the reason why the typically realistic conjectures, such as that of ether, were eliminated in favour of more abstract conceptions. At the present time the domination of positivism appears to have come to an end, and a new era may be opening for realism.

A correct understanding of the true history of relativity has recently produced several surprises, the first one being the realisation that such important scientists as Lorentz [2], Poincaré [3], Larmor [4], Fitzgerald [5] took a fully realistic approach to relativistic physics, though they did not necessarily consider this to be in contradiction with Einstein's theory. For example Poincaré, often described as a conventionalist, repeatedly stated that ether was a necessary ingredient of physics. He did so, for example, in the same page of his famous 1904 St. Louis paper [6], where the first precise modern formulation of the relativity principle is given. It has also become fully evident that Einstein came back to the idea that a physical vacuum must exist, which he called "ether" in several papers from 1918 to 1955 [7]. He confessed to Popper that the greatest mistake of his scientific life was the acceptance of positivistic philosophy in his youth [8].

It has been firmly established that certain fundamental ingredients of the two relativistic theories are basically arbitrary, the main one being the introduction of the so-called "Einstein clock synchronisation." This conclusion surfaced at first in the works of philosophers and historians of physics (Reichenbach [9], Grünbaum [10], Jammer [11]), and then influenced the works of physicists as well (Sexl [12], Sjödin [13], Cavalleri [14], Ungar [15], *etc.*). But Einstein clock synchronisation is based on the assumed invariance of the one-way speed of light. Since a statement whose conventional nature has been recognized cannot be a necessary consequence of a true property of nature, it follows that invariance of *one-way* speed of light is not a law of nature! Accordingly, the general-relativistic invariance of the ds^2 should also be considered as a basically human choice, rather than a property of the physical world [16].

In the fifties and early sixties, Herbert Dingle, professor of History and Philosophy of Science in London, fought a battle against some features of the relativity theory, in particular against the asymmetrical aging present in the twin paradox argument. He believed that the slowing down of moving clocks was pure fantasy. This idea has of course been demolished by direct experimental evidence, collected after his time [17]. Nevertheless, his work has left posterity a rare jewel: the syllogism bearing his name. Given that a syllogism is a technical model of perfect deduction, its consequences are absolutely necessary for any person accepting rational thinking in science.

Dingle's syllogism is the following [18]:

1. (Main premise) According to the philosophy of relativism, if two bodies (for example two identical clocks) separate and reunite, there is no observable phenomenon that will show in an absolute sense that one rather than the other has moved.
2. (Minor premise) If upon reunion, one clock were retarded by a quantity depending on its relative motion, and the other not, that phenomenon would show that the first clock had moved (in an observer independent "absolute" sense) and not the second.
3. (Conclusion) Hence, if the postulate of relativity were universally true, as required by the philosophy of relativism, the clocks must be retarded equally or not at all: in either case, their readings will concord upon reunion if they agreed at separation. If a difference between the two readings were to show up, the postulate of relativity cannot be always true.

Today it can be said that the asymmetrical behaviour of the two clocks is empirically certain (muons in cosmic rays, experiment with the CERN muon storage ring [19], experiments with linear beams of unstable particles, Hafele and Keating experiment [20]). Therefore, as a consequence of point 3. above, the postulate of relativity must somehow be negated. Actually, in recent years it seems to be almost normally accepted in the scientific milieu that the "theory of relativity" is just a name, not to be taken too literally. The total relativism which the theory could seem to embody is now perceived to be only an illusion. One can conclude that not all is relative in relativity, because this theory also contains some features that are observer independent, *i.e.* features which are absolute! As Dingle wrote: "It should be obvious that if there is an absolute effect which is a function of velocity, then the velocity must be absolute. No manipulation of formulae or devising of ingenious experiments can alter that simple fact." [18]

From the new point of view that the theory relativity does not embody a complete relativism the so called "twin paradox" is not a real paradox of the theory, but only a huge problem for the few remaining believers in the philosophy of relativism. In fact, the twin paradox is discussed in many works, and they can be divided into two groups: (a) Those which recognize the velocity of the travelling twin as the cause of the slowing down of his biological processes; (b) Those which instead seek to attribute the same effect to the accelerations felt by the traveler at departure, arrival, and the instant in which the direction of velocity is reversed. Obviously, the followers of the second line of thought try to save the perfect symmetry between the rectilinear uniform motions required by the relativity principle, but their position is really impossible to save, as was shown by Builder [21]. His argument is very simple: in physics one can recognise the cause of a phenomenon by varying it, and verifying that corresponding variations of the effect exist. In short, in the case of the twins, if the traveler doubles the length of the paths described with rectilinear uniform motion and travels in them with the same velocity, leaving unchanged the accelerations, he will find that his age difference from the stationary twin is also doubled: therefore velocity, and not acceleration is responsible for asymmetrical ageing. Accelerations as such have no effect on clocks, as shown very convincingly also by the CERN experiment [19], where accelerations as large as 10^{18} g did not have any effect on muon lifetimes.

For the reasons cited here, the new trends in relativistic research are based on: (1) Overcoming of positivistic limitations to the conceptions to be used in scientific research; (2) Awareness of the limited applicability of the relativity principle itself; (3) Conventionality of the invariance of the one way velocity of light; (4) Probable existence in nature of absolute velocities; (5) Possibility of re-introducing the luminiferous ether. This highly interesting situation has brought new life in a field that many considered finally settled. Particularly remarkable has been the revival of the Lorentzian approach defended by Jánossy [22], Erlichson [23], Prokhovnik [24], Bell [25] and Brandes [26], the study of synchronisation procedures different from the usual one [16], the new discussion of Thomas rotation [27]. Very interesting perspectives are now opening up. At the present time there are several physicists active in the foundations of relativity, and every second year an international conference devoted to these matters is organised in London by the British Society for the Philosophy of Science [28].

The most general space-time transformations leading both to invariance of the two way velocity of light (not necessarily of the one way velocity!) and to the usual time dilation effects have been obtained [16]. Sets of such transformations differ from one another only for the value of a coefficient e expressing the dependence on space of the transformation of time. At our conference the present writer reported that there is necessarily an unacceptable discontinuity between the physics of accelerated frames and the physics of inertial frames, unless $e = 0$. In this way we obtain something different from the Lorentz transformations. A new theory of space, time and motion is clearly starting to emerge. Evidence that the standard theory has great difficulties in explaining the Sagnac effect has been presented [29].

In atomic physics there is has been an indication of superluminal crossing of potential barriers [30]. It has, however, been suggested that this could be a spurious effect: the photonic particles in the first arriving part of the wave packet cross the barrier, those in the final part are reflected. In this way, every photon moves with the speed of light, while the "centre of energy" of the wave packet crossing the barrier looks as if it had propagated superluminally. Astrophysical evidence has been reported of superluminal propagations in jets emerging from galactic nuclei and in active clouds emitted from quasars. These can often be explained away if quasars are indeed associated with nearby galaxies and their redshifts are not due to expansion [31]. There remain the M87 ejections (blue knots propagating at a velocity 5-6 c!) whose distance does not depend on redshift, but was obtained from Cepheids, planetary nebulae, apparent size of galaxy, *etc*. This distance is somewhere around 50 million light years. The M87 evidence was reported immediately after our conference. The so called "leading model" could perhaps get rid of this superluminality as well [32].

The Global Positioning System (GPS) is a set of 24 satellites moving around the Earth in such a way that from any point of our planet at any instant at least four of them are visible. It was built for military applications and is said to allow a ballistic precision such that an intercontinental missile can enter from the chosen window of a building 20 000 Km away. All these satellites have atomic clocks on board. Timing and distances are continuously transmitted to the ground. Some very partial results from the GPS were reported at our conference, the startling news being that there is some evidence of a very weak absorption of the solar gravitational field from the mass of the Earth. Rumors are circulating of evidence for other unexpected physical effects which are difficult to explain with the standard theories (relativity, special and general). The highly frustrating situation is that a military secrecy protects the data, so that only very few privileged physicists have access to them.

A critical reconsideration of the published experiments on Bell's inequality (especially of the Orsay experiments) was been reported at this conference [33], with the conclusion that their meaning is rather doubtful as a consequence of several logical and practical ambiguities which emerge when one considers the experiments carefully. It appears that much more detailed investigations are needed before any conclusion against the validity of local realism in nature can eventually be reached. This new result agrees with older studies of the role of the so called "additional assumptions" in the deduction of Bell type inequalities [34]. In another talk the possibility has been discussed of transmitting superluminal signals by using EPR pairs of photons and unusual reflections on nonlinear crystals [35]. Recent work suggests that a solution of the EPR paradox could come from research on the two neutral kaons arising in the decay of ϕ mesons, *e.g.* produced at rest in the laboratory in e^+e^- collisions at a ϕ factory accelerator [36]. The reported discrepancy between local realistic and quantum theoretical predictions for EPR correlated neutral kaon pairs is numerically very impressive. The full validity of quantum theory could be at stake in these new researches.

References

[1]   A. Einstein, *Ernst Mach, Phys. Zeits.* **17**, 103 (1916); reprinted in: E Mach, *Die Mechanik in ihrer Entwicklung: Historisch-Kritisch Dargestellt*, R. Wahsner and H.H. Borzeszkowski, eds., Akademie Verlag, Berlin (1988), pp. 683-689.

[2]   E. Zahar, *Brit. J. Phil. Sci.* **24**, 95 and 223 (1973).

[3]   H. Poincaré, *Les rapports de la matière et de l'éther, Jour. Phys. théor. appl.* **2**, 347 (1912).

[4]   J. Larmor, *Aether and Matter*, Cambridge Univ. Press (1900).

[5]   J. S. Bell, *George Francis FitzGerald*, Physics World, September 1992, pp. 31-34.

[6]   Henri Poincaré, *The Monist*, **15**, 1 (1905).

[7]   L. Kostro, An Outline of the History of Einstein's Relativistic Ether Conception, in: *Studies in the History of General Relativity*, J. Eisenstaedt & A.J. Kox, eds., Birkhäuser, Boston (1992).

[8]   K. R. Popper, *Unended Quest. An Intellectual Autobiography*, Fontana/Collins, Glasgow (1978), pp. 96-97.

[9]   H. Reichenbach, *The Philosophy of Space & Time*, Dover, New York (1958).

[10] A. Grünbaum, *Philosophical Problems of Space And Time*, Reidel Dordrecht (1973).

[11] M. Jammer, Some foundational problems in the special theory of relativity, in: *Problems in the Foundations of Physics*, G. Toraldo di Francia ed., North Holland, Amsterdam (1979), pp. 202-236.

[12] R. Mansouri and R. Sexl, *General Relat. and Grav.*, **8**, 497 (1977).

[13] T. Sjödin, *Nuovo Cim.* **51** B, 229 (1979).

[14] G. Cavalleri, *Nuovo Cim.* **104** B, 545 (1989).

[15] A.A. Ungar, *Found. Phys.* **21**, 691 (1991).

[16] F. Selleri, *Found. Phys.* **26**, 641 (1996); *Found. Phys. Lett.* **9**, 43 (1996).

[17] H.C. Hayden, *Galilean Electrodynamics*, **2**, 63 (1991).

[18] H. Dingle, *Nature*, **179**, 866 and 1242 (1957); H. Dingle, Introduction, in: Henri Bergson, *Duration and Simultaneity*, pp. xv-xlii, The Library of Liberal Arts, Indianapolis (1965).

[19] J. Bailey, K. Borer, F. Combley, H. Drumm, F. Krienen, F. Lange, E. Picasso, W. von Ruden, F.J.M. Farley, J.H. Field, W. Flegel and P.M. Hattersley, *Nature*, **268**, 301 (1977).

[20] J.C. Hafele and R.E. Keating, *Science*, **177**, 166 (1972).

[21] G. Builder, *Austral. Jour. Phys.*, **11**, 279 (1958); *ibid.* **11**, 457 (1958).

[22] L. Jánossy, *Theory of Relativity Based on Physical Reality*, Akadémiai Kiadó, Budapest (1971).

[23] H. Erlichson, *Am. J. Phys.*, **41**, 1068 (1973).

[24] S. Prokhovnik, *Light in Einstein's Universe*, Kluwer, Dordrecht (1985).

[25] J. S. Bell, How to teach special relativity, in: *Speakable and Unspeakable in Quantum Mechanics*, Cambridge Univ. Press (1987).

[26] J. Brandes, *Die Relativistischen Paradoxien und Thesen zu Raum und Zeit*, VRI, Karlsbad (1995).

[27] C.I. Mocanu, *Galilean Electrodynamics*, **2**, 67 (1991).

[28] M.C. Duffy, ed., *Physical Interpretations of Relativity Theory*, British Society for the Philosophy of Science, London (1996).

[29] A.G. Kelly (HDS Energy, Celbridge): Synchronisation of clock-stations & the Sagnac effect, in this book.

[30] G. Nimtz, A. Enders and H. Spieker, *J. Phys. I France*, **4**, 565 (1994).

[31] H. Arp, *Quasars, Redshifts and Controversies*, Interstellar Media, Berkeley (1987).

[32] H. Arp, private communication.

[33] C. H. Thompson, Behind the Scenes at the EPR Magic Show, in this book.

[34] V.L. Lepore and F. Selleri, *Found. Phys. Letters*, **3**, 203 (1990).

[35] A. Garuccio, Entangled states and the compatibility between quantum mechanics and relativity, in this book.

[36] F. Selleri, *Phys. Rev. A* **56**, 3493 (1997).

Franco Selleri
Bari, May 1998

Velocity of Light

An Explanation of the Sagnac Effect Based on the Special Theory of Relativity, the de Broglie/Bohm Interpretation of Quantum Mechanics, and a Non-Zero Rest Mass for the Photon

Patrick Fleming
Dublin 3
E-mail: flemingp@nsai.ie

If a beam of light (photons) is split by means of a combined beam splitter/interferometer and sent in opposite directions around the circumference of a stationary disc using mirrors or optical fibres, an interference pattern is observed on the interferometer. The disc is capable of being rotated, and the apparatus is fixed in the laboratory. If the disc is now rotated the interference fringe is shifted on the interferometer relative to the stationary disc position. If the disc is now rotated in the other direction the fringe moves to the other side of the stationary disc fringe position. The effect was first observed by the French scientist G. Sagnac in 1910, and is named after him.

The effect is seen irrespective of whether the observer rotates with the disc on its periphery, or is stationary in the laboratory. Subsequent tests have established that the effect is also observed with neutrons [1], and electrons [2].

Over the intervening years many explanations of the phenomenon have been suggested *e.g.*, Anandan [3] gives an explanation based on Special Relativity, and Selleri [4] gives an explanation in terms of inertial transformations. Kelly [5] and this Conference, concludes that the speed of light is not, in all circumstances, independent of the speed of its source. A comprehensive list is given in [6].

I wish to put forward an explanation based on the Special Theory of Relativity, the de Broglie/Bohm interpretation of quantum mechanics, and a massive photon.

The Copenhagen interpretation of quantum mechanics describes particles as either a particle or a wave depending on the mode of observation. It cannot accept the simultaneous presence of particle and wave. The de

Broglie/Bohm model states that all particles, including photons, are always accompanied by a (pilot) wave.

If the tangential velocity of the disc is V_1 and the velocity of the particle (the special situation of the photon is discussed later) is V_2 then, when the particle and the disc are moving in the same direction, the velocity of the particle is $V_2 - V_1$ (say V_3) relative to an observer on the periphery of the disc (in practice a photographic plate). The Lorentz transformation of the addition/subtraction of velocities is not shown for reasons of simplicity of presentation. As V_2 is far greater than V_1 the transformation does not affect the analysis. When the particle and disc are moving in opposite directions the relative velocity of the particle is $V_2 + V_1$ (say V_4). V_3 and V_4 are relativistic. The Special Theory of Relativity states that time for the two particles will be dilated to different extents according to the formula:

$$\gamma = \frac{1}{\sqrt{1 - \dfrac{v^2}{c^2}}}$$

where γ is the gamma factor, v is the velocity of the particle relative to the observer; c is the Einstein assumption of a unique limiting velocity for all phenomena, achieved by a massless particle.

In dealing with the Sagnac effect there are three aspects to be explained:

i)   the fringe shift;
ii)   the direction of the shift in relation to the rotation of the disc;
iii) the fact that the same fringe shift is seen on board the rotating disc and in the laboratory.

Time dilation is greater for the particle travelling against the rotation of the disc (V_4). Therefore, to an observer on the disc, for a given time, this particle will have travelled a distance greater than the other particle causing a fringe shift. The shift is in a direction against the rotation of the disc. This explains the displacement of the particle circumferentially. One is using the de Broglie/Bohm interpretation of quantum mechanics and its model of particles accompanied by a wave. The displaced particle carries its wave with it, causing the fringe and its shift.

All these velocities appear the same to a fixed observer in the laboratory. Therefore, he sees the same fringe shift.

This analysis shows that the photon is behaving exactly like electrons and neutrons in respect of fringe shift. It would seem that one may add to its velocity, and that it obeys the conventional laws of addition and subtraction, and is not, therefore, absolute. It never reaches c, and must, therefore, have a rest mass.

A rest mass for the photon has been suggested by many authors. Vigier [7] states that Einstein, Schrödinger and de Broglie suggested a rest mass of $\sim 10^{-65}$ gr. Goldhaber et al. [8] discuss an upper limit of $\sim 10^{-44}$ gr. Barrow et al.

[9] also discuss limits and some of the implications. The *Particle Data Group* [10] give an upper bound of 3×10^{-33} MeV/c^2.

This time dilation is the same effect as that observed in the CERN experiments of 1976 when muons were accelerated to a speed approaching c, in a circular orbit, in an accelerator ring. This produced the same effect as the Sagnac rotating disc, the ring being fixed in the same frame of reference as the laboratory. Bailey *et al.* [11] report that only the effects of special relativity are relevant even under an acceleration of 10^{18} g. This increased their half-life from 2.2 μs to an observed 64.5 μs, *i.e.* by a factor of 29.3. The speed of the muons in the accelerator ring was 0.9994. Substituting this in the above formula gives a γ factor of 28.9, a near perfect agreement between theory and experimental result. The effect must be used by CERN engineers in designing their particle accelerators.

When similar experiments are carried out on the surface of the earth (which, of course, can be considered as rotating disc at a particular latitude) the same effect is noted. Michelson and Gale [12] carried out an experiment on the effect of the earth's rotation on the velocity of light. They recorded the difference in time taken for the light signals to travel clockwise and anti-clockwise. They got a fringe shift of 0.230 on an interferometer, indicating a time difference.

Saburi *et al.* [13] sent electromagnetic signals around the Earth between standard clock stations. The results showed that the signals travelled slower eastwards than westwards. One predicts that if the tests were carried out in a north-south direction, with the particles not being affected by the rotation of the earth, one would not see a time difference or fringe effect.

Bilger *et al.* [14] carried out tests, using a ring-laser fixed to the earth. The objective was to determine the effect on the laser light of the rotational effect of the earth. The tests were carried out in New Zealand, in the Southern Hemisphere. The light was sent in opposing directions around a circuit of 0.75 m^2. A fringe shift was observed, but in the opposite direction to that of tests carried out in the Northern Hemisphere.

The above analysis, if correct, indicates the validity of the fundamental physical assumption of the de Broglie/Bohm theory of the objective (co)existence of the quantum wave and the particle it guides. It also indicates a non-zero mass for the photon.

References

[1]   Werner, 1979, *Phys. Rev. Lett.* **42** No. 17, 1103-1106

[2]   Hasselbach, F., Nicklaus, M., 1993 Sagnac experiment with electrons: Observation of the rotational phase shift of electron waves in vacuum, *Phys. Rev. A.* Vol. 48, No. 1, 143-151.

[3]   Anandan, J. 1981. Sagnac effect in relativistic and non-relativistic physics. *Physical Review D*, Vol. 24, No. 2, 338-346

[4]   Selleri, F., 1996, Noninvariant one -way velocity of light. (to be published).

[5]   Kelly, A.G., 1996, A New Theory on the Behaviour of Light, The Institution of Engineers of Ireland, *Monograph No. 2.*

[6]   Post, E.J., 1967, Sagnac effect, *Rev. Mod. Phys.*, Vol 39, No. 2, 475-494.

[7]   Vigier, J.P., 1997, Relativistic interpretation (with non-zero photon mass) of the small aether drift velocity detected by Michelson, Morley and Miller. *Apeiron*, Vol 2, Nos.2-3, 71-76.

[8]   Goldhaber, A.S., *et al.*, 1971, Terrestrial and Extraterrestrial Limits on The Photon Mass, *Rev. Mod. Phys.*, Vol 43, No. 2, 277-296.

[9]   Barrow, J.D., *et al.*, 1984, New light on heavy light, *Nature*, Vol, 307, 14-15.

[10]   Particle Data Group, "Review of Particle Properties," *Phys. Rev. D* 50, 1173-1826 (1994).

[11]   Saburi, Y., Yamamoto, M., Harada, K., 1976, *IEEE Trans*, **IM25** No. 4, 473-477

[12]   Bailey, J., *et al.*, 1977, Measurements of relativistic time dilatation for positive and negative muons in a circular orbit, *Nature*, Vol. 268, 301-305.

[13]   Michelson, A.A., Gale, H.G., 1925, The effect of the earth's rotation on the velocity of light, *Astroph. J.*, Vol. LXI No. 3, Part II, 140-145.

[14]   Bilger, H.R., Stedman, G.E., Screiber, W., Schneider, M., 1995 *IEEE Trans.*, **44** IM No. 2, 468-70

On Synchronisation of Clocks in Free Fall Around a Central Body

F. Goy
Dipartimento di Fisica
Universitá di Bari
Via G. Amendola, 173
I-70126 Bari, Italy E-mail: goy@axpba1.ba.infn.it

The conventional nature of synchronisation is discussed in inertial frames, where it is found that theories using different synchronisations are experimentally equivalent to special relativity. On the other hand, in accelerated systems only a theory maintaining an absolute simultaneity is consistent with the natural behaviour of clocks. The principle of equivalence is discussed, and it is found that any synchronisation can be used locally in a freely falling frame. Whatever the synchronisation chosen, the first derivatives of the metric tensor disapear and a geodesic is locally a straight line. But it is shown that only a synchronisation maintaining absolute simultaneity makes it possible to define time consistently on circular orbits of a Schwarzschild metric.

Keywords: special and general relativity, synchronisation, one-way velocity of light, ether, principle of equivalence.

1. Introduction

In the last few decades there has been a revival of so-called "relativistic ether theories." This revival is partly due to the parametrised test theory of special relativity by Mansouri and Sexl [1], which unlike the test theory of Robertson [2], makes explicit allowance for the problem of synchronisation of distant clocks within an inertial frame. Even though it is of vital importance for the definition of time in special relativity, most modern textbooks on relativity treat the question of synchronisation of clocks only briefly, or do not even mention it. The problem of synchronisation of distant clocks arose at the end of the 19th century from the decline of Newtonian mechanics, in which time was absolute and was defined without any reference to experience, and in particular clock synchronisation procedures. The nature of Newtonian time, transcending any experimental definition, was severely criticized by

Mach. However, for the synchronisation procedure one had to take into account that no instantaneous action at distance exists in nature. In his 1905 [3] article in which he expounded the theory of relativity, Einstein, influenced by Mach's epistemological conceptions, gave an operational definition of time:

> "It might appear possible to overcome all the difficulties attending the definition of "time" by substituing "the position of the small hand of my watch" for "time." And in fact such a definition is satisfactory when we are concerned with defining a time exclusively for the place where the watch is located; but is no longer satisfactory when we have to connect in time series of events occuring at different places, or—what comes to the same thing— to evaluate the times of events occurring at places remote from the watch."

Further, he notes:

> "If at the point A of space there is a clock, an observer at A can determine the time values of events in the immediate proximity of A by finding the positions of the hands which are simultaneous with these events. If there is at the point B of space another clock in all respects resembling the one at A, it is possible for an observer at B to determine the time values of events in the immediate neighbourhood of B. But it is not possible without further assumption to compare, in respect of time, an event at A with an event at B. We have so far defined only an "A time" and a "B time." We have not defined a common "time" for A and B, for the latter cannot be defined at all unless we establish by definition that the "time" required by light to travel from A to B equals the "time" it requires to travel from B to A. Let a ray of light start at the "A time" t_A from A towards B, let it at the "B time" t_B be reflected at B in the direction of A, and arrive again at A at the "A time" t'_A.

> In accordance with definition, the two clocks synchronize if

$$t_B - t_A = t'_A - t_B \tag{1}$$

> We assume that this definition of synchronism is free from contradictions, and possible for any number of points; and that the following relations are universally valid:-

> 1. If the clock at B synchronizes with the clock at A, the clock at A synchronizes with the clock at B.

> 2. If the clock at A synchronizes with the clock at B and also with the clock at C, the clocks at B and C also synchronize with each other."

As Einstein himself stresses, the time required by light to travel from A to B and from B to A is equal by definition. This means that the one-way velocity of light is given by a convention, and not by experiment. What is known with great precision is the (mean) two-way velocity of light, which obviously can be measured with only one clock and a mirror. This is known with a precision of $\Delta c/c = 10^{-9}$ [4] and has always been found to be constant in any direction throughout the year, despite the Earth's motion. The one-way velocity of light, on the other hand, cannot be determined experimentally. Let us imagine that someone tries to measure it: he might send a light ray from a clock located at A to a clock located at B, at a distance d from A, and would obtain the one-way velocity of light from A to B by dividing the distance d by the difference between the time of arrival in B and the time of departure from A. But in order to compute this time difference, he first needs clocks which are synchronised, by means of light rays—whose one-way velocity is postulated. Thus the concepts of simultaneity and one-way velocity of light are bound together logically in a circular way.

One may, of course, wonder whether other conventions which are not in contradiction with experiment are possible. First we rewrite equation (1) such that the "B time" is defined as a function of the "A time." That is:

$$t_B = t_A + \tfrac{1}{2}(t'_A - t_A) \tag{2}$$

Reichenbach commented [5]:

> "This definition is essential for the special theory of relativity, but is not epistemologically necessary. If we were to follow an arbitrary rule restricted only to the form
>
> $$t_B = t_A + \varepsilon(t'_A - t_A), \quad 0 < \varepsilon < 1 \tag{3}$$
>
> it would likewise be adequate and could not be called false. If the special theory of relativity prefers the first definition. i.e., sets ε equal to $\tfrac{1}{2}$, it does so on the ground that this definition leads to simpler relations."

Among the "conventionalists", who agree that one can choose ε freely, are Winnie [6], Grünbaum [7], Jammer [8], Mansouri and Sexl [1], Sjödin [9], Cavalleri and Bernasconi [10], Ungar [11], Vetharaniam and Stedman [12] and Anderson and Stedman [13]. Clearly, different values of ε correspond to different values of the one way-speed of light.

A slightly different position was developed in the parametric test theory of special relativity by Mansouri and Sexl [1]. Following these authors, we assume that there is *at least one* inertial frame in which light behaves isotropically. We call it the priviledged frame Σ and denote space and time coordinates in this frame by the letters: (x_0, y_0, z_0, t_0). In Σ, clocks are synchronised with Einstein's procedure. We also consider another system S moving with uniform velocity $v < c$ along the x_0-axis in the positive direction. In S, the coordinates are written with lower case letters (x, y, z, t). Under rather general assumptions as to initial and symmetry conditions on the two systems (S and Σ are endowed with orthonormal axes, which coincide at time $t_0 = 0$, [1,14]) the assumption that *the two-way velocity of light is c and furthermore that the time dilation factor has its relativistic value*, one can derive the following transformation:

$$x = \frac{1}{\sqrt{1-\beta^2}}(x_0 - vt_0)$$
$$y = y_0$$
$$z = z_0 \tag{4}$$
$$t = s(x_0 - vt_0) + \sqrt{1-\beta^2}\,t_0,$$

where $\beta = v/c$. The parameter s, which characterizes the synchronisation in the S frame, remains unknown. Einstein's synchronisation in S involves: $s = -v/c^2\sqrt{1-\beta^2}$ and (4) becomes a Lorentz boost. For a general s, the inverse one-way velocity of light is given by [15]:

$$\frac{1}{c_\to(\Theta)} = \frac{1}{c} + \left(\frac{\beta}{c} + s\sqrt{1-\beta^2}\right)\cos\Theta, \tag{5}$$

where Θ is the angle between the x-axis and the light ray in S. $c_\to(\Theta)$ is in general dependent on the direction. A simple case is $s = 0$. From (4), this means that at $t_0 = 0$ in Σ we set all clocks in S at $t = 0$ (external synchronisation), or that we synchronise the clocks by means of light rays with velocity $c_\to(\Theta) = c/1+\beta\cos\Theta$ (internal synchronisation). We obtain the transformation:

$$x = \frac{1}{\sqrt{1-\beta^2}}(x_0 - vt_0)$$
$$y = y_0$$
$$z = z_0 \tag{6}$$
$$t = \sqrt{1-\beta^2}\,t_0.$$

This transformation maintains an absolute simultaneity (as in Σ) between all inertial frames. It should be stressed that, unlike the parameters of length contraction and time dilation, *this parameter s cannot be tested*, but its value must be assigned in accordance with the synchronisation choosen in the experimental setup. This means, as regards experimental results, that theories using different s are equivalent. Of course, they may predict different values of physical quantities (for example the one-way speed of light). The difference lies not in nature itself, but in the convention used for the synchronisation of clocks. In other words, two transformations (4) with different s represent the same transformation but relative to different time coordinates. For a recent and comprehensive discussion of this subject, see [16]. A striking consequence of (4) is that the negative result of the Michelson-Morley experiment does not rule out an ether. Only an ether with Galilean transformations is excluded, because the Galilean transformations do not lead to an invariant two-way velocity of light in a moving system.

Strictly speaking, the conventionality of clock synchronisation has only been shown to hold in inertial frames. The derivation of equation (4) is done in inertial frames and is based on the assumption that the two-way velocity of light is constant in all directions. This last assumption is no longer true in accelerated systems. However, special relativity is not used just in inertial frames. Many textbooks give examples of calculations done in accelerated systems, using infinitesimal Lorentz transformations. Such calculations use an additional assumption: the so-called *Clock Hypothesis*, which states that, seen from an inertial frame, the rate of an accelerated ideal clock is identical to a clock in the instantaneously comoving inertial frame. In other words, the rate of a clock is not influenced by acceleration *per se*. This hypothesis, first used implicitely by Einstein in his 1905 article, was superbly confirmed in the famous time decay experiment on muons at CERN, where the muons had an acceleration of $10^{18}g$, but where their time decay was due only to their velocity [17]. We stress here the logical independence of this assumption from the structure of special relativity as well as from the assumptions necessary to derive (4). The author's opinion is that the *Clock Hypothesis*, added to special relativity in order to extend it to accelerated systems, leads to logical contradictions when the question of synchronisation is brought up. This idea has also been expressed by Selleri [18]. The following example (see [19]) demonstrates this point: imagine that two distant clocks are secured to an inertial frame (say a train at rest) and synchronised using Einstein's synchronisation. We call this rest frame Σ. The train accelerates for a certain period. After that, the acceleration stops and the train again has inertial motion (sytem S). During acceleration, the clocks are subjected to exactly the same influences, so they have the same rate at all times, and remain synchronous relative to Σ. Due to the relativity of simultaneity in special relativity, where an Einstein's procedure is applied to the synchronisation of clocks in all inertial frames, they are no longer Einstein synchronous in S. *So the Clock Hypothesis is inconsistent with the clock setting of relativity.* On the other hand, the *Clock Hypothesis* has been tested with a high degree of accuracy [20] and cannot be rejected; consequently, we must reject the clock setting of special relativity. The only theory which is consistent with the *Clock Hypothesis* is based on transformations (4) with $s = 0$.

This is an ether theory. The fact that only an ether theory is consistent with accelerated motion provides strong evidence that an ether exists, but does not inevitably imply that our velocity relative to the ether is measurable. The author's opinion is that it cannot be measured, because (6) represents another *coordinatisation* of the Lorentz transformation (obtained by clock resynchronisation). In principle, this prevents any detection of uniform motion through the ether. By changing the coordinate system, one cannot obtain a physics in which new physical phenomena appear. But we can obtain a more consistent description of these phenomena.

In all the above considerations, space-time was flat and no gravitational forces were present. In the following, we want to treat the question of synchronisation of clocks in the framework of general relativity, were special relativity is only valid locally. In section 2, we calculate the equations of motion for circular orbits in a Schwarzschild metric. In section 3, we treat the problem of synchronisation of clocks on these orbits, and discuss the compatibility of different synchronisations with the principle of equivalence.

2. Circular orbits in a Schwarzschild metric

In a reference system R with coordinates S, $(x^0, x^1, x^2, x^3) = (ct, r, \varphi, \theta)$ (θ is the azimuthal angle) the spherical symmetric solution of Einstein's equations in vacuum, with the boundary condition that the metric becomes Minkowskian at infinity is the Schwarzschild metric:

$$ds^2 = -\left(1 - \frac{\alpha}{r}\right)(dx^0)^2 + \left(1 - \frac{\alpha}{r}\right)^{-1} dr^2 + r^2(\sin^2\theta \, d\varphi^2 + d\theta^2), \qquad (7)$$

where $\alpha = 2GM/c^2$ is the Schwarzschild radius of the field of total energy Mc^2 and G the gravitational constant. In the following we will consider only geodesics of test particles of mass m with $r > \alpha$, so that we are not concerned here with the breakdown of the coordinate system at $r = \alpha$. A Lagrangian function can be written as:

$$\mathcal{L} = -m \, c \sqrt{g_{ij} \frac{dx^i}{d\tau} \frac{dx^j}{d\tau}} \qquad (8)$$

and the Lagrange equations by

$$\frac{\partial \mathcal{L}}{\partial x^i} = \frac{d}{d\tau} \frac{\partial \mathcal{L}}{\partial \left(\frac{dx^i}{d\tau}\right)}, \quad i = 0, \dots, 3. \qquad (9)$$

The variables x^0, θ, φ are cyclic and their conjugate momentum is conserved. Without loss of generality we can take $\theta = \pi/2$, i.e., equatorial orbits only. The energy E and angular momentum L per unit of mass are conserved quantities:

$$L = r^2 \frac{d\varphi}{d\tau}$$
$$E = c \frac{dx^0}{d\tau}\left(1 - \frac{\alpha}{r}\right). \qquad (10)$$

From (9) and (10) the equation for the variable r can be written

$$\frac{dr}{d\tau} = \frac{E^2}{c^2} - \left(1 - \frac{\alpha}{r}\right)\left(c^2 + \frac{L^2}{r^2}\right) = \frac{1}{c^2}\left[E^2 - V^2(r)\right] \qquad (11)$$

where $V(r)$ is an effective potential. This effective potential has a local minimum; thus we have stable circular orbits. From (10), we then find for these circular orbits:

$$\frac{dr}{d\tau} = 0 \Rightarrow r = cst$$
$$\frac{d\varphi}{d\tau} = \frac{L}{r^2} \Rightarrow \varphi(\tau) = \varphi(\tau = 0) + cst_1 \tau$$
$$\frac{dt}{d\tau} = \frac{E}{c^2\left(1 - \frac{\alpha}{r}\right)} \Rightarrow \tau(t) = \tau(t = 0) + cst_2 t, \qquad (12)$$

where $cst_1 = \frac{c}{r}\sqrt{\dfrac{\alpha}{2r-3\alpha}}$ and $cst_2 = \sqrt{2-\frac{3\alpha}{r}}$.

3. Two clocks in orbit

We now consider a clock $\mathcal{A}$ in event-point $A(x^0{}_A, r_A, \varphi_A)$, and do all calculations in $1+2$ dimensional space-time, since we treat equatorial orbits only. On a circular orbit, its velocity is given by $U = (c,0,\omega)\sqrt{1-\alpha/r_A - r_A^2\omega^2/c^2}$. We have $U^i U^j g_j = -c^2$ and $\omega = d\varphi/dt$, and is given by the Kepler law $\omega^2 = GM/r_A^3$ for circular orbits [21].

The principle of equivalence assures us that we can find a system of reference $\overset{o}{R}$, with a coordinate system $\overset{o}{S}$ such that at event-point A, $\overset{o}{g}_{ij}(A) = \eta_j$ and $\left(\partial \overset{o}{g}_{ij}/\partial x^k\right)(A)=0$, where $\eta_j = \mathrm{diag}(-1,1,1)$. In particular, it is possible to choose a set of three mutually orthogonal unit vectors $e^i{}_{(a)}$ such that $e^i{}_{(0)} = U^i/c$ and $e_{(1)}$ and $e^i{}_{(2)}$ fulfil the orthonormality conditions: $g_{ik}e^i{}_{(a)}e^{ik}{}_{(b)} = \eta_{ab}$. Indices without parenthesis of $e^i{}_{(a)}$ are lowered with g_{ik}, while indices with parenthesis are raised with η^{ib}. We can choose $e_{(1)}$ radial and $e_{(2)}$ tangential to the orbit:

$$e_{(1)} = \left(0, \sqrt{1-\frac{\alpha}{r_A}}, 0\right)$$

$$e_{(2)} = \frac{1}{\sqrt{1-\alpha/r_A - r_A^2\omega^2/c^2}}\left(\frac{r_A\omega}{c\sqrt{1-\alpha/r_A}}, 0, \frac{\sqrt{1-\alpha/r_A}}{r_A}\right).$$

(13)

The following transformation from coordinate system S to $\overset{o}{S}$ is such that the metric tensor in the new coordinates is Minkowskian and its first derivatives disappear at point A [§9.6][22]:

$$\overset{oi}{x} = e^i_r\left(x^r - x^r_A\right) + \tfrac{1}{2}e^{(i)}_r\Gamma^r_{st}(A)\left(x^s - x^t_A\right), \quad i=0,1,2.$$

(14)

In the case of (7), the Christofell symbols Γ at A are given by:

$$\Gamma^1_{00} = \frac{1}{2}\frac{\alpha(1-\alpha/r_A)}{r_A^2} \qquad \Gamma^0_{01} = \frac{1}{2}\frac{\alpha}{\alpha(1-\alpha/r_A)r_A^2} \qquad \Gamma^2_{12} = \frac{1}{r_A}$$

$$\Gamma^1_{11} = -\frac{1}{2}\frac{\alpha}{2r_A^2(1-\alpha/r_A)} \qquad \Gamma^1_{22} = -r(1-\alpha/r_A).$$

(15)

We obtain for the transformation between S and S :

$$\overset{o}{x}{}^0 = \frac{1}{\sqrt{1-\alpha/r_A - r_A^2\omega^2/c^2}}\left[\begin{array}{l}\left(1-\dfrac{\alpha}{r_A}\right)(x^0 - x_A^0) - \dfrac{\omega r_A^2}{c}(\varphi - \varphi_A) \\[2mm] +\dfrac{1}{2}\dfrac{\alpha}{r_A^2}(x^0 - x_A^0)(r - r_A) + \dfrac{1}{4}\dfrac{\alpha\sqrt{1-\alpha/r_A}}{r_A^2}(r - r_A)\end{array}\right]$$

$$\overset{o}{x}{}^2 = \frac{1}{\sqrt{1-\alpha/r_A}}(r - r_A) + \frac{1}{4}\frac{\alpha\sqrt{1-\alpha/r_A}}{r_A^2}(x^0 - x_A^0)^2$$

$$-\frac{1}{4}\frac{\alpha}{r_A^2(1-\alpha/r_A)^{3/2}}(r - r_A)^2 - \frac{1}{2}\sqrt{1-\alpha/r_A}\,(\varphi - \varphi_A)^2 \tag{16}$$

$$\overset{o}{x}{}^2 = \frac{\sqrt{1-\alpha/r_A}}{\sqrt{1-\alpha/r_A - r_A^2\omega^2/c^2}}\left[\begin{array}{l}-\dfrac{\omega r_A}{c}(x^0 - x_A^0) + r_A(\varphi - \varphi_A) \\[2mm] -\dfrac{1}{2}\dfrac{\omega\alpha}{cr_A(1-\alpha/r_A)}(x^0 - x_A^0)(r - r_A) + (r - r_A)(\varphi - \varphi_A)\end{array}\right]$$

This transformation looks like the Lorentz transformation at first order, in particular, two distant events which are simultaneous in $\overset{o}{S}$ are not simultaneous in S. We now imagine that a clock $\mathcal{B}$ is located at B $(x^0_A + dx^0,\ r_A,\ \varphi_A + d\varphi)$ and we want to synchronise it with $\mathcal{A}$ at A using Einstein's procedure. Since the metric is Minkowskian in $\overset{o}{S}$, the velocity of light is c in this (local) frame. The two clocks will be Einstein synchronised when: $\overset{o}{x}{}^0_A = \overset{o}{x}{}^0_B = 0$. Using (16) we obtain that the infinitesimal time difference in S dx^0 between these events is given by:

$$dx^0 = \frac{\omega r_A^2 d\varphi}{c(1-\alpha/r_A)} \tag{17}$$

We generalise this synchronisation procedure all along the circular orbit. This means that we synchronise $\mathcal{A}$ in (r_A, φ_A), with $\mathcal{B}$ in $(r_A, \varphi_B = \varphi_A + d\varphi)$, and then $\mathcal{B}$ with $\mathcal{C}$ located at $(r_A, \varphi_C = \varphi_B + d\varphi)$, etc. If we do a whole round trip, we find a time lag Δx^0 given by:

$$\Delta x^0 = \oint \frac{\omega r_A^2 d\varphi}{c(1-\alpha/r_A)} = \frac{2\pi\omega r_A^2}{c(1-\alpha/r_A)} \tag{18}$$

This means that A is not synchronisable with itself, when we extend the synchronisation procedure spatially out of a local domain; this is clearly absurd. The problem occurs because dx^0 is not a total differential in r and φ, thus the synchronisation procedure is path-dependent. In general, one can say that if $\mathcal{A}$ is synchronized with $\mathcal{B}$, then $\mathcal{B}$ does not synchronise with $\mathcal{A}$. The same remark is valid for the transitivity of the relation "is synchronous with" in the case of three clocks $\mathcal{A}$, $\mathcal{B}$ and $\mathcal{C}$

According to Einstein in the citation quoted above, the definition of synchronism given by (1) which is free from contradictions in the case of inertial frames in flat space is no longer free from contradictions when we want to define time globally in a curved space. One might think that this difficulty is insuperable, and that it is not possible to:

1. find a local inertial system such that the equivalence principle is respected
2. define time in this system in such a way that extending the synchronisation procedure out of a local domain is self consistent: "is synchronised with" is an equivalence relation.

A similar problem occurs in the case of a rotating disk in flat space. It has been shown that only the transformation (6) allows a consistent definition of time on the rim of a rotating disk, while an Einstein synchronisation leads to the impossibility of defining time without contradictions on the rim of this disk [23].

Guided by the experimental equivalence of relativistic ether theories and special relativity, we are looking for another synchronisation of clocks in $\overset{o}{R}$ such that the conditions 1 and 2 above are fullfilled. The spatial part of transformation (16) is not changed by a resynchronisation of clocks, and we can again choose the vectors $e^{(1)}$, and $e^{(2)}$ as they can be read out from (16). We are looking for a transformation from coordinate system S to local coordinate system $\hat{S}$ such that the time transformation does not depend on the space variables at first order. This means that $e^{(0)}$ is of the type $e^{(0)} = (y,0,0)$. In order to find y, we postulate that the sychronisation only is different in $\hat{S}$ and $\overset{o}{S}$. In other words, the rate of a clock at rest at the origin of $\hat{S}$ and $\overset{o}{S}$ is the same when seen from S. From (16) we easily calculate that:

$\delta \overset{o}{x}^0 = \sqrt{1-\alpha/r_A - \omega^2 r_A^2/c^2}\, \delta x^0$, where $\delta \overset{o}{x}^0$ is the coordinate time difference between two ticks of the clock in $\overset{o}{S}$ and δx^0 is the same quantity in S. We find that $y = \sqrt{1-\alpha/r_A - \omega^2 r_A^2/c^2}$. Thus the transformation of the time coordinate from S to $\hat{S}$ is now given by:

$$\hat{x}^0 = \sqrt{1-\alpha/r_A - \omega^2 r_A^2/c^2}\,(x^0 - x_A^0) + \frac{1}{2}\frac{\alpha\sqrt{1-\alpha/r_A - \omega^2 r_A^2/c^2}}{(1-\alpha/r_A)r_A^2}(x^0 - x_A^0)(r - r_A) \quad (19)$$

1. Are we sure that $\hat{S}$ is a local inertial system of coordinates? Yes. The proof is indeed the same as it would be for $\overset{o}{S}$. From (14) and using the fact that $e_{(r)}{}^i e^r{}_j = \delta_j$, we have:

$$e_{(r)}^i \hat{x}^r = \left(x^i - x_A^i\right) + \frac{1}{2}\Gamma_{st}^i(A)\left(x^s - x_A^s\right)\left(x^t - x_A^t\right) \quad i=0,1,2. \quad (20)$$

Differentiating two times with respect to $\hat{x}^k$ and $\hat{x}^l$ gives:

$$0 = \frac{\partial^2 x^i}{\partial \hat{x}^l \partial \hat{x}^k} + \Gamma^i_{st}(A)\left[\frac{\partial^2 x^s}{\partial \hat{x}^l \partial \hat{x}^k}(x^t - x^t_A) + \frac{\partial x^s}{\partial \hat{x}^k}\frac{\partial x^t}{\partial \hat{x}^l}\right] \tag{21}$$

Thus at point A:

$$0 = \frac{\partial^2 x^i}{\partial \hat{x}^l \partial \hat{x}^k} + \Gamma^i_{st}(A)\left[\frac{\partial x^s}{\partial \hat{x}^k}\frac{\partial x^t}{\partial \hat{x}^l}\right] \tag{22}$$

Because of the law of transformation of Christoffel symbols, this mean that: $\hat{\Gamma}^i_{kl}(A) = 0$. So in $\hat{S}$ at A, a geodesic becomes a straight line:

$$\frac{d^2 \hat{x}^k}{d\lambda^2} + \hat{\Gamma}^k_{il}\frac{d\hat{x}^l}{d\lambda}\frac{d\hat{x}^i}{d\lambda} = \frac{d^2 \hat{x}^k}{d\lambda^2} = 0 \tag{23}$$

2. Can time be defined consistently on the whole circular orbit? Yes. We treat again the problem of synchronising a clock $\mathcal{A}$ at A (x^0_A, r_A, φ_A) and a clock $\mathcal{B}$ at B $(x^0_A + dx^0, r_A, \varphi_A + d\varphi)$ The two clocks are synchronised in the system of coordinates $\hat{S}$ if $\hat{x}^0_A = \hat{x}^0_B = 0$. Then the time difference dx^0 between these events in S calculated with (19) gives: $dx^0 = 0$. A similiar calculation as in (18) shows that $\Delta x^0 = 0$ for a whole round trip. Thus the time can be defined consistently on the orbit with such a synchronisation.

The metric in system $\hat{S}$ at A is given by $e^i_{(a)}\, g_{ij} e^j_{(b)} = \hat{\eta}_{ab}$. We find

$$\hat{\eta}_{ab} = \begin{pmatrix} -1 & 0 & \frac{r_A \omega}{c\sqrt{1-\alpha/r_A}} \\ 0 & 1 & 0 \\ \frac{r_A \omega}{c\sqrt{1-\alpha/r_A}} & 0 & 1 - \frac{r_A^2 \omega^2}{c^2(1-\alpha/r_A)} \end{pmatrix} \tag{24}$$

In the case where the vector potential $\hat{\eta}_{o\alpha}$; $\alpha = 1,2$ is different from zero, the spatial part of the metric is given by the space-space coefficients of the metric as well as by $\hat{\gamma}_{\alpha\beta} = \hat{\eta}_{\alpha\beta} - \frac{\hat{\eta}_{o\alpha}\hat{\eta}_{o\beta}}{\hat{\eta}_{oo}}$. In our case we have $\hat{\gamma}_{\alpha\beta} = \delta_{\alpha\beta}$. Thus the spatial system of coordinates is orthonormal. The velocity of light $c(\Theta)$ is found by solving the equation $ds^2 = \hat{\eta}_{ab}d\hat{x}^a d\hat{x}^b = 0$. We find that:

$$c(\Theta) = \frac{c}{1 + \frac{r_A \omega \cos\Theta}{c\sqrt{1-\alpha/r_A}}} \tag{25}$$

where Θ is the angle between the light ray and the $\hat{x}^2$–axis

4. Remarks

1. The transformation of the time variable can easily be generalised to all synchronisations with a parameter s like in (4):

$$x^0(s) = \sqrt{1 - \alpha/r_A - \omega^2 r_A^2/c^2}\left(x^0 - x^0_A\right) + s\left[r_A(\varphi - \varphi_A) - \frac{r_A \omega}{c}(x^0 - x^0_A)\right] + O(x^i - x^i_A)^2 \tag{26}$$

The transformation (19) is given by $s = 0$ and the transformation (19) by $s = -\dfrac{\omega r_A}{c\sqrt{1 - \alpha/r_A - \omega^2 r_A^2/c^2}}$ A similar argument as in section 3 shows that only $s = 0$ lead to $\Delta x^0 = 0$ for a whole round trip of synchronisation around the orbit.

2. The inertial coordinate systems $\overset{o}{S}$ and $\hat{S}$ are different coordinatisations of the same reference frame $\overset{o}{R}$. The transformation from $\overset{o}{S}$ to $\hat{S}$ does not involve time in the transformation of space variables, and thus is what Møller [p. 267, 316][22] calls a linear gauge transformation.

3. If a clock $\mathcal{A}$ at A (x^0_A, r_A, φ_A) and a clock $\mathcal{B}$ at B $(x^0_A + dx^0, r_A, \varphi_A + d\varphi)$ are Einstein synchronised in the system $\overset{o}{S}$ of section 3 [*i.e* dx^0 is given by (17)], they remain Einstein's synchronised during their trip around the orbit. From the equation of motion (12) one sees that they will be at point $\widetilde{A}$ and $\widetilde{B}$ at a later time with coordinates in S: $\left(x^0_{\widetilde{A}}, r_{\widetilde{A}}, \varphi_{\widetilde{A}}\right)$ and $\left(x^0_{\widetilde{B}} + dx^0, r_{\widetilde{A}}, \varphi_{\widetilde{A}} + d\varphi\right)$. We can take a local inertial system at $\widetilde{A}$ and from (16) one sees that: $\widetilde{x}^0_{\widetilde{A}} = \widetilde{x}^0_{\widetilde{B}} = 0$.

5. Conclusion

In flat space, a whole set of theories equivalent to special relativity can be constructed. These theories are obtained by adopting another convention on the synchronisation of clocks. In accelerated systems, only the theory maintaining an absolute simultaneity is logically consistent with the natural behaviour of clocks.

In general relativity, the principle of equivalence tells us that at every space-time point one can choose a local coordinate system such that the metric is Minkowskian and its first derivatives disapear. Thus, the laws of special relativity are locally valid in general relativity. In this local frame, we can choose another synchronisation of clocks different from Einstein's. The frame is the same but the coordinatisation is different. All these coordinatisations are locally equivallent. The transformation between them is a linear gauge transformation. The spatial part of the metric is orthonormal and the derivates of the space-time metric disapear at the point in question. Thus, a freely falling body has uniform motion in a straight line, and theses local coordinate systems are locally inertial.

An Einstein synchronisation leads to a contradictory definition of time when extended out of a local domain. It was shown in this article that in the case of circular orbits, only a transformation maintaining absolute simultaneity is able to define time globally and consistently on the orbit. An observer moving around a central body, who does not want to adopt a contradictory definition of time (when extended spatially out of his local domain) must then conlude that the velocity of light is not constant.

Acknowledgement

I wish to thank the Physics Departement of Bari University for hospitality, and Prof. F. Selleri for his kind suggestions and criticisms.

References

[1]   R. Mansouri and R. U. Sexl, *Gen. Rel. Grav.* **8**, 497-513 (1977); **8**, 515-524 (1977); **8**, 809-813 (1977).

[2]   H. P. Robertson, *Rev. Mod. Phys.* **21**, 378-382 (1949).

[3]   A. Einstein, *Ann. Phys.* **17**, 891-921 (1905). Translated in The Principle of Relativity (Dover, USA, 1923).

[4]   H. E. Bates, *Am. J. Phys.* **56**, 682-687 (1988).

[5]   H. Reichenbach, *The Philosophy of Space and Time* (Dover, New-York, 1958).

[6]   J. A. Winnie, *Phil. Mag.* **37**, 81-99 and 223-238 (1970).

[7]   A. Grünbaum, *Philosophical Problems of Space and Time* (Reidel, Dodrecht, 1973).

[8]   M. Jammer, *Some Fundamental Problems in the Special Theory of Relativity*, in: Problems in the Foundation of Physics (G. Toraldo di Francia ed., North Holland, Amsterdam, 1979).

[9]   T. Sjödin, *Nuov. Cim.* **51B**, 229-245 (1979).

[10] G. Cavalleri and C. Bernasconi, *Nuov. Cim.* **104B**, 545-561 (1989).

[11] A. A. Ungar, *Found. Phys.* **6**, 691-726 (1991).

[12] I. Vetharaniam and G. E. Stedman, *Found. Phys. Lett.* **4**, 275-281 (1991).

[13] R. Anderson and G. E. Stedman, *Found. Phys. Lett.* **5**, 199-220 (1992); *Found. Phys. Lett.* **7**, 273-283 (1994).

[14] F. Selleri, *Phys. Essays* **8**, 342-349 (1995).

[15] F. Selleri, in *Frontiers of Fundamental Physics*, Ed. by M. Barone and F. Selleri (Plenum Press, New-York, 1994), pp. 181-192.

[16] I. Vetharaniam and G. E. Stedman, *Phys. Lett.* A **183**, 349-354 (1993).

[17] J. Bailey *et al.*, *Nature* **268**, 301-304 (1977).

[18] F. Selleri, *Found. Phys.* **26**, 641-664 (1996).

[19] S. R. Mainwaring and G. E. Stedman, *Phys. Rev.* A **47**, 3611-3619 (1993).

[20] A. M. Eisle, *Helv. Phys. Act.* **60**, 1024-1037 (1987).

[21] C. W. Misner, K. S. Thorne and J. A. Wheeler, *Gravitation* (W. H. Freeman and Company, New-York, 1973).

[22] C. Møller, *The Theory of Relativity*, 2nd ed. (Oxford University Press, Oxford, 1972).

[23] F. Goy and F. Selleri, *Found. Phys. Lett.* **10**, 17-29 (1997).

Remarks on Clock Synchronization

Andrzej Horzela
Department of Theoretical Physics
H. Niewodniczanski Institute of Nuclear Physics
ul. Radzikowskiego 152, 31 342 Kraków, Poland

Introduction

Space-time coordinates, like any other physical quantity, should be given by their own operational definitions. This means that we should point out a physical process which can be used to measure them by comparison with standard phenomena defining the system of units. In particular, in order to be able to measure time at distant points in a unique way we must synchronize all clocks present in the system. The choice of synchronization method defines the model of space-time obtained and significantly influences its properties, physical as well as mathematical, including the structure of the space-time symmetry group [1,2].

Properties of coordinate systems

A general definition of the space-time coordinates may be obtained provided we can specify [1]:

1.   the class of observers,
2.   the class of elementary events,
3.   the class of signals used to communicate between observers and elementary events,
4.   the interaction of the elementary events with signals used for communication.

The set of concepts listed above generalizes Einstein's fundamental assumptions of special relativity theory [3]. Within the latter the classes of observers and elementary events coincide and each of them consists of an identical set of clocks equipped with light emitters and detectors and reflecting mirrors. Communication between observers and events is accomplished by light pulses travelling in space with universal, constant and isotropic velocity. The light pulses interact only with mirrors which are supposed to be ideal and the reflection is described by the laws of geometrical optics. To achieve synchronization of all clocks Einstein's procedure needs only one clock placed at the origin of the reference frame. If at the instant of time T_1 it emits a light pulse towards an event A and if it detects at the instant of time T_2 the same pulse coming back after reflection at A then the instant of time T where reflection has occurred equals

$$T = \frac{1}{2}(T_1 + T_2) \tag{1}$$

while the distance between the origin and the event A is given by

$$X = \frac{c}{2}(T_2 - T_1) \tag{2}$$

where c denotes the velocity of light. In Einstein's synchronization there is no distinction between one-way and two-way velocity of light. They are equal by definition and, as a consequence, no preferred reference frame is permitted to exist. If we change the coefficient in (1) and replace this formula, according to [4], by

$$T = T_1 + \varepsilon(T_2 - T_1) \tag{3}$$

we arrive at models of space-time in which signals obeying frame dependent velocity are used in order to synchronize distant clocks. The example of such an approach are models which distinguish among one-way velocities of light and allow an ether to exist and to be the only medium where the electromagnetic waves propagate isotropically with the velocity c [2]. Also, within Newtonian mechanics it is possible to construct a synchronization procedure based on (3) where observers communicate using signals obeying all the laws of Galilean physics, in particular velocity addition rule [1].

Any recipe postulated to give the time shown by a distant clock should be completed by a prescription which enables us to calculate the distance how far the investigated event is located from the observer. It is obvious that it should be expressed in terms of time $(T - T_1)$ spent by the signal in its travel to the event, as well as in terms of the time of duration of the return trip, $(T_2 - T)$. If it will be unique, we should have

$$X = f_{\rightarrow}(T - T_1) = f_{\leftarrow}(T_2 - T) \tag{4}$$

which may be rewritten as

$$X = f_{\rightarrow}[\varepsilon(T_2 - T_1)] = f_{\leftarrow}[(1 - \varepsilon)(T_2 - T_1)] \tag{5}$$

with functions $f_{\rightarrow}$ and $f_{\leftarrow}$ in general different each from another and depending on the details of the model. They become the same when $\varepsilon = \frac{1}{2}$ which means the situation when the motion towards the event and backwards does not change its properties. It is true for (1) and (2) valid, that is for the standard special relativity.

Does light move uniformly?

In the following it is our aim to continue the analysis performed in [5] where we proposed clock synchronization with the use of signals moving with constant acceleration, and studied possible consequences of such an assumption. The next step in this direction consists in considering a more general situation where (1) holds, but on the right hand side of (5), instead of the linear expression used in special relativity, we are going to investigate a nonlinear, strongly increasing function. We assume this function to be the same in both directions and form-invariant with respect to the choice of a class of reference frames called inertial frames. Introducing distance in such a way, we allow the communication signal to move nonuniformly, always in the forward direction and with all properties of its motion independent on the reference frame as long as it is an inertial one. Within such a model the reflection can not be described by geometrical optics. To be consistent with the assumption above we must exclude classical reflection which preserves instant velocity, and replace it by the process when the mirror at first absorbs the incoming signal and then emits a new one, always with the same initial velocity. Within such a process it is also necessary to consider the possibility of delayed emission of the signal, which, as it is shown in [6], does not influence the form of space-time transformations.

We will derive generalized Lorentz transformation rules between inertial frames for new T and X basing our considerations on the principles of Bondi's k-coefficient method [7]. For the times T_1 and T_2 which are the only directly measurable quantities in radiolocation method their transformation rules when one passes from one inertial reference frame to another read

$$T'_1 = kT_1 \tag{6}$$

$$T'_2 = \frac{1}{k}T_2 \tag{7}$$

Because the function f in (5) is strongly monotonic we can invert the relation (5) and next use it, together with (1), in order to express T_1 and T_2 uniquely in terms of T and $f^{-1}(X)$. In this case, the transformation rules (6) and (7) contain enough information to write down the transformation rules for T and $f^{-1}(X)$ with the form invariance of the latter explicitly taken into account. They are

$$T' = \frac{(k^2+1)T - (k^2-1)f^{-1}(X)}{2k} \tag{8}$$

$$f^{-1}(X') = \frac{(k^2+1)f^{-1}(X) - (k^2-1)T}{2k} \tag{9}$$

which, if needed, may be rewritten as

$$X' = f\left(\frac{(k^2+1)f^{-1}(X) - (k^2-1)T}{2k}\right) \tag{10}$$

The physical meaning of reference frames connected by transformations (6)–(7) or (8)–(9),(10) may be found if one looks for the motion of the origin of the primed frame $X' = 0$ observed in the unprimed frame. If the physically obvious relation

$$f(0) = f^{-1}(0) = 0 \qquad (11)$$

is satisfied then we immediately obtain its trajectory

$$X(t) = f\left(\frac{k^2 - 1}{k^2 + 1} t\right) \qquad (12)$$

which means that within the proposed model all inertial reference frames move nonuniformly with respect to another. The characteristics of their motion are defined by those of synchronizing signal scaled by factors $\left(k^2 - 1/k^2 + 1\right)^N$ respectively. Absolute values of all these factors are less than 1 which means that characteristics of the synchronizing signal take the biggest values allowed in the model. This generalizes the case of special relativity theory where inertial frames move uniformly with the velocity v connected to k of (6) and (7) by

$$\frac{v}{c} = \frac{k^2 - 1}{k^2 + 1} \qquad (13)$$

The transformation rules in the form (8) and (9) possess the same structure as the usual Lorentz transformations, and they become these transformations for $f(X)$ as an identity function. We can therefore ask what are the forms of the time dilation and the length contraction in the model proposed, and compare them with the standard ones following from the ordinary Lorentz transformations. If (11) holds then the formula (8) implies the same relation between the time intervals measured by moving and resting clocks as the special relativity does. It reads

$$\Delta T' = \frac{k^2 + 1}{2k} \Delta T \qquad (14)$$

The analogy of the Lorentz-Fitzgerald contraction of moving rods may be derived from (8) and (9) according to the standard construction. The difference of time intervals passed between emission and detection of signals seen in a moving reference frame as reflected simultaneously at the ends of the rod is

$$f^{-1}(X_2') - f^{-1}(X_1') = \frac{2k}{k^2 + 1}\left[f^{-1}(X_2) - f^{-1}(X_1)\right] \qquad (15)$$

with primed quantities denoting results of measurements obtained by a moving observer. It is clear that

$$f^{-1}(X'_2) - f^{-1}(X'_1) \le f^{-1}(X_2) - f^{-1}(X_1) \qquad (16)$$

because of the factor $2k/(k^2 + 1)$ which implies that moving observers see rods being shorter. However (15) does not determine $L' = X'_2 - X'_1$ uniquely if distances are not proportional to time intervals. This points out the case of special relativity theory, where in order to calculate length defined as above, we do not need to know an instant of time when the observation has been done because a light pulse has a velocity c wherever it reaches a rod. For linear f's there is always

$$L' = \frac{2k}{k^2 + 1} L \tag{17}$$

which we understand as a consequence of a particularly chosen definition of length being additional assumption in the radiolocation method.

Conclusions

Einstein's synchronization which leads to the definition of time shown by a distant clock in a form (1) has been discussed from many years and we understand now that it is a consequence of *a priori* assumed invariance of one-way velocity of light. This is an attractive hypothesis because it enables us to synchronize all clocks with respect to only one chosen clock, but in fact it never has been checked as precisely as other fundamental assumptions of physics. The main problem is that in order to make such a measurement we must not perform an experiment which uses distant clocks being synchronized in advance according to Einstein's method and we should know the distance between them given independently from (2) which assumes the properties of the light motion. Such an experiment is not easy to imagine, and is sometimes even considered to be impossible, but there exist physical phenomena which known explanations forbid from replacing one-way velocity of light by an accurately measured two-way velocity and suggest that the former is frame dependent.

Our consideration shows that not only the clock synchronization formula (1) may be treated as a matter of convention. Also coordinatization of space dimensions may be different from (2) provided the definitions used are independent from the choice of the reference frame. It is possible to do this in agreement with the transformation rules (6) and (7) which give mutual relations between inertial reference frames and reflect the physically observed Doppler effect. The proposed approach enlarges the class of reference frames considered as inertial ones and connects their definition to the properties of the signal used for synchronization. The parameters of its motion give limits for allowed velocities, accelerations and other properties of any motion generated by higher order derivatives, which agrees with physical intuition and our experience from special relativity. Elimination of the definition of inertial reference frames as those which move only uniformly may be also useful in considerations where the presence of the gravitational field must be taken into account, or it is necessary to pass to rotating reference frames.

Acknowledgments

The author is very grateful to the Organizing Committee of the conference "Relativistic Physics and Some of Its Applications" for the support which made possible his participation in the conference. He also expresses his gratitude to Prof. E. Kapuscik for discussions and fruitful remarks as well as for information on results presented in reference [6].

References

[1]   E. Kapuscik, *Acta Phys. Pol.* **B17**, 569 (1986).

[2]   F. Selleri, Theories equivalent to special relativity, in *Frontiers of Fundamental Physics*, pp. 181-192, M. Barone & F. Selleri eds. (Plenum Press, London/New York, 1994); also *Found. Phys.* **26**, 641 (1996).

[3]   A. Einstein, *Ann. Phys.* (Germany) **17**, 891 (1905).

[4]   H. Reichenbach, *The Philosophy of Space & Time*, (Dover Publ., New York, 1958).

[5]   A. Horzela, E. Kapuscik, J. Kempczynski, *Phys. Essays* **6**, 314 (1992).

[6]   V. S. Barashenkov, E. Kapuscik, M. V. Liablin, Nature of Relativistic Effects and Delayed Clock Synchronization, this volume.

[7]   H. Bondi, *Brandeis Lectures*, 1964, vol. 1, (Englewood Cliffs, NJ: Prentice Hall), p. 386.

Synchronisation of Clock-Stations and the Sagnac Effect

A. G. Kelly
HDS Energy Ltd.
Celbridge, Co. Kildare
Ireland 1

It is shown that the Sagnac correction, as applied to time comparisons upon the Earth, does not derive from the normal Relativistic corrections. It is proposed that the reason given for the application of the Sagnac correction, and the circumstances appropriate to its application, require amendment.

Key words:: Clock synchronisation; Sagnac effect; Relativistic corrections.

Standards for the synchronisation of clock-stations upon the Earth are to be found in the 1990 publication of the CCIR (International Radio Consultative Committee: International Telecommunication Union) [1]. Similar rules are in the 1980 publication of the CCDS (*Comité Consultatif Pour la Définition de la Seconde: Bureau International des Poids et Mesures*) [2]. Two methods are used to synchronise clocks at different clock stations. The first method is physically to transport a clock from one site to the other, and thereby to compare the times recorded at the two clock stations. The second method is to send an electromagnetic signal, from one site to the other.

Three corrections to be applied, as listed in the above publications, are as follows:-

(a)   to take account of the Special Relativistic velocity effect, caused by carrying a portable clock at speed aboard an aeroplane, from one site to the other.

(b)   under General Relativity, to allow for height above sea level.

(c)   a correction described as being for the *rotation of the earth.*

Correction (a) is quantified as $v^2/2c^2$. This is the slowing of time as calculated under the Special Theory of Relativity. A clock transported from one site to another will have such a correction applied, because of the ground speed v of the aeroplane; c is the velocity of light. Correction (b) is quantified as $g(\phi)h/c^2$ where g is the total acceleration at sea level (gravitational cum centrifugal) at a latitude of ϕ, and h is the height over sea level. Correction (c) is

Open Questions in Relativistic Physics
Edited by Franco Selleri (Apeiron, Montreal, 1998)

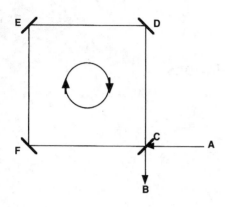

Figure 1 - Sagnac Test

quantified as $2A_E\,\omega/c^2$, where A_E is the equatorial projection of the area enclosed by the path of travel of the clock being transported from one site to another (or of the electromagnetic signal) and the lines connecting the two clock-sites to the centre of the Earth; ω is the angular velocity of the Earth. As the area AE is swept, it is taken as positive when the projection of the path of the clock (or signal), on to the equatorial plane, is Eastward.

Both reports include all three terms under the umbrella description of being "*of the first order of general relativity.*" The first two corrections are clearly the result of the Special Theory and the General Theory of Relativity respectively. But, what is the third? This paper examines the precise meaning and derivation of the third correction.

To understand the meaning of the third term, we must study the Sagnac effect. Sagnac (1914) showed that light took different times to traverse a path, in opposite directions, upon a spinning disc [3]. Figure 1 shows a schematic representation of the test that was done by Sagnac. A source at A sends light to a half-silvered mirror at C. Some of the light goes from C to D, E, F and C and is reflected to a photographic plate at B. Some of the light goes the other way around. The whole apparatus (including A and B), can turn with an angular velocity of ω.When the apparatus is set spinning, a fringe shift occurs at an interferometer, indicating a difference (dt) in the time taken by the light to traverse the path in opposite directions. For this difference in time Sagnac derived the formula

$$\delta t = \frac{4A\omega}{c^2} \qquad (1)$$

where A is the area enclosed by the path of the light signals, and ω is the angular velocity of spin in R/s.

Sagnac also showed that the centre of rotation can be away from the geometric centre of the apparatus, without affecting the results, and that the shape of the circuit was immaterial. He also proved that the tilting of the mirrors, as they spin, caused an insignificant alteration in the overall effect.

In order to derive the Sagnac equation, consider the theoretical circular model shown in Figure 2. Light is emitted at S; a portion of the signal goes clockwise (denoted by the inner line), and some goes anti clockwise, around a circular disc of radius r. The light source at S and the photographic recorder, also situated at S, rotate with the disc. The disc is rotating with an angular velocity ω in a clockwise direction. The anti clockwise beam is going against the rotation of the equipment, and will return to Point S when it has moved to S'. The second beam, travelling clockwise, will return when S has moved to S." As viewed by an observer on the spinning platform, the light signals return to the same point, but at different times.

Taking t_o as the time observed when the disc is stationary, *i.e.* the path length divided by the speed of light

$$t_o = \frac{2\pi r}{c} \tag{2}$$

Let $\delta s'$ be the distance SS' and $\delta s''$ be the distance SS." Let t' be the time for the light to go from S to S' in the anti clockwise direction.

$$t' = \frac{2\pi r - \delta s'}{c} \tag{3}$$

But, t' is also the time taken for the disc to move the distance $\delta s'$ in the clockwise direction. Therefore $t' = \delta s'/v$, and $\delta s' = t'v$; $\delta s' = (2\pi r - \delta s')v/c$; $\delta s'/v = 2\pi r/(c+v)$, and

$$t' = \frac{2\pi r}{(c+v)} \tag{4}$$

Similar calculations give the time (t'') for the light to go from S to S" in a clockwise direction,

$$t'' = \frac{2\pi r}{(c-v)} \tag{5}$$

Subtracting equation (4) from (5), the difference (δt) between the times for the light to go clockwise (t'') and anti clockwise (t') is

$$\delta t = 2\pi r \left[\frac{1}{(c-v)} - \frac{1}{(c+v)} \right] = \frac{4\pi r v}{(c^2 - v^2)} \tag{6}$$

This is the same as equation (1), because v^2 is negligible.

From the point of view of the observer in the fixed laboratory the disc moves a distance $\delta s'$ while the light completes a distance of $2\pi r - \delta s'$ around in the other direction from S to S'. Equation (3) describes the time interval, as it would be discerned by the observer in the laboratory. From the point of view of the moving observer, upon the spinning disc, the light has, relative to that observer, completed one revolution of the disc ($2\pi r$) at velocities of $c \pm v$ in the two opposing directions. Equations (4) and (5) describe this.

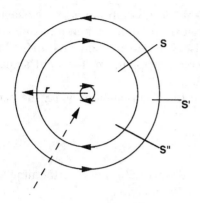

Figure 2 - Circular Sagnac Test: Whole apparatus turning at ω clockwise

In the above calculation, the light is assumed to travel at a constant velocity of c in relation to the fixed laboratory. But, the fringe shift measured solely aboard the spinning disc, and which is a record of the time difference for the light beams to complete a circuit in opposing directions, corresponds exactly to the time difference in equation (1). How can this be? The only possible explanation is that the time in the fixed laboratory, and that upon the spinning disc are precisely the same. This fact is at the core of the postulates being put forward in this paper.

The Sagnac effect shows that the velocity of the light is not affected by the movement of the source of that light (Point S); this accords with Special Relativity theory. It also shows that the light travels at the velocity c solely relative to the laboratory. Assuming that the light travels at the velocity c, relative to the laboratory, gives the correct result. The light does not adapt to the movement of the disc.

To get a fringe shift of one fringe, the velocity of Point S in Figure 2, relative to the laboratory, has to be about 13 m/s per meter of radius. This is a very low velocity. Fringe shift is got from time difference by multiplying by c/λ. Where, for example, $\lambda = 5500 \times 10^{-10}$ m, this gives $v = 13$ m/s, per meter of radius, from $1 = (4A\omega)/(c\lambda) = (4\pi rv)/(3 \times 108 \times 5{,}500 \times 10^{-10})$. In equation (4), as v approaches c, t' becomes $t_o/2$, and the speed relative to the observer is now 2c. In equation (5), as the speed v approaches c, t'' becomes infinite, because the light and the Point S are travelling in the same direction, and the time for the light signal to gain one complete circuit on the Point S is infinite; the speed of the light, relative to the observer, becomes zero.

Dufour & Prunier (1942) repeated the Sagnac test, and got the same result [4]. They then did a variation of that test. A practical example of a case where the signal is not solely in the plane of the disc is their test, in which the path of the light was partly on the spinning disc, and partly in the fixed laboratory. The light signal was introduced (Figure 3) from C out

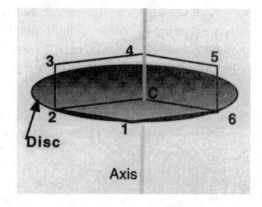

Figure 3 - Dufour & Prunier Test

to Point 1, and sent from there in opposite directions.

As shown schematically, the light went firstly on a path on the spinning disc (Point 1 to Point 2), then went vertically up to a mirror fixed to the laboratory overhead the disc (Point 3). It then traversed linear paths 3 to 4 to 5 in the fixed laboratory, and came vertically back down to the disc at Point 6, whereupon it finished the trajectory on the disc back to the starting point at Point 1. The reverse beam went the other way. The plane of the path, of the portion that was fixed in the laboratory, was parallel to the plane of the disc. Lines 3-4 and 4-5 are directly overhead 2-C and C-6. The two short connections, 2-3 and 5-6 (shown exaggerated here for clarity) were 10 cm each. The mirrors at 2 & 6 rotated with the disc. The fringe shifts were the same as in their repeat of a test with the light path solely upon the spinning disc (on the circuit 1-2-C-6-1). This test by Dufour & Prunier confirms that the light does not adapt to the movement of the disc, and that it is travelling relative to the fixed laboratory.

A young German student Harress (1911) had done a test on the refraction of light [5]. This test was later shown by von Laue (1920) to have produced the Sagnac effect, but Harress was not aware of this [6]. Harress had both the photographic equipment and the light source fixed in the laboratory, whereas Sagnac had both on the spinning disc. This shows that the photographic record of the fringe shift and/or the origin of the light may be made on or off the disc, without affecting the result; this is because it is the behaviour of the light relative to the spinning disc that is being measured. Dufour & Prunier also did tests with the light source fixed in the laboratory and with the photographic plate fixed in the laboratory; the results were the same as in a traditional Sagnac test. The fringe shift occurs, whether there is any observer (camera) present on the disc, or in the fixed laboratory. There is a slight Doppler effect in the case where the photographic equipment is in the fixed laboratory, because the disc is moving past the viewing lens. Post (1967) discusses the magnitude of the distortion introduced, and correctly dismisses the effect as too small to have any observable effect, being "v/c times smaller than the effect one wants to observe." [7].

Michelson & Gale (1925) showed that electromagnetic signals sent around the Earth did not travel at the same speed in the East-West direction [8]. They constructed a large rectangular piping system fixed to the Earth, and sent light signals in opposite directions around the circuit. The signals did not arrive back at the same time, as evidenced by the resulting fringe shift. That test was a Sagnac test on a disc of radius equal to that of the Earth at the Latitude concerned, and rotating at the angular velocity of the Earth. The results were within 3% of the forecast and were also in the correct direction (signal retarded in the direction of the spin of the Earth). Tests by Bilger et al. (1995) using a ring-laser, confirmed the Sagnac effect to better than one part in 10^{20}. This was a Michelson & Gale type test with the ring laser fixed to the Earth; the retardation of the signal was also in the direction of the spin of the

Earth (as this was done in the Southern hemisphere, the retardation was in the opposite sense to the Michelson & Gale test)[9].

Saburi *et al.* (1976) transported a clock from Washington (USA) to Tokyo (Japan), and compared the difference in the time displayed by the two clocks on the arrival of the transported clock, with the time relayed from one station to the other, *via* an electromagnetic signal.[10] The two sites were almost at the same latitude. They calculated from the Sagnac effect that there should be a difference of +0.333 ms (Japan ahead of Washington, DC, because of the direction of rotation of the Earth). The Sagnac correction, on its own, applied solely to the electromagnetic signal (and not to the time displayed by the clock that was physically transported from one site to the other), bridged the gap to a very close agreement with the test results (to –0.02 μs). The Relativistic effects applied solely to the portable clock, which was physically transported from one site to the other, amounted to +0.08 μs. The uncertainty of the reading being recorded by the portable clock was ±0.2 μs. This test could nowadays be repeated to greater accuracy.

Special Relativity has no role in trying to explain the Sagnac effect. Post (1967) states that the Sagnac effect and the Special Relativity effect are of very different orders of magnitude. He says that the alteration to be applied to the Sagnac effect under Special Relativity is a v^2/c^2 effect which is *"indistinguishable with presently available equipment"* and *"is still one order smaller than the Doppler correction, which occurs when observing fringe shifts."* Post derives the Sagnac formula as given above in equation (1) and then applies the Special Relativity γ factor to that formula; in this he distinguishes clearly between the two. Post says that *"for all practical purposes we may accept as adequate for the time interval in the stationary as well as in the rotating frame, the formula"* as in equation (1). This confirms that the difference in the time recorded in a Sagnac test is the same in the laboratory and upon the spinning disc. Post also says that *"the time interval between the consecutive positions of the beam splitter is observed in the stationary frame and is therefore dilated by a factor γ ."* Here again Post distinguishes between the Sagnac effect and the Relativistic time dilation.

The basis of timekeeping by the CCIR is time at the non-rotating centre of the Earth. It defines that the *"TAI is a coordinate time scale defined at a geocentric datum line."* The unit of time is defined as *"one SI second as obtained on the geoid in rotation."* The time scale and the unit of time are not measured at the same place; the unit of time is based upon the spinning Earth, which has motion in relation to the geocentre where the time scale is measured. The CCIR report recommends that *"for terrestrial use a topocentric frame be chosen."* It continues *"when a clock B is synchronised with a clock A (both clocks being stationary on the Earth) by a radio signal travelling from A to B, these two clocks differ in coordinate time by"* the Sagnac effect. These statements make it clear that the time upon the rotating Earth is viewed as differing from that at the geocentre. This assumption is in contradiction of the analysis in this paper, and of the conclusions of Post [7].

The CCIR report states that *"the time of a clock carried eastward around the earth at infinitely low speed at h = 0 at the equator will differ from a clock remaining at rest by −207.4 ns."* That amount is the Sagnac one-way effect. The significance of the $h = 0$ is that there would be no effect under the General Theory of Relativity. The infinitely low speed eliminates any effect from the Theory of Special Relativity. The CCIR report here assumes that when a clock is physically transported around the globe, a Sagnac-type correction has to be applied. Because the area is taken as *"positive if the path is traversed in a clockwise sense as viewed from the South Pole,"* a clock transported around the Earth in a Westward direction would gain time by +207.4 ns, relative to the stationary clock. Consider two clocks that are sent, in opposite directions, around the globe at the equator at the same time; when they have completed one revolution each, there would be a supposed time difference of 414.8 ns between them, and they would each differ from a clock that remained at the starting place by 207.4 ns. They have had no effect from Special Relativity (velocity infinitely slow) or from General Relativity (at sea level). We then would have the strange situation where we have three clocks at the same spot on the Earth recording different times; we could repeat the circumnavigation as often as we wish and get clocks, at the same spot, which have had zero corrections under normal Relativity theory, recording times which are different from each other by larger and larger amounts. All times here are coordinate times as earlier defined.

Both the CCIR and CCDS reports make it clear that considering time upon the Earth, from the point of view of "a geocentric non-rotating local inertial frame," requires no Sagnac correction. But, when considering time upon the rotating Earth, they apply a Sagnac correction.. Langevin (1937) proposed that, to explain the Sagnac effect, one had to assume that either (a) the velocity of the signal was $c \pm v$ in the two directions or, (b) the time aboard the spinning disc was altered by $2A\omega/c^2$ [11]. The CCIR and CCDS reports assume that (b) is true. As we saw above, it is (a) that is the correct explanation.

Special Relativistic time dilation does nor contribute very much towards the Sagnac effect. Taking an example, where the surface velocity of the Earth at a particular latitude is $v = 300$ m/s and a portable clock is transported at, say, $x = 10$ m/s (the CCIR defines the transportation as"slowly"). In this case the difference between the $v^2/2c^2$ and $(v+x)^2/2c^2$, which is $(2vx+x^2)/2c^2$, gives a difference of 4×10^{-14} s/s. An electromagnetic signal circumnavigates the Earth in about 0.1 s. The Sagnac one-way difference $2A\omega/c^2$ for a light signal to circumnavigate the Earth is about 2×10^{-7} s/s, as calculated in the CCIR Report. The ratio of the two is thus 10^7. Thus, the two effects are not at all of the same magnitude. This agrees with the analysis by Post [7]. Another basic difference between the Relativistic and Sagnac effects, as calculated for movements measured upon the spinning Earth, is that the former is non-directional, whereas the latter is ± depending upon the direction of sending the signal West or East respectively, and zero in a North-South direction.

In the CCIR analysis, the starting point is time at the *"local non-rotating geocentric reference frame."* This is done *"to account for relativistic effects in a self-consistent manner."* If we assume that the speed of light upon the rotating Earth must be the constant value c, then perforce we must vary the time upon the Earth, by the Sagnac formula, as compared with time measured from the geocentric reference frame. Special Relativity theory is designed specifically to alter the time upon the moving object in direct accord with the requirement that the speed of light must be the constant value c.

The application of the γ factor correction, under Special Relativity, to time on the spinning Earth, as compared to time at the centre of the Earth, ensures that

a) the speed of light is c as calculated in all directions upon the spinning Earth
b) time upon the Earth is consequently calculated to vary by precisely the amount necessary to agree with the constant value for the speed of light.

If this were not so, the velocity of the electromagnetic signals upon the Earth would remain unchanged as $c \pm v$ in the opposing directions. This method arrives at a solution that conforms with Relativity theory. Is this method justifiable?

It is convenient to start with time at a geocentric datum line. This conforms with the fact that electromagnetic signals do not adapt to the spin of the Earth; this datum corresponds to the 'laboratory' in a Sagnac bench test. The speed of the signals can confidently be taken as c as measured in that frame of reference. The orbital movement of the Earth around the Sun. and its other movements in the Universe, can be ignored, and assumed to have no effect upon the results being calculated.

Allan *et al.*, compare the Sagnac correction as applied to (a) slowly moving portable clocks upon the Earth and (b) electromagnetic signals, used for clock synchronisation [12]. They state that *"the Sagnac effect has the same form and magnitude whether slowly moving portable clocks or electromagnetic signals are used to complete the circuit."* They say that the Sagnac correction applies in both cases, and that it has the same magnitude. In case (a) they define the Sagnac effect as *"being due to a difference between the second-order Doppler shift (time dilation) of the portable clock and that of the master clock whose motion is due to the Earth's motion"* as *"viewed from a local nonrotating geocentric frame."* Petit & Wolf also state that the correction $2A\omega/c^2$ is applied equally *"if the two clocks are compared by using portable clocks or electromagnetic signals in the rotating frame of the Earth."* If we take the time measured at the geocentric datum line as t_o, and the time upon the spinning Earth as t', Special Relativity Theory requires that $t_o = t' \gamma$. Applying a correction of $v^2/2c^2$ to the time taken for two clocks, which move at speeds of v relative to the ground, to circumnavigate the Earth in opposing directions, as viewed from the geocentre, gives the following result.

The moving clocks have the speeds of $\omega r + v$ and $\omega r - v$ in the opposing directions, relative to the geocentric time frame; r is the radius of the Earth, and ω its angular velocity. The time dilations of the two clocks are $\int \frac{1}{2}\left[(\omega r + v)/c\right]^2 dt$, and $\int \frac{1}{2}\left[(\omega r - v)/c\right]^2 dt$ respectively. The difference between these two time dilations is therefore $\int \left[(2\omega rv)/c^2\right] dt$.

When the two clocks have gone right around the equator (a distance of $2\pi r$) the $\int cdt = 2\pi r$, and the difference between the time dilations is $(4\pi r2\omega)/(c^2)$, which is the same as equation (1) (Burt, 1973) [13]. The result is independent of v, so the speed of transportation of the clocks will not affect this result. A similar analysis using electromagnetic signals to circumnavigate the globe Eastward and Westward (that is substituting c for v in the above equations) also gives the same result. In this way this analysis gives the Sagnac formula as a supposed correction for the difference in the time taken by two electromagnetic signals sent in opposing directions around the globe, or for the time correction to be applied to clocks that are physically transported around the globe in an East-West direction. This is the correction published in the CCIR and CCDS reports. It also shows that the application of the γ factor to time as measured upon a moving object agrees with the speed of light being measured upon that object as c in all directions. All of this scheme is consistent.

There is one problem. The Sagnac tests, done with ever increased accuracy down the years, show a difference in the time taken by electromagnetic signals to circumnavigate any spinning disc (including a cross-section of the Earth) and consequently a difference in the speed of the signal in the opposing directions. No difference in time for activities upon the spinning disc is required, when viewed from the stationary laboratory. This difference in the speed of the signal contradicts a basic assumption of the scheme of synchronisation that is used.

It is assumed by the CCIR that the time upon the spinning Earth is altered by the γ factor of Special Relativity in all calculations carried out on time durations upon the Earth, from the viewpoint of the geocentric non-rotating system. If no difference in time was measured in a Sagnac test, then the speed of the signal would have been measured as c in the opposing directions, upon the spinning disc. It can be argued that the rotating disc is not an Inertial Frame, and that therefore the matter is not relevant. As larger and larger discs are considered, we approach the situation where the movement is tantamount to that in a straight line at constant velocity. In this case the matter applies to an Inertial frame. There does not seem to be any plausible solution which shows the fringe shift measured aboard the spinning disc to be caused by other than a difference in the speed of light relative to that disc.

The application of the CCIR correction to time upon the spinning Earth gives a correct answer, whenever electromagnetic signals are used to compare the time being recorded at two clock stations upon the Earth. However, where the physical transportation of a clock around the globe is concerned, it introduces an error.. The CCIR report works out an example where the three specified corrections are applied to the frequency of a clock that is physically transported from one site to another. However, as discussed earlier, Saburi *et al.* showed that the physical transportation of a clock does not require the application of any Sagnac correction to the time being recorded upon that travelling clock [10]. They also confirmed that it is the electromagnetic signal speed Eastward and Westward that varies, and that requires a Sagnac correction to its speed of transmission.

Allan *et al.* (1985) did a Sagnac-type test between standard time-keeping stations in USA, Germany and Japan [12]. These tests confirm the Sagnac effect, as applied to electromagnetic signals, sent right around the Earth in opposing directions, to an accuracy of 1% over a period of 3 months. There were no further corrections made to the results (got by sending electromagnetic signals between the clock-stations) on the basis of Special Relativity or General Relativity; in this case, where electromagnetic signals are used to synchronise the clock stations, no measurable effect under Special Relativity or General Relativity are to be expected. Saburi *et al.* state that *"in a comparison experiment via a satellite, it is considered that the effect of gravitational potential on the light path is small and cancelled out by the two-way method, and that other relativistic effects are negligibly small."*

In the CCIR report, the corrections to be applied are listed as three *viz.* *"the corrections for difference in gravitational potential and velocity and for the rotation of the Earth."* In describing these corrections the report names them as *"corrections of the first order of general relativity."* We now see that the third one is the Sagnac effect ($2A\omega/c^2$). By naming the Sagnac correction as a separate item from the other two factors, the CCIR report tacitly accepts that it is not a Special Relativity or a General Relativity effect. This paper shows that no such Sagnac correction should be applied to the case where a clock is physically transported from one site to another. However, in all cases of synchronising clocks by electromagnetic signal comparison, the Sagnac correction is properly quantified in the CCIR report, and thus the timekeeping authorities are applying it correctly, even if they assume that it is derived from the Theory of General Relativity. The Sagnac correction is nowadays automatically applied to all electromagnetic signals used in the synchronisation of clock stations. The CCIR report gives an incorrect value for the angular velocity of the Earth (7.992 R/s instead of 7.292 R/s); this error was not carried forward into the calculations given in examples in the report.

Winkler (1991), in a paper on the subject of the synchronisation of clocks around the world, ascribed the Sagnac effect to the General Theory of

Relativity [14]. He explained the effect by saying that *"accelerations have an effect on timekeeping and on the propagation of light."* He also stated that *"on a rotating system, the velocity of light must be added to (or subtracted from) the speed due to rotation, an effect that produces a time difference for two rays that travel in opposite directions around a closed path."* Here he has accepted that the velocity of the signal is different in the opposing directions, and that the signals take different times to complete the circuit, relative to an observer upon the rotating Earth.

Other publications also purport to show that the Sagnac effect is part of the General Relativity Theory. An example is the paper by Petit & Wolf (1994), which begins by assuming that the light travels relative to the stationary frame (in their case the *"geocentric 'non-rotating frame'"*) [15]. They assume that the light velocity relative to the spinning object is not c. They take it as *"c + s where s represents the time taken for the signal to travel the extra path due to the motion of b in the non-rotating frame during transmission"*: *"b"* is the clock moving on a rotating disc. This is the same as the analysis of the Sagnac effect, given earlier in this paper, where the extra distance travelled by the Point S in Figure 2, while the signal is travelling around the circuit, yields a speed of the light of $c - v$ in one direction. But, they then assume that the time aboard the spinning Earth alters by the equivalent of the Sagnac effect; this is, as seen above, not sustainable.

Two clocks upon the Earth at the same Latitude have no relative motion in respect to each other, as considered in a geocentric Earth-fixed system. It is only when we attempt to compare the time being kept by the two clocks that we have to employ either an electromagnetic signal or a physical transportation of a comparison clock. The time keeping of those two clocks does not alter because of the measuring process. The Sagnac correction has to be applied to the time taken by the electromagnetic signal to get from one clock site to the other. No corrections apply to the time being kept by the clocks in relation to each other. By shifting the time base to the geocentre, the CCIR introduce a supposed Sagnac effect alteration to the time difference measured between the two clocks when transporting a portable clock or sending an electromagnetic signal between the two sites.

There are various reasons that can be advanced to answer the apparent contradictions between Relativistic theory and the Sagnac effect. One could say that it is correct to state that the Sagnac effect is not relativistic; but it comes out naturally if one writes the equations of time transfer, from the geocentric frame to the spinning Earth, in the context of general relativity, with some very small additional terms that are genuinely relativistic. It can be claimed that Newtonian Mechanics are not relativistic, but that General Relativity includes all terms of Newtonian theories of motion plus additional corrections. So we could claim that it is not wrong to say that the Sagnac effect is also relativistic in the sense that it also appears in the solution in a general relativity theory. Such an argument would agree that the Sagnac effect is a first order effect that cannot have any explanation purely by Relativistic theory.

It could be debated that we have to (a) adopt a relativistic model, because the classical treatment leads to contradictions with experiment, and (b) have a convention for the meaning of clock comparison. As a model, we use Einstein's General Relativity because this theory is the simplest which, up to now, agrees with all observed facts. The convention for clock comparison is based on the convention of coordinate simultaneity; the readings of the clocks take place at the same value of some specified coordinate time (geocentric in metrology on the Earth). The question, it could be claimed, is not to distinguish in the theory of clock comparison some classical terms, some terms due to Special Relativity, and some gravitational terms. General Relativity, it can be said, is a self-contained theory and provides all the terms we need, as a consequence of its basic postulates. The separation of the various terms is a consequence of the choice of coordinates we have made and of the low level of approximation which is accepted.

General Relativity theory is required to make corrections to the time keeping of the clocks. It includes the corrections for height over sea level, and also the corrections under Special Relativity (velocity effects). The setting of the atomic clocks that are to be placed aboard a satellite are made, in advance of launching the satellite, to allow for both of those corrections. These corrections anticipate the increased reading that will emerge as a result of height over sea level, and also the decreased reading that will emerge because of the higher velocity of the satellite as compared with the velocity of the surface of the Earth. The clocks are set before launch, and will then be correct in keeping time the same as upon the surface of the Earth, when they are in orbit. These alterations are appreciable, and are a precise confirmation of these two corrections. Without making these corrections, the clock on the satellite would not keep an unaltered time, as compared with a clock upon the surface of the Earth.

However, there is another correction to be made and that is the Sagnac correction, whenever one has to compare the time upon such a satellite with the time being recorded by a clock on another satellite or upon the Earth. It is this quite separate correction that is the dichotomic problem being addressed here.

Some publications try to avoid the problem of the Sagnac effect by declaring that the Theory of Special Relativity is not applicable to a rotating Frame of Reference. But, some precise explanation of the effect is required. It is not sufficient to say that the Sagnac effect is not explained by Special Relativity theory, and to leave the matter at that. Einstein (1905) seems to have accepted, in his first paper on Relativity Theory, that movement on a circular path had the same result as movement in a straight line, when considering the question of measurement of distance or time.[16]. Having derived his formula for straight line movements, he said *"it is at once apparent that this result still holds good if the clock moves from A to B in any polygonal line"* and "*if we assume that the result proved for a polygonal line is also valid for a continuously curved line, we arrive at this result: If one of two synchronous clocks at A is moved in a closed curve with constant velocity until it returns to A, the journey lasting t seconds, then by the clock which has remained at rest the travelled clock on its arrival at A will be $\frac{1}{2}tv^2/c^2$ second slow.*" An observer riding upon the moving clock B will not be measuring time in an Inertial Frame with respect to clock A, but in a Rotating Frame of Reference. The argument that Special Relativity Theory is not applicable to movement in a circuit, such as that of circumnavigation of the Earth, is thus open to different interpretations. Even though the effect Einstein described is much smaller than the Sagnac effect (as shown above), it is the application, from a straight path to a curved path, that is of interest here.

The Sagnac corrections applied by the CCIR and the CCDS is not a Relativistic correction. It is not a continuing correction, such as are the Relativistic corrections. It is necessary when comparing the time being recorded at different clock stations, because the velocity of electromagnetic signals, travelling in an East-West direction, as measured upon the Earth, is not constant, but $c \pm v$, where v is the spin velocity of the surface of the Earth at the particular Latitude. The outstanding problem is to devise a theory that will fit both the Relativistic corrections, that are vindicated in everyday use, and the Sagnac correction.

The Sagnac effect is proof that light travels at a constant velocity, in relation to the fixed laboratory, and does not adapt to the movement of a spinning disc. This requires that time aboard a spinning disc is the same as time in the fixed laboratory. The Sagnac correction is being correctly applied to the sending of electromagnetic signals between standard clock stations on the Earth; the reason given (relativistic correction) is incorrect. It is proposed that the Sagnac correction should not be applied to the physical transportation of clocks between sites, as is presently done in the CCIR and CCDS rules; it is solely the Relativistic corrections, due to velocity of travel and height over sea level, that should be applied in such a case.

The CCIR report concludes by saying that "*additional definitions and conventions are under consideration.*" These are awaited with interest. An amendment to relativistic theory to accommodate the true application of the Sagnac correction would give a more precise solution to the problem of clock synchronisation.

References

[1] *CCIR* Internat. Telecom. Union Annex to Vol. 7 1990, No 439-5, 150-4

[2] *CCDS* Bureau Internat. Poids et Mesures, 1980, 9th Sess., 14-17

[3] Sagnac M G *J. de Phys.* 1914, 4, 177-95

[4] Dufour A & Prunier F *J. de Phys.* 1942, 3, No 9, 153-61

[5] Harress F Thesis (Unpublished) Jena 1911

[6] von Laue M *Ann. der Phys.* 1920, 62, 448-63

[7] Post E J *Rev. Mod. Phys.* 1967, 39, No 2, 475-93

[8]   Michelson A A & Gale H G *Astroph. J.*,1925, 61, 137-45

[9]   Bilger H R *et al. IEEE Trans.* 1995, 44 No 2, 468-70

[10] Saburi Y *et al. IEEE Trans.* 1976, IM25, 473-7

[11] Langevin P *Compt. Rend.* 1937, 205, 304-6

[12] Allan D W *et al. Science* 1985, 228, 69-70

[13] Burt E. G. C. *Nature Phys. Sci.* 1973, 242, 94-5

[14] Winkler G M R *et al. Meterologia*, 1970, 6, No 4, 126-33

[15] Petit G & Wolf P *Astron Astrophys.* 1994, 286, 971-4

[16] Einstein A, Lorentz H *et al. The Principle of Relativity* (Metheun, 1923)

Is Simultaneity Relative or Absolute?

Joseph Lévy
4, Square Anatole France
91250 Saint-Germain les Corbeil
France

Due to their conviction that the laws of nature must be identical in any inertial frame, the physicists of the beginning of the twentieth century were led to extend the relativity of Galileo to the electromagnetism of Maxwell, but this seemed to imply the abandonment of universal time and absolute simultaneity.

In a previous paper* we have criticised, *from a logical viewpoint*, the criteria intended to demonstrate the relativity of simultaneity and we have proposed replacing them by other criteria. According to these, the relativity of simultaneity was called into question.

We propose here a rigorous *experimental* method intended to verify the simultaneity of two events. By means of this device, we demonstrate that one can define an absolute simultaneity. The method also permits an exact synchronization of clocks. We then demonstrate that the relativity of time of Einstein's theory must be discarded. On the other hand, the relativity principle appears as an approximation only, valid for bodies moving at low speeds relative to one another.

On the contrary, the slowing of the clocks moving with respect to the ether can be maintained. But this does not mean that Lorentz's theory can be retained without change.

I. Position of the Problem

One of the concepts which has most drastically changed our vision of the world since the origin of philosophical thinking is the idea of "relativity of time." Until the beginning of this century, time was considered as absolute, flowing uniformly, and identical for all observers.

Certainly, if one supposes that different stars are inhabited, one could suppose that the people living there use units of time different from our own,

---

*   J. Lévy, Some important questions regarding Lorentz Poincaré's theory and Einstein's relativity II. *Proceedings of the P.I.R.T. Conference*, supplementary papers, Imperial College London, 6-9 September 1996.

Open Questions in Relativistic Physics
Edited by Franco Selleri (Apeiron, Montreal, 1998)

adjusted to the rhythm of their central sun. Nevertheless, time by itself would not be affected. An appropriate conversion would be enough to come to an agreement.

Until the end of the nineteenth century this conception was the object of a large consensus. Better still, the idea that it could have been called into question looked deprived of any meaning. It was not before the advent of Maxwell's electromagnetic theory that the problem began to be addressed. In effect, by using the Galilean transformations, (although these were universally accepted), one could realize that Maxwell's equations, which crown this theory, were not invariant under a change of inertial reference frame. In other words, the electromagnetic laws were different for an observer on earth than for another travelling inside a rocket, or for an inhabitant of another solar system.

The physicists could not content themselves with such a disparity. Knowing that their obsession was the discovery of the universal laws of nature, this result conflicted with their firm convictions. But, on the other hand, to call into question the Galilean relationships seemed to imply a heartbreaking revision of the concept of universal time.

The end of the nineteenth century and the beginning of the twentieth century were periods of deep reappraisal. The notions which appeared most firmly established seemed to collapse. The experiment of Michelson and Morley came at the right moment to increase this perplexity.

Different physicists, all around the world, set themselves the task of putting in order the sum of the newly appeared disparate notions. Voigt, Larmor, Lorentz, Poincaré, in turn, showed great concern about them. Lorentz was particularly shocked when he knew the null result of the Michelson experiment, which called into question the concept of ether. In order to save this, he formulated (at the same time as Fitzgerald) the hypothesis of a contraction of the lengths moving through the ether. Alas, the hypothesis could never be verified experimentally; the experiments of Rayleigh and Brace, those of Chase and Tomashek, those of Trouton and Rankine, and those of Wood, all proved negative. Lorentz was then compelled to formulate other hypotheses in order to explain such negative results: at first, the variation of mass with speed, but this was not sufficient. It was necessary to postulate the existence of a local time needed by the consistency of the theory.

The sum of these *ad hoc* hypotheses finally ended in the formulation of a set of equations, improved and modified by Poincaré, and named by him "Lorentz transformations." This was the first breach against universal time. Nevertheless, the latter was not really abolished since, for Lorentz, the idea of an absolute space (absolute inertial frame) was not called into question, and the time measured in this privileged frame was exclusively considered as the real time. The local time was described as fictitious, which implied that the measurement of the time is distorted. That is the reason why it is more appropriate to define the process as 'slowing of clocks' rather than relativity of time.

It is at that moment that Poincaré took part in the debate. Poincaré acknowledged the Lorentz transformations but, at the same time, showing great concern about the fact that the laws of nature ought to be universal, he believed that they should take the same form in any inertial frame. The final form he gave to them, seemed to satisfy this requirement: in effect, the Lorentz transformations assume a group structure. He then formulated his relativity principle. Although he was convinced that a preferred inertial system supporting the ether should exist, Poincaré was persuaded of the impossibility of distinguishing it from the other inertial frames.

What Poincaré did not realize is that, if one takes for granted the postulates of Lorentz (existence of an ether at rest in the absolute inertial frame, contraction of moving lengths absolute and not reciprocal), the inertial transformations take the form of those of Lorentz exclusively in the absolute frame. They take a different form in all other inertial frames[†] and, as a consequence, the whole of these inertial transformations do not really constitute a group. From a strictly mathematical viewpoint Poincaré was right, but the point of view of physics is different, since we must make allowance for the systematic errors carried out during the measurements inherent in Lorentz's theory.

The approach of Einstein proved very different. Einstein also showed great concern about the requirement of the universal character of the laws of nature. His relativity principle is based on hypotheses different from those of Lorentz and Poincaré. Since it assumes a total reciprocity of the observations (contraction observational and reciprocal of moving lengths), it is not affected by the same difficulties as that of Poincaré.

Einstein's transformations constitute a fully fledged group. In other words, Einstein succeeded in deriving a set of transformations that maintain invariant the laws of nature. His system was universally adopted because, at the time when the relativity theory was published, the universality of the laws of nature appeared to be an unquestionable requirement.

At first sight, Einstein's transformations look completely identical to those of Lorentz, but when we look at them more accurately, we note two essential differences: first, the constant C appears universal whereas, for Lorentz, the speed of light is constant exclusively in the privileged inertial frame. On the other hand, contrary to Lorentz's approach, there is no

---

[†] J. Lévy, Relativity and Cosmic Substratum. *Precirculated proceedings of the P.I.R.T. Conference* p. 231 Imperial College London 6-9 September 1996.
Some important questions regarding Lorentz-Poincaré's theory and Einstein's relativity I. *Proceedings of the PIRT* 1996 Late papers p. 158.
Relativité et substratum cosmique, a book of 230 p. Dist. Lavoisier, 14 rue de Provigny 94 236 Cachan Cedex France Tél.: 0147406700.

absolute time and fictitious times; the different local times assume the same status.

It is this aspect of Einstein's transformations that we propose talking about here.

The question asked is the following: what is the price to pay for maintaining the universality of the laws of nature, and should they be kept? This question, which could not be asked some decades before without raising a general outcry, is seriously envisaged today by numerous physicists.

In other words, is the relativity principle really unquestionable? For certain physicists[‡] it has no absolute meaning but only a conventional character, in connection with a method of synchronization of clocks questionable and relative.

The relativity principle seems to require the abandonment of the universal time in favour of local times having identical properties. We will try to see, in what follows, if these two concepts are mutually compatible, and if Einstein's local time is in agreement with logic.

In order to make us understand the relativity of time, Einstein takes the classical example of the train and the two flashes of lightning that we have studied in detail in a previous paper[§]. Let us recall it briefly here: two flashes of lightning break at the two ends of a railway platform at the very instant when the two ends of the train meet them. The two flashes reflect against mirrors and then run in opposite direction towards the middle of the platform. According to Einstein, the two flashes are considered simultaneous if they reach the middle of the platform at the same instant. The same definition is also valid for the train, but this one moves towards one of the flashes and moves away from the other and, consequently, the middle of the train will be reached at different instants by the two flashes. Einstein concludes that two events which are simultaneous for an observer, are not simultaneous for another moving with respect to the first. From which the relativity of simultaneity.

In our previous article[**], we demonstrated that the assertion was incorrect, because it implied a confusion between the instantaneous flashes breaking at the ends of the platform, and the light issuing from them. We have also defined differently the simultaneity after correction of the errors of judgement generated by the non instantaneous translation of light.

[‡]   F. Selleri, Inertial systems and the transformations of space and time, *Phys Essays* 8, 342, 1995, section 3. & Le principe de relativité et la nature du temps, *Fusion*, Paris, 66, 50, June 1997.

[§]   J. Lévy, Some important questions regarding Lorentz Poincaré's theory and Einstein's relativity II. *Proceedings of the P.I.R.T. Conference*, supplementary papers, Imperial College London, 6-9 September 1996.

[**] *Ibid.*

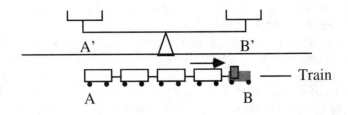

Figure 1

II. A New Experimental Test of Simultaneity

We now propose another experimental test of simultaneity deprived of the defects of that of Einstein. Of course, it is a matter of a thought experiment that may be difficult to carry out, but this does not deprive it of its character of logical foundation of thinking.

Let us consider a precision balance (of great sensitivity) longer than an ordinary one, and suppose that two rubber spheres fall down and bounce instantaneously on its pans. If the beam of the balance remains steady during the experiment, one will be authorized to conclude that the spheres have met the pans at the same instant.[tt]

Let us suppose now that a train passes alongside the beam of the balance, and that the two ends of the train meet the pans at the very instant when the spheres bounce. Let us designate by A the rear of the train, by B its front, and by A' and B' the places of the corresponding pans (figure 1).

After the spheres have bounced, a recording device situated at the middle of the train, will receive the light issuing from B' before that issuing from A'. From this, one could conclude that two events simultaneous for the earth, are not simultaneous for the train.

Nevertheless, an observer inside the train will realize that the beam of the balance has not moved. In consequence, he will deduce that the two spheres have fallen down at the same instant. There is no doubt that this criterion of simultaneity is better than that of Einstein, because it allows an instantaneous appreciation which does not need the mediation provided by the photons. It permits the absolute simultaneity independent of the motion of the observer, to be rediscovered.

Moreover the method should permit the clocks to be exactly synchronized. For that it could suffice placing two clocks in proximity to the two pans of the balance; if the spheres bounce without making the beam move, then it is the same time at the two ends of the balance: for example 8

[tt]   Such a schematic device could be replaced by a more sophisticated one implying beams of electrons and coincidence circuits. We entrust the engineers with the task of imagining an appropriate device.

o'clock. It will also be possible to synchronize the clocks of the train identically. It will suffice for that to make them indicate 8 o'clock when they pass in front of A' and B'.

Now, with the help of several identical closely related balances, it is theoretically possible to synchronize clocks distributed on the whole surface of the Globe.

Let us now consider a train equipped at its two ends with clocks synchronized with the clocks situated at A' and B'. Suppose now that the train continues on its way and, after a certain time, meets two other clocks synchronous with A' and B', aligned with them, and separated from one another with the same distance. The question asked now is to know if the clocks of the train will be synchronous with these two new clocks.

According to the relativity principle, this should be the case, because there is no reason to favor the terrestrial frame rather than that of the train.

This can be easily understood with the help of the following reasoning:

Let A and B be the clocks of the train, A' and B' the terrestrial clocks met first and A" and B" the terrestrial clocks met secondly.

Let us recall that A' and B' and A" and B" have been synchronized beforehand. On the other hand A and B are synchronized with A' and B' when they pass in front of A' and B'. So, at this instant, the six clocks are synchronous.

If one supposes that the earth frame is an inertial frame (which is only approximately true, but that we will consider absolutely true for the purpose in hand) the frame of the train and the earth frame are equivalent. Therefore, when A and B meet A" and B" they will indicate the same time.

This fact demonstrates that the relativity principle is not compatible with the relativity of time (contrary to what Einstein's postulates suggest), and that the relativity principle also excludes the slowing down of moving clocks.

Now, knowing that clock retardation is an experimental fact, the relativity principle cannot be maintained. This result calls into question all the derivations of the inertial transformations which assume the relativity principle, or are in agreement with it, including our own.[‡‡] (However, the arguments in the same paper, according to which the speed of light must be different from a limiting velocity, remain unchanged.)

Conversely, the existence of a privileged inertial frame could generate a dissymmetry responsible for a slowing of a pair of clocks relative to the other. So the notion of local time (slowing of clocks), proposed by Lorentz, looks possible. But it is a matter of a physical effect concerning the clocks rather than an effect regarding the time itself. This probably explains the experiments of Hafele and Keating, and those regarding the pions.[§§]

[‡‡] J. Levy, Invariance of light speed: Reality or fiction? *Phys. Essays* 6:241, 1993.

[§§] B. Rossi, D.B. Hall, *Phys. Rev.*, 59, 223, 1941.

D.H. Frish, J.H. Smith, *Am J. Phys*, 31, 342, 1963

J. Hafele, R. Keating, *Science*, 177, 166, 1972.

But, as demonstrated in previous papers[***], the existence of a privileged frame is not compatible with the relativity principle (Contrary to the opinion of Poincaré).

The conclusion of this study is that we are compelled to make a choice: we either consider the relativity principle as true and abandon the relativity of time, or we take for granted the slowing of moving clocks and renounce the relativity principle.

It is easy to see that the first option is contradictory to logic. The second option implies that the laws of nature are different in the different inertial frames. This option looks more likely today. Nevertheless, although in agreement with the theory of Lorentz, it does not imply a total support for it for the reasons previously mentioned[†††].

Acknowledgements

I would like to thank Professor Franco Selleri for having given me the opportunity to participate in the debate raised around Special Relativity, at the occasion of this Conference.

Note added in proofs:

In fact, the relativity principle remains approximately true for bodies moving with respect to one another, and with respect to the ether frame, at low speeds ($v/c << 1$). For this reason, the conclusions of Galilei can be retained as a good approximation.

[***] J. Lévy, Relativity and Cosmic Substratum. *Precirculated proceedings of the P.I.R.T. Conference* p. 231 Imperial College London 6-9 September 1996.
Some important questions regarding Lorentz-Poincaré's theory and Einstein's relativity I. *Proceedings of the PIRT* 1996 Late papers p. 158.
Relativité et substratum cosmique, a book of 230 p. Dist. Lavoisier, 14 rue de Provigny 94 236 Cachan Cedex France Tél.: 0147406700.
[†††] *Ibid.*

Reception of Light Signals in Galilean Space-Time

Adolphe Martin
2235 Brebeuf, Apt. 3
Longueuil, Quebec, Canada J4J 3P9

By interpreting Relativity in Galilean space and time, it was found that the time of light reception by an observer moving relative to a source is a different event from the reception of the same light by an observer at rest. The Einstein viewpoint considers these two events to be the same, thus introducing the Special Relativity paradoxes. Using only Einstein's own coordinates and speed parameter as given by the Lorentz equations but introducing his Doppler factor and reasonning as a Galilean, we arrive at a different viewpoint from Einstein's with light speeds c, relative to the source, and c', relative to the observer, different from Einstein's constant c_o. This viewpoint becomes an isomorphism of a truly Galilean viewpoint with relative velocity V given by $v/c_o = \tanh(V/c_o)$ and light speeds $C = C' + V$ and $C' = C - V$, where angles remain the same as in Einstein space-time, and all corresponding distances have the same ratio $\sinh B/B$, the Galilean coordinates being given by both the Lorentz equations and Galilean transformation. This eliminates the Special Relativity paradoxes, though respecting the relativity principle and invariance of space and time of Galilean Relativity.

Keywords: Special Relativity, Galilean Relativity, Light speeds.

Introduction

A study was started as early as 1960 to determine the possibility of interpreting Special Relativity in invariant Galilean space and time. By using the Doppler factor and reasoning as a Galilean, the reception of light by an observer moving relative to a source was found to be a different event from the reception of light by an observer at rest relative to the source. As a result, entirely different viewpoints from Einstein's were obtained for Einstein speed v and proper speed v/L, L being the Lorentz factor:

$$L = \sqrt{1 - \frac{v^2}{c_o^2}}$$

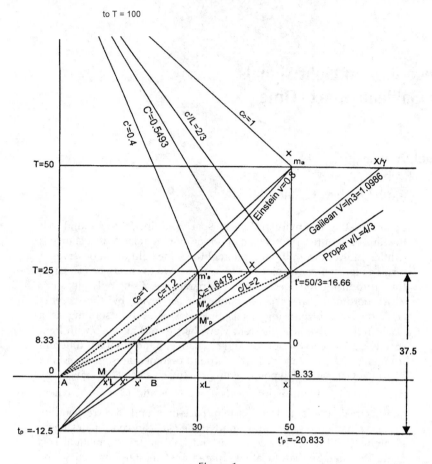

Figure 1

In 1970, the speed V was introduced as a purely Galilean speed, by making the celerity B of the Einsteinians equal to V/c_o. Equations for Galilean light speeds C relative to the source and C' relative to the observer were obtained before December 1981.[4] These light speeds and V add vectorially, as they should in Galilean space and time: $C = C' + V$ and $C' = C - V$. Four related viewpoints on Relativity are thus possible.[3] Einstein viewpoint with speed v and three Galilean viewpoints with Einstein speed v, proper speed v/L and Galilean speed V.

The present paper is restricted to the arguments leading to a differentiation between Einstein and Galilean viewpoints at speed v to emphasize the fact that the reception of light by the moving observer is not the same event as the reception of light by the equivalent fixed observer. Again for ease of comprehension, the arguments are restricted to one space dimension and the time dimension.

For ease of understanding, Einstein values are shown in lower case while Galilean values are shown in UPPER CASE.

Einstein viewpoint

Einstein considers two postulates [2]:

   a)   The principle of relativity.
   b)   The principle of constancy of light speed.

In Figure 1, light speed $c_o = 1$ is represented by lines at $45°$ from time axis, speed zero is parallel to time axis. The example is for an observer moving at $0.8c_o$ passing at M at a distance 10 ahead of a source of light emitted at time $t = 0$. Time $t = 50$ of reception of light by m is calculated by Einstein and all others by:

$$t = \frac{AM}{c_o - v} = \frac{10}{1 - 0.8} = 50, \quad x = c_o t = 50. \tag{1}$$

The distance 50 covered by light also corresponds to the position m of a fixed object receiving light at time 50. Light would be reflected back at same speed toward A at time $t_r = 100$.

$$t_r = 100; \quad \frac{t_r - t_e}{2} = 50; \quad t_p = \frac{AM}{v} = -12.5. \tag{2}$$

This corresponds to a radar shot: dividing the time interval taken by light to go and come back by 2, distance 50 of object m is obtained. In practice, to localize the trajectory of a moving object a minimum of two radar shots is required. Time of passage of M at world line of A is $t_p - 12.5$, equation (2). This corresponds to a second radar shot with zero time interval: trajectory and speed are then determined.

Galilean viewpoint

The Galilean viewpoint also considers two postulates:

   a)   The principle of relativity.
   b)   The principle of space and time invariance.

A Galilean frame at rest with the source is same as Einstein rest frame. A moving Galilean frame utilizes the same scales as the rest frame.

Time dilation

The experimentally proven Doppler factor value, given by the Einstein formula but inverted for wavelengths, is used to determine the time of light reception by the moving observer:

$$\beta = \frac{v}{c_o} = 0.8, \quad D_\lambda = \frac{\sqrt{1 - \beta^2}}{1 - \beta \cos \theta}, \quad D_\lambda = \frac{1 + \beta \cos \theta'}{\sqrt{1 - \beta^2}} \tag{3}$$

$$\cos(0) = 1, \quad D_o = \sqrt{\frac{1+\beta}{1-\beta}} = 3 \text{ in one space dimension'}$$

$$\cos(180) = -1, \quad D_{180} = \sqrt{\frac{1-\beta}{1+\beta}} = \frac{1}{3} \text{ in one space dimension'}$$

Subscripts 0 and 180 refer to the angle in degrees between relative velocity and light velocity. The light wavelength at observer M, moving away from A is 3 times the wavelength at the source, and the frequency is decreased 3 times (as shown in Fig. 1).

If the source sent light toward M from time of passage t_p to time $t = 0$ during 12.5 time units, the moving observer receives the same number of waves during a 3 times longer interval, or 37.5 units. The reception would cease at time: $T = 37.5 - 12.5 = 25$.

According to the relativity principle, an observer at A also receives light back from M at a 3 times reduced frequency and longer time. Emission from M lasts 37.5 time units; the reception at A lasts 3 times 37.5 or 112.5 units measured from t_p to $t_r = 112.5 - 12.5 = 100$, exactly the return time of a radar shot as given by Einstein and observed!

Since the Doppler effect is a fact of experience, all viewpoints have to respect it, including Einsteinians. No matter what speed, equivalent to Einstein's v, is assigned to the moving observer passing at $(0,t_p)$, the observer receives light emitted at A(0,0) at time $T = 25$, in order to have the same Doppler factor.

Similarly light reflected back to A from the moving observer is received at time $t_r = 100$, the observed return time. These are experimental facts to be respected by all.

At velocity $0.8c_o$ the Lorentz factor L is 0.6.

$$\beta = \frac{v}{c_o}, \quad L = \sqrt{1 - \frac{v^2}{c_o^2}} = 0.6 . \qquad (4)$$

Therefore, with $x = 50, t = 50$

$$x' = \frac{(x - vt)}{L} = 16.66, \quad t' = \frac{(t - vx/c_o^2)}{L} = 16.66 .$$

and with $x = 0, t_p = -12.5$

$$x' = \frac{x - vt}{L} = 16.66, \quad t'_p = \frac{(t - vx/c_o^2)}{L} = -20.833 ,$$

and $t' - t'_p = 37.5$, $T - t_p = t' - t'_p = 0.6(t - t_p) = L(t - t_p)$.

Using coordinates of m (50,50) in the Lorentz equations, $x' = t' = 16.66$ is obtained. Similarly, using coordinates of t_p (0, −12.5) gives the same $x' = 16.66$ and $t_p' = -20.83$. Notice that time intervals $(t' - t_p')$ and $(T - t_p)$ are strictly equal to 37.5—proof that there is *no time dilation*, since the Galilean time interval $(T - t_p)$ is invariant.

Time intervals $(T - t_p)$ and $(t' - t_p')$ are smaller than $(t - t_p)$ by exactly the Lorentz factor. Einsteinians then say: the moving frame time is dilated by the inverse of the Lorentz factor. However, it only appears so. In fact, the reception of light by an observer moving away from the source preceeds the reception by the observer at rest whose coordinates are given by the Lorentz equations.

If the origin of time is taken at t_p in both reference frames, to agree with the canonical Lorentz equations to make x, t, x', $t' = 0$ then t_p and $t_p' = 0$. Einstein time t', proper time (τ) and the Galilean invariant time T (same in both reference frames in relative motion) become numerically equal, showing again there is *no time dilation*.

$$T = t' = t\sqrt{1 - \frac{v^2}{c_o^2}} = \tau \qquad (5)$$

This is extremely important, as it proves beyond a doubt, using only the Lorentz equations with the Einstein coordinates and his Doppler factor, that there is *no time dilation*.

Space contraction

Einstein says t' is the time of light reception on the moving observer clock. It is then positioned on the right side of Figure 1, at light reception time $T = 25$ and $t_p' = -20.833$ at $t_p = -12.5$. Zero of $t' = 0$ is at $T - t' = 25 - 16.66 = 8.33$. This corresponds exactly to the time required by the moving observer starting at $x'L$ to reach x':

$$T - t' = \frac{x' - x'L}{v} = 8.33 \qquad (6)$$

Now x' is the coordinate of the moving observer in his own rest frame. At time $t' = 0$, the origin of the moving frame coincides with origin of the source frame at time $t = 8.33$.

Figure 1 shows that at time T, the moving observer is at distance 30 (exactly smaller by the Lorentz factor 0.6 than $x = 50$) where Einstein says he receives the light. The position of the moving observer at $T = 25$ is then

$$AM + vT = xL = x' + vt' = (c_o + v)t' = 1.8 \times 16.66 \qquad (7)$$

We see why, for Einsteinians, assuming that events m' and m coincide in the source frame, the moving frame appears contracted by the Lorentz factor L. Looking at the Lorentz equations in Galilean form, we recognize, in Figure 1, distances $AM = x'L$ and of $m' = xL$ given by

$$x'L = x - \beta c_o t, \qquad xL = x' + \beta c_o t'$$
$$c_o t'L = c_o t - \beta x, \qquad c_o tL = c_o t' + \beta x' \qquad (8)$$

and when $x = c_o t$, $x' = c_o t'$ in one space dimension.

The difficulties with Special Relativity stem entirely from the Einsteinian assumption that the reception of a light flash by an observer at rest relative to the source is the same event as reception of the signal by a moving observer whose coordinates are related by the Lorentz equations. In Galilean space-time the two events, definitively different, are (x,t) and (xL,T) in the source and fixed observer frame, and $(x',t' + 25)$ and (x',t') respectively in the moving observer frame. Thus $x'L$ and xL are not contractions of x' and x, but positions of the moving observer in the source frame at times of light emission $t = 0$ and reception $T = 25$ respectively.

We immediately see that there is *no space contraction*. The postulate of space and time invariance is respected. We are living in Galilean space and time!

Velocity addition

Einsteinians use equation (1) to calculate t, which is very different from Einstein's own formula for addition of velocities—a glaring inconsistency! This would mean that light from A takes 50 time units to travel a distance 10 in the moving frame at a relative speed 0.2, not 1, contrary to Einstein's second postulate. To respect his postulate in the moving frame, Einstein has to postpone light emission to 8.33 corresponding to his time $t' = 0$ and bring the reception back from $t = 50$ to $t' = 16.66$ corresponding to Galilean time $T = 25$. Only then can he write, following Lorentz, equation (8).

Costa de Beauregard [1] wrote the Lorentz equations in hyperbolic form:

$$x' = x \cosh B - c_o t \sinh B \qquad x = x' \cosh B + c_o t' \sinh B$$
$$c_o t' = c_o \cosh B - x \sinh B \qquad c_o t = c_o t' \cosh B + x' \sinh B \qquad (9)$$

These formulae give the same numerical results as the canonical Lorentz equations. The relationships of the hyperbolic functions to Einstein values is given by

$$\beta = \frac{v}{c_o} = \tanh B \qquad \cosh B = \frac{1}{\sqrt{1-\beta^2}} = \frac{1}{L} \qquad \sinh B = \frac{\beta}{\sqrt{1-\beta^2}} = \frac{\beta}{L} \qquad (10)$$

The equations of Galilean time T in relation to Einstein values or to hyperbolic values are given by

$$T = t\left[1 - \cos\theta\left(\frac{1}{\beta} - \frac{L}{\beta}\right)\right] = t'\left[1 + \cos\theta'\left(\frac{1}{\beta} - \frac{L}{\beta}\right)\right]$$
$$T = t\left[1 - \cos\theta\left(\frac{\cosh B - 1}{\sinh B}\right)\right] = t'\left[1 + \cos\theta'\left(\frac{\cosh B - 1}{\sinh B}\right)\right] = T' \qquad (11)$$

The hyperbolic argument B is called *celerity* by Einsteinians. Jacques Trempe [4,5] assumed that $B = V/c_o$, V being the Galilean velocity corresponding to Einstein's v. These speeds are related by $v/c_o = \tanh V/c_o$. When $v = c_o$, V = infinite.

In Galilean space-time, due to the invariance of space and time, velocities add vectorially in three dimension space, or velocity components add arithmetically along each space dimension. Galilean velocity V meets these requirements. In Einstein–Minkowski space-time, the velocity addition formula is really the addition of hyperbolic tangents of B, while proper velocities add as hyperbolic sines of B.

$$\beta = \frac{\beta_1 + \beta_2}{1 + \beta_1 \beta_2} \qquad \tanh(B_1 + B_2) = \frac{\tanh B_1 + \tanh B_2}{1 + \tanh B_1 \tanh B_2}$$

$$\frac{\beta}{L} = \frac{v}{c_o L} = \sinh(B_1 + B_2) = \sinh B_1 \cosh B_2 + \cosh B_1 \sinh B_2 \qquad (12)$$

Therefore, Einstein velocity v and proper velocity v/L cannot be Galilean, since they do not add according to the Galilean prescription, as V does.

Light velocity

This Galilean viewpoint uses the same coordinates used in Einstein space-time. One must be careful not to confuse this viewpoint using Einstein velocity v in Galilean space-time where light speeds c and c' are variable, with the Einstein viewpoint in Einstein–Minkowski space-time where light speed c_o is constant. This is not an idle remark, as the great majority of critics of Einstein, claiming to be Galileans, use the speeds $c_o + v$ and $c_o - v$ in their arguments. These speeds, when divided by c_o, are hyperbolic tangents of speed parameter additions, and thus are not Galilean speeds[6]. The variable speeds c and c' are truly Galilean in one space dimension and, if considered as vectors, in three space dimensions. Hence, $\mathbf{c'} + \mathbf{v} = \mathbf{c}$ and $\mathbf{c} - \mathbf{v} = \mathbf{c'}$ are vectors mapping Galilean vectors $\mathbf{C'} + \mathbf{V} = \mathbf{C}$ and $\mathbf{C'} - \mathbf{V} = \mathbf{C}_,$. $\mathbf{C}$ and $\mathbf{C'}$ being light speeds relative to source and to observer respectively. The Galilean light speeds are obtained by equating the Lorentz equations for Galilean coordinates to the Galilean transformations, with $T = T'$:

$$X' = X \cosh B - CT \sinh B = X - VT \qquad X = X' \cosh B + C'T' \sinh B = X' + VT' \quad (13)$$

$$C'T' = CT \cosh B - X \sinh B \qquad CT = C'T' \cosh B + X' \sinh B \qquad (14)$$

$$C = \frac{c_o B}{\sinh B - \cos \theta (\cosh B - 1)} \qquad C' = \frac{c_o B}{\sinh B + \cos \theta' (\cosh B - 1)} \qquad (15)$$

Moreover, the ratio of light speed c or C relative to the source to the light speed c' or C' relative to the observer is equal to the Doppler factor, even in three space dimensions, for all angles between light and relative speeds.

$$\frac{c}{c'} = \frac{C}{C'} = \frac{(c/L)}{(c'/L)} = D \qquad (16)$$

According to Einstein, light emitted by the moving observer at M (10,0) should arrive with speed c_o at A at time 10, while light emitted at A (0,0) should arrive at the moving observer at time 50. This is a flagrant breach of the relativity principle! On the contrary, in Galilean space-time, with light velocities dependent on the source–observer relative velocity, light emitted at A and moving M at instant $t = 0$ is received respectively at moving M and fixed A at the same time $T = 25$, in full compliance with the relativity principle. Similarly, a light flash reflected from them at $T = 25$ is received by both at the same return time $t_r = 100$.

The ratio of all distances or lengths between Galilean and Einstein values are $B/\sinh B$:

$$\frac{X}{x} = \frac{Y}{y} = \frac{Z}{z} = \frac{X'}{x'} = \frac{Y'}{y'} = \frac{Z'}{z'} = \frac{CT}{c_o t} = \frac{C'T}{c_o t'} = \frac{VT}{(v/L)T} = \frac{B}{\sinh B} \quad (17)$$

Since angles are the same in both space-times, and the Galilean coordinates are also related by the Lorentz equations, the Einstein–Minkowski space-time with speed v or v/L, as viewed by a Galilean, is a mathematical mapping of Galilean space-time with speed V! This makes it possible to solve problems in one (*e.g.* Einstein-Minkowski) space-time and convert the results into the other (*e.g.* Galilean.

Consequently, the Galilean viewpoint is also in accordance with the experimentally proven Doppler effect, the vectorial composition of velocities and the relativity principle.

Twins paradox

The Galilean viewpoint eliminates the Einsteinian concepts of space contraction and time dilation. Furthermore, it removes the twin paradox. In this example, by placing event m' at m, with Einstein, a sizeable interval of time (equal to 2 times 25 or 50) is forgotten in the Einsteinian calculation of the moving observer's age. This brings it down to 75, for a total time span of 125, the age of the stay-at-home twin, in a ratio of 0.6, the Lorentz factor, according to the time dilation concept.

In Galilean space-time, the two twins age at the same rate; both would be 125 at the return. This may be a disappointment for science fiction authors. However, trips would be shorter in time since Galilean velocities greater than c_o are attainable for all Einstein velocities above $0.7616\,c_o$ or at Doppler $D > 2.71828$!

Hence, for $B = 1\ D = e = 2.71828$

$$\beta = \frac{D^2 - 1}{D^2 + 1} = \frac{e^2 - 1}{e^2 + 1} = 0.7616 \qquad v \geq 0.7616\,c_o \quad (18)$$

Moreover, there is only one Universe with one universal time. Time travel is absolutely physically impossible as it would entail the existence of an infinity of simultaneous past and future universes.

References

[1]   Costa de Beauregard, O., 1968. *Relativité et quanta. Les grandes théories de la Physique Moderne*, Masson et Cie Éditeurs, Paris.
[2]   Einstein, A., 1905.On the electrodynamics of moving bodies. *Annalen Der Physik* 17(1905), Doc. 23, *The Collected Papers of Albert Einstein*, Vol. 2, Princeton University Press.
[3]   Martin, Adolphe, 1994. Light signals in Galilean Relativity. *Apeiron* 18(1994)

[4]   Trempe, Jacques, 1981. Einstein aurait–il pris des vessies pour des lanternes? *Spectre*, decembre 1981, Montreal.

[5]   Trempe, Jacques, 1990. Laws of light propagation in Galilean space-time, *Apeiron* 8:1,Montreal.

[6]   Trempe, Jacques, 1992. Light kinematics in Galilean space-time, *Physics Essays* 5:1

Experiments on the Velocity c

J.Ramalho Croca
Departamento de Física
Faculdade de Ciências, Universidade de Lisboa
Campo Grande, Ed C1, 1700 Lisboa Portugal
email: croca@fc.ul.pt

1. Introduction

One of the cornerstones of contemporary physics is the invariance of the velocity of light. As a natural consequence of this postulate comes the fact that the velocity of light does not depend, in any way, on the velocity of the emitting source.

Unfortunately, the experimental evidence confirming this corollary of the basic postulate of the relativity is not very accurate. It comes mainly from astronomical sources, such as star doublets, see Fig.1.

In 1913 De Sitter, as quoted by Wesley[1], showed that if the velocity of light depended on the velocity of the source, then stellar binaries revolving on a mass center would be very irregular and in certain cases should show ghosts. The light from the approaching star would arrive before light from the receding star. It is known that no such phantom stars are seen.

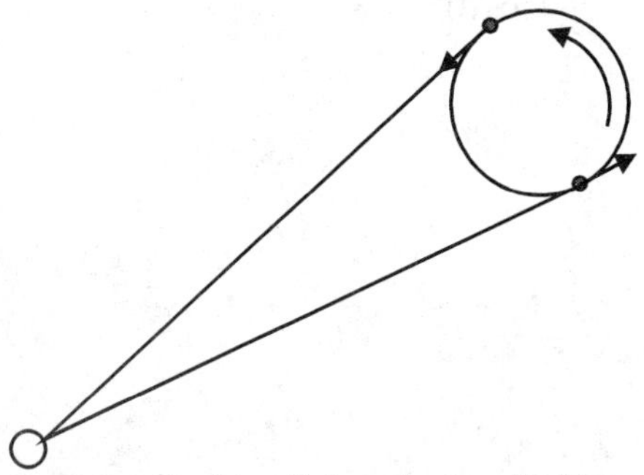

There are still other possibilities discussed in the book of Wesley[1]. Yet most of them are astronomical scale experiments where the errors of measurement are so great that they are not even quantified.

Here a laboratory scale experiment is proposed to remedy this situation. It is a one-way experiment where the light travels a single independent path.

The absolute temporal precision of the measurement is of subfemtosecond order, that is of

Fig.1. Star doublet. If the velocity of light depended on the velocity of the source, then the orbit of the star doublet would be very irregular.

about of 10^{-17} s. This high temporal precision will allow the detection of an hypothetical change in the one-way velocity of the light, due to the movement of the source, of the order of one meter per second or even better. The temporal precision of the measurement is possible thanks to recently developed technology in the production of pulsed laser beams, associated with interferometric detection techniques of second or fourth order.

2. Second Order Interference

It known that when two coherent gaussian pulses overlap in an interferometer a steady, in time, interference pattern is observed.

In this interferometer, the mixing region of which is shown in Fig.2, interference is observed when the phase shifting device P_s moves slightly, changing the relative phase difference between the two overlapping beams.

Consider incident gaussian pulses of the form

$$V = \int_{-\infty}^{+\infty}\int_{-\infty}^{+\infty} g(k,\omega)e[i(kx - \omega t)]dk\,d\omega, \tag{1}$$

with

$$g(k,\omega) = e\left(-\frac{(k-k_0)^2}{2\sigma_k^2} - \frac{(\omega-\omega_0)^2}{2\sigma_\omega^2}\right), \tag{2}$$

which by substitution and integration of expression (1) becomes

$$V = Ae\left[-\frac{(x-ct)^2}{2\sigma^2}\right]e[i(k_0 x - \omega_0 t)]. \tag{3}$$

The generic intensity at the output port of the interferometer is given by

$$I =< |V_1 + V_2|^2 >=< |V_1|^2 > + < |V_2|^2 > + < |V_1 V_2^*| > + < |V_1^* V_2| >, \tag{4}$$

which for the gaussian pulse of the form given by (3) transforms, for each output port, into

$$\begin{cases} I_1 = 2I_0(1 + e\left[-\frac{\varepsilon^2}{2\sigma^2}\right]\cos\delta) \\ I_2 = 2I_0(1 - e\left[-\frac{\varepsilon^2}{2\sigma^2}\right]\cos\delta) \end{cases}, \tag{5}$$

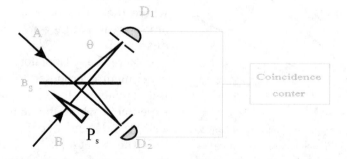

Fig.2. Overlapping region of a second order interferometer. P_s phase shifting device.

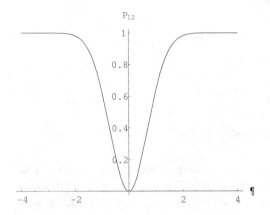

Fig.3. Plot for the joint probability detection in the case of coherent overlapping.

where ε represents the time delay between the two pulses, σ the length of the pulses and δ the minute relative phaseshift.

The joint probability detection $P_{12} = I_1 I_2$ is given by

$$P_{12} = 1 - e\left(-\frac{\varepsilon^2}{2\sigma^2}\right)\cos^2\delta,$$

which, for a relative phase shift $\delta = 0,...,2n\pi$ becomes

$$P_{12} = 1 - e\left(-\frac{\varepsilon^2}{2\sigma^2}\right), \qquad (6)$$

a plot of which is shown in Fig.3

As can be seen, from the plot, when the two pulses arrive at the same time, meaning that $\varepsilon = 0$, the normalised coincidence detection goes to zero. In this case coincidence never occurs, the photons go either to one output port or to the other. If the time delay between them is much greater than the length of the pulse, $\varepsilon >> \sigma$, then all happens as if each pulse arrives alone at the overlapping region. In this situation the pulse is split into two parts, each going to a different detector, and hence the probability of coincidence is one. For the case of partial overlapping, the probability of coincidence lies between zero and one. What is important in this measure is that from the actual value of the joint probability detection, given by expression (6), one can deduce the time delay between the two pulses.

3. Fourth Order Interference

When two overlapping waves come from mutually incoherent sources, no interference is to be seen in a usual second order interferometer. Nevertheless, even in this case it is possible to detect intensity correlation in

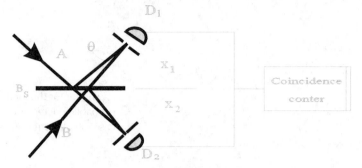

Fig.4. Fourth order interferometer.

the field resulting from the superposition of two incoherent waves. This phenomenon, usually known as fourth order interference, since it relates four fields, was experimentally discovered by Brown and Twiss[2] and is now widely applied in stellar interferometry.

The mathematical treatment of the effect can be found in many works that now are classic, such as those of Paul[3], Mandel[4] and others. An interferometer of this type is shown in Fig.4, where the interference is observed when the detectors scan the distance $x_1 - x_2$. The joint probability detection for the experimental set-up is given by

$$P_{12} \propto \left\langle (I_A + I_B)^2 \right\rangle - 2 \left\langle I_A I_B \right\rangle \cos \delta \qquad (7)$$

which can be written

$$P_{12} \propto 1 - \frac{2 < I_A I_B >}{< I_A^2 > + < I_B^2 > + 2 < I_A I_B >} \cos \delta \qquad (8)$$

where $\delta = 2\pi (x_1 + x_2)/L$, and the separation between the fringes $L = \lambda/\sin\theta \cong \lambda/\theta$. By substitution in expression (8) of the gaussian pulses, of the form described previously, the joint probability detection for this case turns out to be

$$P_{12} = 1 - \frac{e\left(-\frac{\varepsilon^2}{2\sigma^2}\right)}{1 + e\left(-\frac{\varepsilon^2}{2\sigma^2}\right)} \cos \delta \qquad (9)$$

which for the minimum coincidence rate, $\delta = 2\pi (x_1 + x_2)/L = 2n\pi$ is

$$P_{12} = 1 - \frac{e^{-\frac{\varepsilon^2}{2\sigma^2}}}{1 + e^{-\frac{\varepsilon^2}{2\sigma^2}}} . \qquad (10)$$

The plot for this case is shown in Fig. 5, together with the one for coherent overlapping, dashed line. The difference between them is that for complete overlapping when the pulses are coherent, the probability of coincidence is zero while in the incoherent case it is one half.

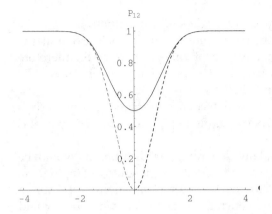

Fig.5. Plot for the probability of coincidence. Solid line incoherent overlapping. Dashed line coherent superposition.

4. Proposed Experiment

The set-up representing the experimental arrangement is shown in Fig.6. It is composed of a laser able to emit short pulses of femtosecond order, a turning wheel plus an interferometer.

The pulsed light from the laser enters the axis of the turning wheel, and is split into two and directed by means of an optical fiber, or some other device, to the points S_1 and S_2, which act as the moving sources.

Based on suggestions from the experimentalists[5] the experiment is best done using the alternate set-up shown in Fig.7.

This alternate device differs from the previous one in the way the moving sources are made. The laser pulse is split outside and the two beams are directed to the turning wheel made of stainless steel, which acts as a moving mirror.

In either case the light from S_1 and S_2 follows two independent parallel paths of the same length and is mixed in the in the interferometer set for the optimal conditions.

If the wheel is not turning, since the two paths are equal, the two pulses arrive in complete overlapping conditions corresponding to the situation

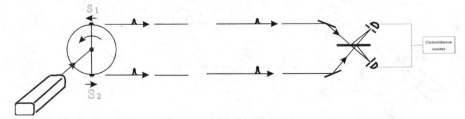

Fig.6. Set-up to test the independence of C on the velocity of the emitting source through single independent paths.

$\varepsilon = 0$ which results in the minimum in the coincidence rate.

When the wheel is set in motion, the sources S_1 and S_2 start moving in opposite directions with a velocity V.

Assuming that the velocity of light depends hypothetically on the velocity of the moving source, them the pulse from S_2 shall arrive at the interferometer a certain time in advance of the pulse from S_1, therefore increasing the joint detection probability. The probability of coincidence P_{12} may increase to reach its maximum when no overlapping occurs.

From the actual value of the joint detection probability P_{12}, one derives the hypothetical relative delay ε between the two pulses when the wheel is turning.

If the arriving pulses are totally coherent, the distribution would be given by the expression (6). In the case of perfect incoherent overlapping, the second expression (10) should fit the data. The real situation is probably a mixture of the two cases. Only the concrete experimental conditions would allow us to determine the right distribution.

The absolute precision, in time, of this experiment depends on the time width of the laser pulse. For steady laser pulses of about 3 fs, currently operating, it is possible, in principle, to determine fractions (say one thousandth) of this value, corresponding to differences in time up to 10^{-18} s.

As is well known, there is a certain controversy surrounding the possibility of measuring the one-way velocity of light[6] due to the problem of clock calibration. Some authors even deny the possibility of measuring the one-way velocity of light. Hence, in order to measure any hypothetical change in the one-way velocity of light due to the motion of the source, it must be assumed, of course, that this property of light is measurable, even if only for the sake of logical coherence.

On this problem my opinion coincides with Selleri who says[7] "*After all, light goes from one point to another in a well-defined way and it would be very strange if its true velocity were forever inaccessible to us.*" Concrete experiments may eventually show that this working hypothesis is wrong.

Since we do not know the law for the change in the velocity of light, due to the motion of the emitting source, for simplicity, we assume that it will increase by a certain hypothetical amount δv when the emitting source moves in direction of the light

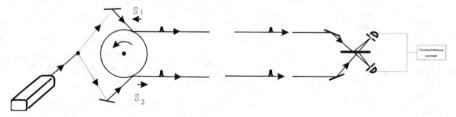

Fig.7.   Alternate set-up to test the independence of c of the velocity of the emitting source.

$$c+ = c + \delta v, \tag{11}$$

where $c+$ represents the velocity of light when the emitting source moves in direction of the light, c is the usual velocity of light and δv is hypothetical amount of velocity needed to add to c in order to attain $c+$.

Symmetrically, we shall assume that when the velocity of the emitting source is contrary to the light its velocity would be given by

$$c^- = c - \delta v. \tag{12}$$

It must be stressed, in order to avoid misinterpretations, that δv is not the velocity of the emitting source. It only represents the hypothetical change in the velocity of light due to the motion of the source.

No attempt is made to propose an expression for the hypothetical dependence of the velocity of light on the velocity of the emitting source.

According to those assumptions we shall write, for the hypothetical time difference between the two paths

$$\varepsilon = t_2 - t_1 = \frac{\ell}{c^-} + \frac{\ell}{c^+} = \frac{\ell}{c - \delta v} - \frac{\ell}{c + \delta v} \approx \frac{2\ell}{\delta v}\left(\frac{\delta v}{c}\right)^2. \tag{13}$$

That is

$$\delta v \approx \frac{\varepsilon c^2}{2\ell}, \tag{14}$$

where t_1, t_2 are the time the laser pulses take to go through paths one and two and δv is the hypothetical change in the velocity of light due to the motion of the sources.

Assuming a moderated minimal time resolution $\varepsilon \approx 10^{-16}$ s and that the experiment is done with an interferometer with arms of about $3m$, one obtains a lower limit of $\delta v \approx 1.5$ m/s

5. Conclusion

It was shown, under reasonable experimental conditions, I believe, that it is possible, in principle, to determine, in one-way independent trip, if the velocity of the light is independent or not on the velocity emitting source to an hypothetical change of about 1.5 m/s in 300 000 km/s (3×10^8 m/s).

References

[1]   J.P. Wesley, *Selected Topics in Advanced Fundamental Physics*, (Benjamin Wesley, Blumberg, 1991)

[2]   R.H. Brown and R.Q. Twiss, *Nature* (London) 177(1956) 27; 178(1956) 1046

[3]   H. Paul, *Rev. Mod. Phys.* 58(1986) 209

[4]   Z.Y. Ou and L. Mandel, *Phys. Rev. Lett.* 62(1989) 2941.

[5]   A. Melo, as suggested this change in the planning of a concrete experiment to be done with J.P. Ribeiro *et al.* at Lisbon.

[6]   F. Selleri, *Found Phys*. Letters, 9(1996) 43.
[7]   F. Selleri, *Found. Phys*. 26(1996) 641.

Inertial Transformations from the Homogeneity of Absolute Space

Ramón Risco-Delgado
Dpto. de Física Aplicada, E. Superior de Ingenieros
Universidad de Sevilla (Spain)
e-mail: ramon@cica.es

A transformation law for the space and time coordinates between two reference systems is deduced from the hypothesis of a privileged frame. No experiment performed up till now can distinguish them from Lorentz transformations.

Introduction

After the work of Mansouri and Sexl [1], it is well known that there exist many theories equivalent to special relativity from the experimental point of view that negate the relativity principle. This is possible because the experiments have only established the constancy of the two-way velocity of light. Accordingly, the one-way velocity of light is a degree of freedom in the space-time transformations that allows one to build a infinite set of transformations, indistinguishable from those of special relativity until now.

The existence of absolute affects, like the acceleration in the clock paradox or the length contraction in prerelativistic physics due to the deformation of the electromagnetic field of moving charges, allows us to change the starting point, and take the homogeneity of absolute space and time, instead of the relativity principle, as the basis of our argument. Homogeneity of space and time is a very plausible characteristic, one on which well established conservation laws are based.

Space-time transformations

If absolute space and time are homogeneous, then the transformation laws between a reference system fixed in this space and time (S_0) and any other inertial one (S) must be linear (for the sake of simplicity we shall only consider one spatial axis):

$$x = a_1 x_0 + a_2 t_0, \quad t = b_1 x_0 + b_2 t_0, \tag{1}$$

supposing that the origins of S_0 and S coincide at time $t = t_0 = 0$. If the systems move in the standard configuration, in such a way that the origin of S is seen from S_0 to move with velocity v, then $a_2 = -a_1 v$.

Now let us postulate the well classical fact that due to the deformation of the electromagnetic fields of moving atomic charges, a *material* body contracts itself just by the usual factor γ^1 [2]:

$$x = \gamma(x_0 - vt_0), \qquad t = b_1 x_0 + b_2 t_0 \qquad (2)$$

The next step will be to introduce the well established experimental result that the two-way velocity of light must be c in all reference systems. In order to do that we shall first compute the velocity transformation law from Eq. (2). Next, we shall apply this velocity transformation law for light going forward and backward with velocity c and $-c$ in the S_0 system respectively. Then we shall require that the two-way velocity of light be c in S as well.

By taking differentials in (2)

$$\begin{aligned} dx &= \gamma(dx_0 - vdt_0) \\ dt &= b_1 dx_0 + b_2 dt_0 \end{aligned} \qquad (3)$$

dividing by one another and calling $u = dx/dt$ and $u_0 = dx_0/dt_0$, we obtain

$$u = \gamma \frac{(u_0 - v)}{b_2 + b_1 u_0}. \qquad (4)$$

If Eq. (4) is applied to light that travels forward (with velocity c) and backward (with velocity $-c$) in S_0, then the corresponding velocities in S are:

$$c_+ = \gamma \frac{c - v}{b_2 + b_1 c}; \quad c_- = \gamma \frac{-c - v}{b_2 - b_1 c}. \qquad (5)$$

When light in S makes a two-way trip, travelling a distance $2d$, then the total time spent in the journey must be the sum of the times in the forward and backward trips. Requiring the two-way velocity of light to be c in S, then

$$\frac{2d}{c} = \frac{d}{c_+} + \frac{d}{-c_-}. \qquad (6)$$

By introducing (5) in (6) it is possible to find a relation between b_1 and b_2:

$$b_2 = \gamma^1 - vb_1, \qquad (7)$$

and the transformations read

$$\begin{aligned} x &= \gamma(x_0 - vt_0) \\ t &= b_1 x_0 + (\gamma^{-1} - vb_1)t_0 \end{aligned}. \qquad (8)$$

The former equations are the most general transformations compatible with experimental knowledge until now.

Clock synchronization

Eqs. (8) still have a parameter to be found. This freedom in the transformations is fixed by the value of the one-way velocity of light, a quantity that has not been measured previously. This fact is translated into different clock synchronization criteria. The only thing we can do for the moment is to follow our physical intuition in order to fix its value. Although in principle the choices can be many, there are two that seem specially justified: the Einstein approach and absolute approach.

Einstein approach

By denying the existence of an absolute space and time, Einstein arrives at the Lorentz transformations. Denying an absolute space means the equivalence of all inertial systems. This implies the validity of Maxwell's equations in all frames and, therefore, the constancy of the one-way velocity of light. In this section we shall obtain the particular element of (8) corresponding to the Lorentz transformations. We shall do so by denying the existence of an absolute space; in particular by imposing on (8) one of the most amazing consequences of this idea, the constancy of the velocity of light.

From (8), if we compute the velocity transformation laws between S_0 and S we obtain

$$u = \gamma \frac{u_0 - v}{\gamma^{-1} + b_1(u_0 - v)}. \tag{9}$$

Let us now impose that when $u_0 \equiv c$, then also $u \equiv c$.

$$c = \gamma \frac{c - v}{\gamma^{-1} + b_1(c - v)}, \tag{10}$$

and from here it is straightforward to obtain b_1:

$$b_1 = -\gamma \frac{v}{c^2}. \tag{11}$$

Obviously if (11) is carried into (8) we obtain Lorentz transformations. These are the only transformations that follow from denying the existence of an absolute space.

Absolute approach

Instead of denying the absolute standard, here we admit its existence. We shall see how, by assuming homogeneity, we can arrive at a set of equations, different from Lorentz transformations, but still compatible with experience.

Let us imagine the following ideal experiment. Two spacecraft, each carrying a clock, are at rest in S_0 at a certain distance from one another. At the same time, as measured from S_0, they start to move with the same acceleration until a certain preassigned time, and then they move with constant velocity. If S_0 is *homogeneous* the two clocks have acquired the same delay with respect to those at rest in S_0. In other words, this delay cannot depend on the spacecraft we chose, if space is homogeneous. This means that in (8) the transformation for time cannot depend on the position x_0. The only way to achieve this is by taking $b_1 = 0$. The transformations are then

$$x = \gamma(x_0 - vt_0)$$
$$t = \gamma^{-1}t_0$$

$$(12)$$

Using a similar approach, these equations were obtained first by Tangherlini [3] and generalized by Selleri [4].

For the moment there is no experimental reason to prefer the Lorentz transformations to (12). On the contrary, the behavior of clocks reported in this section seems quite plausible. The existence of a crucial test is still an open question.

I acknowledge Prof. Selleri for encouragement. I also thank *La Revista Espanola de Física* for the permission to reproduce here a revised version of the manuscript published there in Spanish in December 1997.

References

[1]   R. Mansouri and R. Sexl, *General Relativity and Gravitation* 8, 497 (1977).
[2]   J. S. Bell, How to Teach Special Relativity, in *Speakable and Unspeakable in Quantum Mechanics* (Cambridge University Press, 1988).
[3]   F. R. Tangherlini, *Nouvo Cim. Suppl.* 20, 1 (1961).
[4]   F. Selleri, *Found. Phys.* 26, 641 (1996).

On a Physical and Mathematical Discontinuity in Relativity Theory

F. Selleri
Università di Bari - Dipartimento di Fisica
INFN - Sezione di Bari
I 70126 Bari, Italy

An isotropic inertial reference frame ("stationary") is considered, in which a circular disk of radius R rotates uniformly. The velocity of light $\widetilde{c}$ relative to the rim of the disk is calculated under *very general* assumptions and found to satisfy $\widetilde{c} \neq c$. This $\widetilde{c} \neq c$ remains the same if R is increased but the peripheral velocity of the disk is kept constant. Since by so doing any small part of the circumference can be considered (for a short time) better and better at rest in a ("moving") inertial system, there is a discontinuity between accelerated reference frames with arbitrarily small acceleration and inertial frames, *if* the velocity of light is assumed to be c in the latter. Elimination of the discontinuity is shown to imply *for inertial systems* a velocity of light $\widetilde{c} \neq c$, necessarily equal to that obtained from recently published "inertial transformations."

1. Space and time on a rotating platform

In this paper it is shown that the relativistic description of inertial reference systems contains a basic difficulty. It is well known that no perfectly inertial frame exists in practice, *e.g.*, because of the Earth's rotation around its axis, orbital motion around the Sun and Galactic rotation. Therefore, all knowledge about inertial frames has necessarily been obtained in frames having small but nonzero acceleration a. For this reason the mathematical limit $a \to 0$ taken in the theoretical schemes should be a smooth limit, and no discontinuities should arise between systems with small acceleration and inertial systems. From this point of view the existing relativistic theory will be shown to be inconsistent.

Consider an inertial reference frame S_0 and assume that it is isotropic. Therefore, the one-way velocity of light relative to S_0 can be taken to have the usual value c in all directions. In relativity, the latter assumption is true in all inertial frames, while in other theories only one such frame exists [1]. A laboratory in which physical experiments are performed is assumed to be at rest in S_0, and in it clocks are assumed to have been synchronised with the Einstein method, *i.e.*, by using light signals and assuming that the one-way velocity of light is the same in all directions.

In this laboratory there is a perfectly circular platform having radius R, which rotates around its axis with angular velocity ω and peripheral velocity $v = \omega R$. On its rim, consider a clock C_Σ and assume it to be set as follows: When a laboratory clock momentarily very near C_Σ shows time $t_0 = 0$, then C_Σ is also set at time $t = 0$. When the platform is not rotating, C_Σ constantly shows the same time as the laboratory clocks. When it rotates, however, motion modifies the rate of C_Σ, and the relationship between the times t and t_0 is taken to have the general form

$$t_0 = t\, F_1(v,a) \tag{1}$$

where F_1 is a function of velocity v, acceleration $a = v^2/R$, and eventually of higher derivatives of position (not shown).

The circumference length is assumed to be L and L, measured in the laboratory S and on the platform, respectively. Since motion can modify length as well, the relationship

$$L_0 = L\, F_2(v,a) \tag{2}$$

is assumed, where F_2 is another function of the said arguments. Notice that the assumed isotropy of the laboratory frame implies that F_1 (F_2) does not depend on the point on the rim of the disk where the clock is placed (where the measurement of length is started). It depends only on velocity, acceleration, *etc.*, and these are the same at all points of the edge of the rotating circular platform.

One is of course far from ignorant about the nature of the functions F_1 and F_2. In the limit of small acceleration and constant velocity, they are expected to become the usual time dilation and length contraction factors, respectively:

$$F_1(v,0) = \frac{1}{\sqrt{1-v^2/c^2}} \quad ; \quad F_2(v,0) = \sqrt{1-v^2/c^2} \tag{3}$$

There are even strong indications, at least in the case of $F_1(v,a)$, that the dependence on a is totally absent [2]. All this is, however, unimportant for present purposes, because the results obtained below hold for all possible functions F_1 and F_2.

2. Velocity of light on rotating platforms

On the rim of the platform, in addition to clock C_Σ there is a light source Σ, placed in a fixed position very near C_Σ. Two light flashes leave Σ at the time t_1 of C_Σ, and are forced to move on a circumference, by "sliding" on the internal surface of a cylindrical mirror placed at rest on the platform, all around it and very near its edge. Mirror apart, the light flashes propagate in the vacuum. The motion of the mirror cannot modify the velocity of light, because the mirror is like a source (a "virtual" one), and the motion of a source never changes the velocity of light signals originating from it. Thus, relative to the laboratory, the light flashes propagate with the usual velocity c.

The description of light propagation given by the laboratory observers is as follows: two light flashes leave Σ at time t_{01}. The first one propagates on a circumference, in the sense opposite to the platform rotation, and comes back to Σ at time t_{02} after a full rotation around the platform. The second one propagates on a circumference, in the same rotational sense of the platform, and comes back to Σ at time t_{03} after a full rotation around the platform. These laboratory times, all relative to events taking place in a fixed point of the platform very near C_Σ, are related to the corresponding platform times via (1):

$$t_{0i} = t_i \, F_1(v,a) \qquad (i = 1,2,3) \qquad (4)$$

Light propagating in the direction opposite to the disk rotation, must cover a distance smaller than L_0 by a quantity $x = \omega R(t_{02} - t_{01})$ equalling the shift of Σ during the time $t_{02} - t_{01}$ taken by light to reach Σ. Therefore

$$L_0 - x = c(t_{02} - t_{01}) \quad ; \quad x = \omega R(t_{02} - t_{01}) \qquad (5)$$

From these equations one gets:

$$t_{02} - t_{01} = \frac{L_0}{c(1+\beta)} \qquad (6)$$

Light propagating in the rotational direction of the disk, must instead cover a distance larger than the disk circumference length L_0 by a quantity $y = \omega R(t_{03} - t_{01})$ equalling the shift of Σ during the time $t_{03} - t_{01}$ taken by light to reach Σ. Therefore

$$L_0 + y = c(t_{03} - t_{01}) \quad ; \quad y = \omega R(t_{03} - t_{01}) \qquad (7)$$

One now gets

$$t_{03} - t_{01} = \frac{L_0}{c(1-\beta)} \qquad (8)$$

By taking the difference between (8) and (6) one sees that the delay between the arrivals of the two light flashes back in Σ is observed in the laboratory to be

$$t_{03} - t_{02} = \frac{2L_0\beta}{c(1-\beta^2)} \qquad (9)$$

Incidentally, this is the well known delay time for the Sagnac effect [3], calculated in the laboratory (our treatment is, however, totally independent of this effect). It is shown next that these relations to some extent fix the velocity of light relative to the disk. In fact (2) and (4) applied to (6) and (8) give

$$(t_2 - t_1) F_1 = \frac{L F_2}{c(1+\beta)} \quad ; \quad (t_3 - t_1) F_1 = \frac{L F_2}{c(1-\beta)} \quad (10)$$

whence

$$\frac{1}{\widetilde{c}(\pi)} = \frac{t_2 - t_1}{L} = \frac{F_2}{F_1 c(1+\beta)} \quad ; \quad \frac{1}{\widetilde{c}(0)} = \frac{t_3 - t_1}{L} = \frac{F_2}{F_1 c(1-\beta)} \quad (11)$$

if $\widetilde{c}(0)$ and $\widetilde{c}(\pi)$ are the light velocities, relative to the rim of the disk, for the flash propagating in the direction of the disk rotation, and in the opposite direction, respectively. From (11) it follows that the velocities of the two light flashes relative to the disk must satisfy

$$\frac{\widetilde{c}(\pi)}{\widetilde{c}(0)} = \frac{1+\beta}{1-\beta} \quad (12)$$

Notice that the unknown functions F_1 and F_2 cancel from the ratio (12). The consequences of (12) will be discussed in the next section.

Now comes an important point: Clearly, Eq. (12) gives us not only the ratio of the two global light velocities for full trips around the platform in the two opposite directions, but *the ratio of the instantaneous velocities as well.* In fact, we have assumed that the "stationary" inertial reference frame in which the centre of the platform is at rest is isotropic, and isotropy of space ensures that the instantaneous velocities of light are the same at all points of the rim of the circular disk, and, therefore, that the average velocities coincide with the instantaneous ones.

The result (12) holds for platforms having different radii, but the same peripheral velocity v. Let a set of circular platforms with radii $R_1, R_2, \ldots R_i, \ldots$ ($R_1 < R_2 < \ldots < R_i < \ldots$) be given, and made to spin with angular velocities $\omega_1, \omega_2, \ldots \omega_i, \ldots$ such that

$$\omega_1 R_1 = \omega_2 R_2 = \ldots = \omega_i R_i = \ldots = v \quad (13)$$

where v is constant. Obviously then, (12) applies to all such platforms with the same β $(\beta = v/c)$. The respective centripetal accelerations are

$$\frac{v^2}{R_1}, \frac{v^2}{R_2}, \ldots \frac{v^2}{R_i}, \ldots \quad (14)$$

and *tend to zero with increasing* R_i. This is so for all higher derivatives of position: if $\vec{r}_0(t_0)$ identifies a point on the rim of the i-th platform seen from the laboratory, one can easily show that

$$\left| \frac{d^n \vec{r}_0}{d t_0^n} \right| = \frac{v^n}{R_i^{n-1}} \quad (n \geq 2) \quad (15)$$

This tends to zero with increasing R_i for all $n \geq 2$. Therefore, a very small part AB of the rim of a platform, having peripheral velocity v and very large radius, for a short time is completely equivalent to a small part of a "co-moving" inertial reference frame (endowed with a velocity only). For all practical purposes the "small part AB of the rim of a platform" will belong to that inertial reference frame. But the velocities of light in the two directions AB and BA must satisfy (12), as shown above. It follows that the velocity of light relative to the co-moving *inertial* frame cannot be c.

3.   Speed of light relative to inertial frames

As shown in [1], one can always choose Cartesian co-ordinate systems in two inertial reference frames S and S_0 and assume:

(1)   that space is homogeneous and isotropic, and that time is also homogeneous;

(2)   that relative to S_0 the velocity of light is the same in all directions, so that Einstein's synchronisation can be used in this frame and the velocity v of S relative to S_0 can be measured;

(3)   that the origins of S and S_0 coincide at $t = t_0 = 0$;

(4)   that planes (x_0, y_0) and (x,y) coincide at all times t_0; that also planes (x_0, z_0) and (x,z) coincide at all times t_0; but that planes (y_0, z_0) and (y,z) coincide at time $t_0 = 0$ only.

It then follows [1] that the transformation laws from S_0 to S are necessarily

$$\begin{cases} x &= f_1(x_0 - vt_0) \\ y &= g_2\, y_0 \\ z &= g_2\, z_0 \\ t &= e_1\, x_0 + e_4 t_0 \end{cases} \tag{16}$$

where the coefficients f_1, g_2, e_4 and e_1 can depend on v. If at this point one assumes the validity of the relativity principle (including invariance of light velocity) these transformations reduce *necessarily* to the Lorentz ones. It was shown in [1] that the most general transformations (16) satisfying the well established experimental conditions of constant *two-way* velocity of light and of time dilation according to the usual relativistic factor, are such that

$$f_1 = \frac{1}{R(\beta)} \quad ; \quad g_2 = 1 \quad ; \quad e_4 = R(\beta) - e_1 \beta c$$

where $\beta = v/c$, and

$$R(\beta) = \sqrt{1 - \beta^2} \tag{17}$$

so that

$$\begin{cases} x = \dfrac{x_0 - \beta ct_0}{R(\beta)} \\ y = y_0 \\ z = z_0 \\ t = R(\beta)t_0 + e_1(x_0 - \beta ct_0) \end{cases} \qquad (18)$$

In (18) only e_1 remains unknown. Length contraction by the factor $R(\beta)$ is also a consequence of (18): This is natural, because in [4] I showed that space-time transformations from a "stationary" isotropic inertial system S_0 to any other inertial system S imply complete equivalence between the three possible pairs of assumptions chosen among the following: A1. Lorentz contraction of bodies moving with respect to S_0; A2. Larmor retardation of clocks moving with respect to S_0; A3. Two-way velocity of light equal to c in all inertial systems and in all directions. The inverse speed of light compatible with (18) was shown to be [1]:

$$\frac{1}{\widetilde{c}(\theta)} = \frac{1}{c} + \left[\frac{\beta}{c} + e_1 R(\beta) \right] \cos\theta \qquad (19)$$

where θ is the angle between the direction of propagation of light and the absolute velocity $\vec{v}$ of S. The transformations (18) represent the complete set of theories "equivalent" to the Special Theory of Relativity (STR): if e_1 is varied, different elements of this set are obtained. The Lorentz transformation is found as a particular case with $e_1 = -\beta / c\, R(\beta)$. Different values of e_1 are obtained from different clock-synchronisation "conventions." In all cases except that of STR, such values exclude the validity of the relativity principle, and imply the existence of a privileged frame [1]. For all these theories only subluminal motions are possible.

In the previous sections we found a ratio of the one-way velocities of light along the rim of the disk, and relative to the disk itself, different from 1 and given by Eq (12). Our principle of local equivalence between the rim of the disk and the "tangent" inertial frame requires (12) to apply in the latter frame as well. Eq. (19) applied to the cases $\theta = 0$ and $\theta = \pi$ yields

$$\frac{1}{\widetilde{c}(0)} = \frac{1}{c} + \left[\frac{\beta}{c} + e_1 R(\beta) \right] \quad ; \quad \frac{1}{\widetilde{c}(\pi)} = \frac{1}{c} - \left[\frac{\beta}{c} + e_1 R(\beta) \right] \qquad (20)$$

This gives

$$\frac{\widetilde{c}(\pi)}{\widetilde{c}(0)} = \frac{1 + \beta + ce_1 R(\beta)}{1 - \beta - ce_1 R(\beta)}$$

which can agree with (12) only if

$$e_1 = 0 \qquad (21)$$

The space-dependent term in the transformation of time is thus seen to disappear from (18). The same result (21) was obtained in [1] by requiring that the Sagnac effect be explained on the rotating disk also, and not only in the laboratory. See also Ref. [5] for a detailed discussion of that effect both from the special and the general relativistic point of view.

4. The inertial transformations

In the previous section it was shown that the condition $e_1 = 0$ has necessarily to be used. This generates transformations different from the Lorentz ones [1]:

$$\begin{cases} x = \dfrac{x_0 - \beta c t_0}{R(\beta)} \\ y = y_0 \\ z = z_0 \\ t = R(\beta) t_0 \end{cases} \qquad (22)$$

The velocity of light predicted by (22) can be found by taking $e_1 = 0$ in (19):

$$\frac{1}{\tilde{c}(\theta)} = \frac{1 + \beta \cos \theta}{c} \qquad (23)$$

The transformation (22) can be inverted and gives:

$$\begin{cases} x_0 = R(\beta) \left[x + \dfrac{\beta c}{R^2(\beta)} t \right] \\ y_0 = y \\ z_0 = z \\ t_0 = \dfrac{1}{R(\beta)} t \end{cases} \qquad (24)$$

Note the formal difference between (22) and (24). The latter implies, for example, that the origin of S_0 ($x_0 = y_0 = z_0 = 0$) is described in S by $y = z = 0$ and by

$$x = -\frac{\beta c}{1 - \beta^2} t \qquad (25)$$

This origin is thus seen to move with speed $\beta c/(1-\beta^2)$, which can exceed c, but cannot be superluminal. In fact, a light pulse seen from S to propagate in the same direction as S_0 has $\theta = \pi$, and thus [using (23)] has speed $\tilde{c}(\pi) = c/(1-\beta)$, which can easily be checked to satisfy

$$\frac{c}{1-\beta} \geq \frac{c\beta}{1-\beta^2}$$

One of the new features of these transformations is, of course, the presence of relative velocities exceeding c. This is not a problem, in practice, because one-way velocities have never been measured. Absolute velocities are, instead, always smaller than c [1]. It is clear from (25) that the velocity of S_0 relative to S is not equal and opposite to that of S relative to S_0. In STR one is used to relative velocities that are always equal and opposite, but this symmetry is a consequence of the synchronisation used, and cannot be expected to hold more generally [1].

Now consider a third inertial system S', moving with velocity βc and its transformation from S_0, which of course is given by (18) with β' replacing β. By eliminating the S_0 variables, one can obtain the transformation between the two moving systems S and S':

$$\begin{cases} x' = \dfrac{R(\beta)}{R(\beta')}\left[x - \dfrac{\beta' - \beta}{R^2(\beta)}ct\right] \\ y' = y \\ z' = z \\ t' = \dfrac{R(\beta')}{R(\beta)}t \end{cases} \tag{26}$$

Equations (22)-(24)-(26) will be called "inertial transformations." In its most general form (26) the inertial transformation depends on two velocities (v and v'). When one of them is zero, either S or S' coincide with the privileged system S_0, and the transformation (26) becomes (22) or (24), respectively.

One feature that characterizes the transformations (22)-(24)-(26) is *absolute simultaneity*: two events taking place at different points of S but at the same t are judged to be simultaneous also in S' (and *vice versa*). The existence of absolute simultaneity does not imply that time is absolute: on the contrary, the β-dependent factor in the transformation of time gives rise to time-dilation phenomena similar to those of STR. *Time dilation* in another sense is, however, also absolute: a clock at rest in S is seen from S_0 to run slower, but a clock at rest in S_0 is seen from S to run *faster*, so that both observers agree that motion relative to S_0 slows down the pace of clocks. The difference with respect to STR exists because a clock T_0 at rest in S_0 must be compared with clocks at rest at different points of S, and the result is, therefore, dependent on the "convention" adopted for synchronising the latter clocks.

Absolute length contraction can also be deduced from (25): all observers agree that motion relative to S_0 leads to contraction. The discrepancy with the STR is due to the different conventions concerning clock synchronisation: the length of a moving rod can only be obtained by marking the *simultaneous* positions of its end points, and therefore depends on the very definition of simultaneity of distant events.

5. Further comments on the discontinuity problem

Our choice of synchronisation (called "absolute" by Mansouri and Sexl [6]) was made by considering rotating platforms. The main result of this paper is Eq. (12): the ratio

$$\rho \equiv \frac{\widetilde{c}(\pi)}{\widetilde{c}(0)}$$

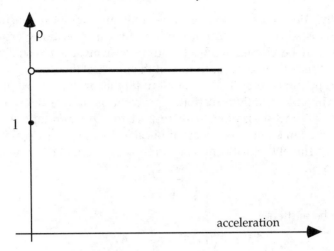

Figure 1. The ratio $\rho = \widetilde{c}(\pi)/\widetilde{c}(0)$ plotted as a function of acceleration for rotating platforms of constant peripheral velocity and increasing radius (decreasing acceleration). The prediction of SRT is 1 (black dot on the ρ axis) and is not continuous with the ρ value for the rotating platforms.

has been calculated along the rims of the platforms and shown, under very general conditions, to have the value (12), which in general is different from unity. Therefore, the velocities of light parallel and anti-parallel to the disk peripheral velocity are not the same. For SRT, this is a serious problem, because a set of platforms with growing radius, but all with the same peripheral velocity, locally approaches better and better an inertial frame. To say that the radius becomes very large with constant velocity is the same as saying that the centripetal acceleration goes to zero with constant velocity. The logical situation is shown in Fig. 1: SRT predicts a discontinuity for ρ at zero acceleration, a sudden jump from the accelerated to the "inertial" reference frames. While all experiments are performed in the real physical world ($a \neq 0$, $\rho = (1+\beta)/(1-\beta)$), our theoretical physics seems to have gone out of the world ($a = 0$, $\rho = 1$)!

The above discontinuity probably is the origin of the synchronisation problems encountered by the Global Positioning System: after all, Earth is also a kind of rotating platform.

It should be stressed that non-invariant velocity of light is required for all (but one!) inertial systems. In fact, given any such system, and a small region of it, it is always possible to conceive a large and rotating circular platform, a small part of which is locally at rest in that region, and the result (12) must then apply. Therefore, the velocity of light is non-isotropic in all inertial reference frames, with the exception of one (S_0) where isotropy can be postulated.

Finally, one must also conclude that the famous synchronisation problem [7] is solved by nature itself: it is not true that the synchronisation procedure can be chosen freely, because the convention usually adopted leads to an unacceptable discontinuity in the physical theory.

It was pointed out by T. Van Flandern [8] that an "orthodox" approach to dealing with the rotating platform problem is to consider a position-dependent desynchronization, with respect to the laboratory clocks, as an objective phenomenon, concretely applicable to the clocks set at different points of the rim of the platform. The Lorentz transformation of time

$$t' = \frac{t - x\beta/c}{\sqrt{1-\beta^2}}$$

can in fact be written

$$t'\sqrt{1-\beta^2} - t = -x\beta/c \tag{27}$$

and can be read as follows: the difference between the time t' of the "moving" frame S' (corrected by a $\sqrt{1-\beta^2}$ factor in order to cancel the time dilation effect) and the time t of the "stationary" frame S has a linear dependence on the x coordinate of S as given by (27). This difference is called "desynchronization."

The "orthodox idea" would be to apply the previous approach to the rotating platform. Assume that rotation is from left to right in Fig. 2. Two flashes of light are emitted at laboratory time $t = 0$ by the source Σ in opposite directions along the platform rim. When the right-moving (left-moving) flash reaches point A (point B) at laboratory time $t_A (t_B)$ after covering a distance $x_A (x_B)$ [measured in the laboratory from Σ along the platform border], it will find a local clock which is desynchronized by an amount $\Delta t_A (\Delta t_B)$ with respect to the laboratory clocks given by

$$\Delta t_A = x_A v/c^2 + \alpha \quad ; \quad \Delta t_B = -x_B v/c^2 + \alpha \tag{28}$$

where α represents whatever desynchronization the clock placed in Σ might have with respect to the laboratory clocks.

It is at once possible to see that Eq. (28) is in sharp contradiction with the rotational invariance assumed above, which leads to the existence of the discontinuity in Fig. 1: if the inertial system S_0 (in which the centre of the perfectly circular platform is at rest) is isotropic, and if the platform is set in motion in a regular way, no difference between clocks on its rim can ever arise, and Eq. (28) represents a logical impossibility. It is impossible to understand why the clock in A should be desynchronized differently from the clock in B, unless this is achieved artificially by some observer.

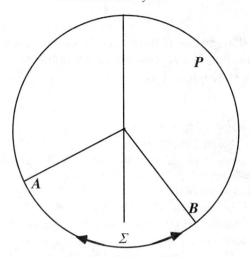

Figure 2. Two flashes of light emitted at laboratory time $t = 0$ by Σ in opposite directions along the platform border overlap in a point P where the local clock on the platform should show two incompatible times.

Furthermore, the whole argument culminating in Fig. 1 was based on the existence of a single clock, *arbitrarily retarded with respect to the laboratory clocks*, so that even if the previous strange desynchronization were assumed true, it would still be possible to use one of the clocks satisfying (28) to repeat the argument and again deduce the existence of the same discontinuity!

It should also be stressed that the experimental evidence is that many little clocks (muons) injected in different positions of the CERN Muon Storage Ring behave in exactly the same way, independently of their position in the ring [2]. And in the present case, it is not possible to conceive a human intervention that desynchronizes the muons. Therefore, not only common sense, but also direct experimentation, shows that a position-dependent desynchronization is out of the question.

Finally, it is possible to show that the desynchronization in (28), which is unable to cure the discontinuity of Fig. 1, as just shown, introduces a further discontinuity in the time shown by clocks set all around the platform rim, so that far from simplifying the problem, it introduces new complications.

From the point of view of the laboratory observer, the space between Σ and the right-moving (left-moving) flash widens at a rate $c - v$ ($c + v$), so that from (28) we get

$$(c - v)t_A = x_A \qquad ; \qquad (c + v)t_B = x_B$$

Therefore

$$\Delta t_A = (c - v)t_A v / c^2 + \alpha \qquad ; \qquad \Delta t_B = -(c + v)t_B v / c^2 + \alpha \qquad (29)$$

There will be a time t_P when the two flashes overlap at a point P, while moving in opposite directions. When this happens

$$t_A = t_B = t_P \tag{30}$$

The problem is that the Eqs. (29) should both be valid for the same clock at the common time t_P, but they are instead incompatible if (30) is satisfied, as their equality is easily shown to reduce to $c = -c$.

References

[1]   F. Selleri, *Found. Phys.* 26, 641 (1996).

[2]   J. Bailey *et al.*, *Nature* 268, 301 (1977) .

[3]   G. Sagnac, *Compt. Rend.* 157, 708 (1913); *ibid.* 1410; P. Langevin, *Compt. Rend.* 173, 831 (1921); 205, 304 (1937); E. J. Post, *Rev. Mod. Phys.* 39, 475 (1967).

[4]   F. Selleri, *Apeiron*, 4, 100 (1997).

[5]   F. Goy and F. Selleri, *Found. Phys. Lett.* 10, 73 (1997).

[6]   R. Mansouri and R. Sexl, *Gen.Relat. Gravit.* 8, 497 (1977).

[7]   A. Einstein, *Relativity, the Special, the General Theory*, (Chicago, 1951); H. Reichenbach, *The Philosophy of Space & Time* (Dover Publ., New York, 1958). M. Jammer, Some fundamental problems in the special theory of relativity, in: *Problems in the Foundations of Physics*, G. Toraldo di Francia, *ed.*, (Società Italiana di Fisica, Bologna, and North Holland, Amsterdam, 1979).

[8]   Tom Van Flandern, private communication.

What the Global Positioning System Tells Us about Relativity

Tom Van Flandern
Univ. of Maryland & Meta Research
P.O. Box 15186, Chevy Chase, MD 20825, USA

1. What is the GPS?

The Global Positioning System (GPS) consists of a network of 24 satellites in roughly 12-hour orbits, each carrying atomic clocks on board. The orbital radius of the satellites is about four Earth-radii (26,600 km). The orbits are nearly circular, with a typical eccentricity of less than 1%. Orbital inclination to the Earth's equator is typically 55 degrees. The satellites have orbital speeds of about 3.9 km/s in a frame centered on the Earth and not rotating with respect to the distant stars. Nominally, the satellites occupy one of six equally spaced orbital planes. Four of them occupy each plane, spread at roughly 90-degree intervals around the Earth in that plane. The precise orbital periods of the satellites are close to 11 hours and 58 minutes so that the ground tracks of the satellites repeat day after day, because the Earth makes one rotation with respect to the stars about every 23 hours and 56 minutes. (Four extra minutes are required for a point on the Earth to return to a position directly under the Sun because the Sun advances about one degree per day with respect to the stars.)

The on-board atomic clocks are good to about 1 nanosecond (ns) in epoch, and about 1 ns/day in rate. Since the speed of light is about one foot per nanosecond, the system is capable of amazing accuracy in locating anything on Earth or in the near-Earth environment. For example, if the satellite clocks are fully synchronized with ground atomic clocks, and we know the time when a signal is sent from a satellite, then the time delay for that signal to reach a ground receiver immediately reveals the distance (to a potential accuracy of about one foot) between satellite and ground receiver. By using four satellites to triangulate and determine clock corrections, the position of a receiver at an unknown location can be determined with comparable precision.

2.   What relativistic effects on GPS atomic clocks might be seen?

General Relativity (GR) predicts that clocks in a stronger gravitational field will tick at a slower rate. Special Relativity (SR) predicts that moving clocks will appear to tick slower than non-moving ones. Remarkably, these two effects cancel each other for clocks located at sea level anywhere on Earth. So if a hypothetical clock at Earth's north or south pole is used as a reference, a clock at Earth's equator would tick slower because of its relative speed due to Earth's spin, but faster because of its greater distance from Earth's center of mass due to the flattening of the Earth. Because Earth's spin rate determines its shape, these two effects are not independent, and it is therefore not entirely coincidental that the effects exactly cancel. The cancellation is not general, however. Clocks at any altitude above sea level do tick faster than clocks at sea level; and clocks on rocket sleds do tick slower than stationary clocks.

For GPS satellites, GR predicts that the atomic clocks at GPS orbital altitudes will tick faster by about 45,900 ns/day because they are in a weaker gravitational field than atomic clocks on Earth's surface. Special Relativity (SR) predicts that atomic clocks moving at GPS orbital speeds will tick slower by about 7,200 ns/day than stationary ground clocks. Rather than have clocks with such large rate differences, the satellite clocks are reset in rate before launch to compensate for these predicted effects. In practice, simply changing the international definition of the number of atomic transitions that constitute a one-second interval accomplishes this goal. Therefore, we observe the clocks running at their offset rates before launch. Then we observe the clocks running after launch and compare their rates with the predictions of relativity, both GR and SR combined. If the predictions are right, we should see the clocks run again at nearly the same rates as ground clocks, despite using an offset definition for the length of one second.

We note that this post-launch rate comparison is independent of frame or observer considerations. Since the ground tracks repeat day after day, the distance from satellite to ground remains essentially unchanged. Yet, any rate difference between satellite and ground clocks continues to build a larger and larger time reading difference as the days go by. Therefore, no confusion can arise due to the satellite clock being located some distance away from the ground clock when we compare their time readings. One only needs to wait long enough and the time difference due to a rate discrepancy will eventually exceed any imaginable error source or ambiguity in such comparisons.

3.   Does the GPS confirm the clock rate changes predicted by GR and SR?

The highest precision GPS receiver data is collected continuously in two frequencies at 1.5-second intervals from all GPS satellites at five Air Force monitor stations distributed around the Earth. An in-depth discussion of the data and its analysis is beyond the scope of this paper. [1] This data shows that the on-board atomic clock rates do indeed agree with ground clock rates to the predicted extent, which varies slightly from nominal because the orbit actually achieved is not always precisely as planned. The accuracy of this comparison is limited mainly because atomic clocks change frequencies by small, semi-random amounts (of order 1 ns/day) at unpredictable times for reasons that are not fully understood. As a consequence, the long-term accuracy of these clocks is poorer than their short-term accuracy.

Therefore, we can assert with confidence that the predictions of relativity are confirmed to high accuracy over time periods of many days. In ground solutions with the data, new corrections for epoch offset and rate for each clock are determined anew typically once each day. These corrections differ by a few ns and a few ns/day, respectively, from similar corrections for other days in the same week. At much later times, unpredictable errors in the clocks build up with time squared, so comparisons with predictions become increasingly uncertain unless these empirical corrections are used. But within each day, the clock corrections remain stable to within about 1 ns in epoch and 1 ns/day in rate.

The initial clock rate errors just after launch would give the best indication of the absolute accuracy of the predictions of relativity because they would be least affected by accumulated random errors in clock rates over time. Unfortunately, these have not yet been studied. But if the errors were significantly greater than the rate variance among the 24 GPS satellites, which is less than 200 ns/day under normal circumstances, it would have been noticed even without a study. So we can state that the clock rate effect predicted by GR is confirmed to within no worse than $\pm200 / 45,900$ or about 0.7%, and that predicted by SR is confirmed to within $\pm200 / 7,200$ or about 3%. This is a very conservative estimate. In an actual study, most of that maximum 200 ns/day variance would almost certainly be accounted for by differences between planned and achieved orbits, and the predictions of relativity would be confirmed with much better precision.

12-hour variations (the orbital period) in clock rates due to small changes in the orbital altitude and speed of the satellites, caused by the small eccentricity of their orbits, are also detected. These are observed to be of the expected size for each GPS satellite's own orbit. For example, for an orbital eccentricity of 0.01, the amplitude of this 12-hour term is 23 ns. Contributions from both altitude and speed changes, while not separable, are clearly both present because the observed amplitude equals the sum of the two predicted amplitudes.

4.   Is the speed of light constant?

Other studies using GPS data have placed far more stringent limits than we will here. However, our goal here is not to set the most stringent limit on possible variations in the speed of light, but rather to determine what the maximum possible variation might be that can remain consistent with the data. The GPS operates by sending atomic clock signals from orbital altitudes to the ground. This takes a mere 0.08 seconds from our human perspective, but a very long (although equivalent) 80,000,000 ns from the perspective of an atomic clock. Because of this precision, the system has shown that the speed of radio signals (identical to the "speed of light") is the same from all satellites to all ground stations at all times of day and in all directions to within ±12 meters per second (m/s). The same numerical value for the speed of light works equally well at any season of the year.

Technical note: Measuring the one-way speed of light requires two clocks, one on each end of the path. If the separation of the clocks is known, then the separation divided by the time interval between transmission and reception is the one-way speed of the signal. But measuring the time interval requires synchronizing the clocks first. If the Einstein prescription for synchronizing clocks is used, then the measured speed must be the speed of light by definition of the Einstein prescription (which assumes the speed of light is the same in all inertial frames). If some other non-equivalent synchronization method is used, then the measured speed of the signal will not be the speed of light. Clearly, the measured signal speed and the synchronization prescription are intimately connected.

Our result here merely points out that the measured speed does not change as a function of time of day or direction of the satellite in its orbit when the clock synchronization correction is kept unchanged over one day. As for seasonal variations, all satellite clocks are "steered" to keep close to the U.S. Naval Observatory Master Clock so as to prevent excessive build up of errors from random rate changes over long time periods. So we cannot make direct comparisons between different seasons, but merely note that the same value of the speed of light works equally well in any season.

5.   What is a "GPS clock"?

Cesium atomic clocks operate by counting hyperfine transitions of cesium atoms that occur roughly 10 billion times per second at a very stable frequency provided by nature. The precise number of such transitions was originally calibrated by astronomers, and is now adopted by international agreement as the definition of one atomic second.

GPS atomic clocks in orbit would run at rates quite different from ground clocks if allowed to do so, and this would complicate usage of the system. So the counter of hyperfine cesium transitions (or the corresponding phenomenon in the case of rubidium atomic clocks) is reset on the ground before launch so that, once in orbit, the clocks will tick off whole seconds at the same average rate as ground clocks. GPS clocks are therefore seen to run slow compared to ground clocks before launch, but run at the same rate as ground clocks after launch when at the correct orbital altitude.

We will refer to a clock whose natural ticking frequency has been pre-corrected in this way as a "GPS clock". This will help in the discussion of SR effects such as the twins paradox. A GPS clock is pre-corrected for relativistic rate changes so that it continues to tick at the same rate as Earth clocks even when traveling at high relative speeds. So a GPS clock carried by the traveling twin can be used to determine local time in the Earth's frame at any point along the journey—a great advantage for resolving paradoxes.

6.   Is acceleration an essential part of resolving the "twins paradox"?

If the traveling twin carries both a natural clock and a GPS clock on board his spacecraft, he can observe the effects predicted by SR without need of any acceleration in the usual twins paradox. That is as it should be because cyclotron experiments have shown that, even at accelerations of 10^{19} g (g = acceleration of gravity at the Earth's surface), clock rates are unaffected. Only speed affects clock rates, but not acceleration *per se*.

Suppose that the traveling twin is born as his spaceship passes by Earth and both of his on-board clocks are synchronized with clocks on Earth. The natural on-board clock ticks more slowly than the GPS on-board clock because the rates differ by the factor gamma that SR predicts for the slowing of all clocks with relative speed v. [gamma = $1/\sqrt{(1 - v^2/c^2)}$] But everywhere the traveling twin goes, as long as his speed relative to the Earth frame does not change, his GPS clock will give identical readings to any Earth-synchronized Earth-frame clock he passes along the way. And his natural clock will read less time elapsed since passing Earth by the factor gamma. His biological processes (including aging), which presumably operate at rates comparable to the ticking of the natural clock, are also slowed by the factor gamma.

Since this rate difference is true at every instant of the journey beginning with the first, there are no surprises if the traveling twin executes a turn-around without change of speed and returns to Earth. He will find on journey's completion what he has observed at every step of the journey: His natural clock and his biological age are slower and younger by the factor gamma than that of his Earth-frame counterparts everywhere along his journey, including at its completion. The same would have been true if he had not turned around, but merely continued ahead. He would be younger than his peers on any planet encountered who claim to have been born at the same time that the traveler was born (*i.e.*, when he passed Earth) according to their Earth-frame perspective.

Clearly, acceleration or the lack thereof has no bearing on the observed results. If acceleration occurs, it is merely to allow a more convenient comparison of clocks by returning to the starting point. But since the traveler can never return to the same point in space-time merely by returning to the same point in space, the results of a round-trip comparison are no different in kind from those made anywhere along the journey. The traveler always judges that his own aging is slower than that in any other frame with a relative motion.

Then why isn't the traveler entitled to claim that he remained at rest and the Earth moved? The traveler is unconditionally moving with respect to the Earth frame and therefore his clocks unconditionally tick slower and he ages less as judged by anyone in the Earth frame. However, if the traveler makes the same judgment, the result will depend on whether he values his natural clock or his GPS clock as the better timekeeper. If he takes readings on the GPS clock to represent Earth time, his inferences will always agree with those of Earth-frame observers. If he instead uses the results of the exchange of light signals to make inferences of what time it is at distant locations, he will conclude that the Earth-bound twin is aging less than himself because of their relative motion. But on the occasion of any acceleration his spaceship undergoes, the traveler will infer a discontinuity in the age of his Earth-bound counterpart, which can be either forward or backward in time depending on which direction the traveler accelerates. At the end of any round trip after any number of such accelerations, the traveler and Earth-bound twins will always agree about who should have aged more.

7.   Does the behavior of GPS clocks confirm Einstein SR?

To answer this, we must make a distinction between Einstein SR and Lorentzian Relativity (LR). Both Lorentz in 1904 and Einstein in 1905 chose to adopt the principle of relativity discussed by Poincaré in 1899, which apparently originated some years earlier in the 19th century. Lorentz also popularized the famous transformations that bear his name, later used by Einstein. However, Lorentz's relativity theory assumed an aether, a preferred frame, and a universal time. Einstein did away with the need for these. But it is important to realize that none of the 11 independent experiments said to confirm the validity of SR experimentally distinguish it from LR—at least not in Einstein's favor.

Several of the experiments bearing on various aspects of SR (see Table 1) gave results consistent with both SR and LR. But Sagnac in 1913, Michelson following the Michelson-Gale confirmation of the Sagnac effect for the rotating Earth in 1925 (not an independent experiment, so not listed in Table 1), and Ives in 1941, all claimed at the time they published that their results were experimental contradictions of Einstein SR because they implied a preferred frame. In hindsight, it can be argued that most of the experiments contain some aspect that makes their interpretation simpler in a preferred frame, consistent with LR. In modern discussions of LR, the preferred frame is not universal, but rather coincides with the local gravity field. Yet, none of these experiments is impossible for SR to explain.

For example, Fresnel showed that light is partially dragged by the local medium, which suggests a certain amount of frame-dependence. Airy found that aberration did not change for a water-filled telescope, and therefore did not arise in the telescope tube. That suggests it must arise elsewhere locally. Michelson-Morley expected the Earth's velocity to affect the speed of light because it affected aberration. But it didn't. If these experimenters had realized that the aether was not a single entity but changed with the local gravity field, they would not have been surprised. It might have helped their understanding to realize that Earth's own Moon does not experience aberration as the distant stars do, but only the much smaller amount appropriate to its small speed through the Earth's gravity field.

Another clue came for De Sitter in 1913, elaborated by Phipps [3], both of whom reminded us that double star components with high relative velocities nonetheless both have the same stellar aberration. This meant that the relative velocity between a light source and an observer was not relevant to stellar aberration. Rather, the relative velocity between local and distant gravity fields determined aberration. In the same year, Sagnac showed non-null results for a Michelson-Morley experiment done on a rotating platform. In the simplest interpretation, this demonstrated that speeds relative to the local gravity field do add to or subtract from the speed of light in the experiment, since the fringes do shift. The Michelson-Gale experiment in 1925 confirmed that the Sagnac result holds true when the rotating platform is the entire Earth's surface.

When Ives and Stilwell showed in 1941 that the frequencies of radiating ions depended on their motion, Ives thought he had disposed once and for all of the notion that only relative velocity mattered. After all, the ions emitted at a particular frequency no matter what frame they were observed from. He was unmoved by arguments to show that SR could explain this too because it seemed clear that nature still needed a preferred frame, the motion relative to which would determine the ion frequencies. Otherwise, how would the ions know how often to radiate? Answers to Ives' dilemma exist, but not with a comparable simplicity. Richard Keating was surprised in 1982 that two atomic clocks traveling in opposite directions around the world, when compared with a third that stayed at home, showed slowing that depended on their

Experiment	Description	Year
Bradley	Discovery of aberration of light	1728
Fresnel	Light suffers drag from the local medium	1817
Airy	Aberration independent of the local medium	1871
Michelson-Morley	Speed of light independent of Earth's orbital motion	1881
De Sitter	Speed of light independent of speed of source	1913
Sagnac	Speed of light depends on rotational speed	1913
Kennedy-Thorndike	Measured time also affected by motion	1932
Ives-Stilwell	Ions radiate at frequencies affected by their motion	1941
Frisch-Smith	Radioactive decay of mesons is slowed by motion	1963
Hafele-Keating	Atomic clock changes depend on Earth's rotation	1982
GPS	**Clocks in all frames continuously synchronized**	1997

Table 1. Independent experiments bearing on Special Relativity

absolute speed through space—the vector sum of the Earth's rotation and airplane speeds—rather than on the relative velocities of the clocks. But he quickly accepted that astronomers always use the Earth's frame for local phenomena, and the solar system barycentric frame for other planetary system phenomena, to get results that agreed with the predictions of Relativity. Being unaware of LR, he did not question the interpretation at any deeper level.

Table 2 summarizes what the various experiments have so say about a preferred frame. These experiments confirm the original aether-formulated relativity principle to high precision. However, the issue of the need for a preferred frame in nature is, charitably, not yet settled. Certainly, experts do not yet agree on its resolution. But of those who have compared both LR and SR to the experiments, most seem convinced that LR more easily explains the behavior of nature.

8.   How does the resolution of the "twins paradox" compare in LR and SR?

In LR, the answer is simple: The Earth frame at the outset, and the dominant local gravity field in general, constitutes a preferred frame. So the high-speed traveler always comes back younger, and there is no true reciprocity of perspective for his or other frames.

In SR, the answer is not so simple; yet an explanation exists. The reciprocity of frames required by SR when Einstein assumed that all inertial frames were equivalent introduces a second effect on "time" in nature that is not reflected in clock rates alone. We might call this effect "time slippage" so we can discuss it. Time slippage represents the difference in time for any remote event as judged by observers (even momentarily coincident ones) in different inertial frames.

For example, we would argue that, if it is 9/1998 here and now, it is also 9/1998 "now" at Alpha Centauri. But an observer here and now with a sufficiently high relative motion (say, 99% of c; gamma = 7) might judge that

Experiment	Type	Notes on Reciprocity
Bradley	Aberration	Moon exempt
Fresnel	Fresnel drag	Existence of aether
Airy	Existence of aether	Water in 'scope ignored
Michelson-Morley	No universal aether	Aether "entrained"?
De Sitter	c independent of source	Double star aberration
Sagnac	c depends on rotation	Local gravity field non-rotating
Kennedy-Thorndike	Clocks slow	Motion w.r.t. local gravity field
Ives-Stilwell	Ions slow	"
Frisch-Smith	Mesons live longer	"
Hafele-Keating	Clocks depend on rotation	Preferred frame indicated
GPS	Universal synchronization	Preferred frame = local gravity

Table 2. Independent experiments bearing on Special Relativity

it is 9/1994 at Alpha Centauri "now" (meaning that he just left there one month of Earth time ago, and it was 8/1994 then). Or he might judge that it is 9/2002 at Alpha Centauri "now" (meaning that he will arrive there in one month of Earth-elapsed time, and will find the time to be 10/2002). These differences of opinion about what time it is at remote locations are illustrations of time slippage effects that appear only in Einstein SR to preserve the frame independence of its predictions.

So as a traveler passes Earth in 8/1994 at a speed of 0.99 c, time slippage effects begin to build up. Seven months later by his natural clock, the traveler arrives at Alpha Centauri. His own GPS clock shows four years of elapsed time, and indeed Alpha Centauri residents who think they are calendar-synchronized with Earth agree that the twin arrives in 9/1998. But the traveler is convinced by Einstein SR that only one month of Earth time has elapsed since he passed Earth and noted the time as 8/1994. The traveler, upon arriving at Alpha Centauri, claims that the time is "now" 9/1994 on Earth. Alpha Centauri residents claim it is "now" 9/1998 on Earth. The difference is the time slippage predicted by SR.

If the traveler orbits Alpha Centauri at a speed of 0.99 c, then whenever he is headed in the direction of Earth his opinion changes to Earth time "now" is 9/2002. And whenever he is again headed away from Earth, Earth time is once again 9/1994. Earth time "now" changes continually, according to SR, because of these time slippage effects needed to retain frame reciprocity. Earth residents—even the ones who died in 1998—are oblivious to their repeated passages into the future and past of the traveling twin, with concomitant deaths and resurrections.

So when the traveler finally does return, he will indeed find that time on Earth is 10/2002, just as his GPS clock shows. He accounts for this as two months of elapsed time on Earth's slow-running clocks during his own 14-month (by his natural clock) journey, plus 8 years of "time slippage" when the traveler changed frames. There is no logical or mathematical inconsistency in this resolution, which is why SR remains a viable theory today.

We are, of course, free to question whether or not this mathematical theory retains a valid basis under the principles of causality. For those of us who answer "yes", LR is unnecessary, and inelegant because it depends on a preferred frame. For those of us who answer "no", LR is then the better descriptor of nature, requiring the sacrifice of symmetry ("covariance") to retain causality.

9.   What physical consequences arise from the differences between LR and SR?

In SR, speed causes changes in time and space themselves, not just in clocks and rulers. Rest mass remains unchanged, but resistance to acceleration increases toward infinity as speed approaches c. There is no absolute time or space in the universe. The time at remote locations depends on what frame one observes from. All frames are equivalent.

In LR, speed relative to the preferred frame (the local gravity field) causes clocks to slow and rulers to contract. Electromagnetic-based forces become increasingly less efficient with increasing speed relative to the preferred frame, and approach zero efficiency as speed approaches c. There are natural, physical reasons why these things should be so. [2] The frame of the local gravity field acts as a preferred frame. Universal time and remote simultaneity exist.

The single most important difference is that, in SR, nothing can propagate faster than c in forward time. In LR, electromagnetic-based forces and clocks would cease to operate at speeds of c or higher. But no problem in principle exists in attaining any speed whatever in forward time using forces such as gravity that retain their efficiency at high speeds.

References

[1]   Alley, C.O. and Van Flandern, T. (1998). "Absolute GPS to Better Than One Meter," preprint not yet submitted for publication.

[2]   Van Flandern, T. (1993). *Dark Matter, Missing Planets and New Comets*, North Atlantic Books, Berkeley, CA.

[3]   Phipps, T. (1989). "Relativity and Aberration," *Amer.J.Phys.* 57, 549-551.

History and Philosophy

Some Almost Unknown Aspects
of Special Relativity Theory

M. Barone
National Scientific Research Centre "Demokritos"
15310 Aghia Paraskevi-Athens-Greece.

Introduction

T. Kuhn in his book *The structure of Scientific Revolutions*, referring to the textbooks tradition says:

> *Characteristically, textbooks of science contain just a bit of history, either in an introductory chapter or often in scattered references to the great heroes of an earlier age. From such references both students and professionals come to feel like participants in a long-standing historical tradition. Yet the textbook-derived tradition in which scientists come to sense their participation is one that, in fact, never existed. For reasons that are both obvious and highly functional, science textbook (and too many of the older histories of science) refer only to that part of the work of past scientists that can easily be viewed as a contribution to the statement and solution of the text paradigm problems. Partly by selection and partly by distortion, the scientists of earlier ages are implicitly represented as having worked upon the same set of fixed problems and in accordance with the same set of fixed canons that the most recent revolution in scientific theory and method has made seem scientific. No wonder that textbooks and the historial tradition they imply have to be rewritten after each scientific revolution. And no wonder that, as they are rewritten, science once again comes to seem largely cumulative.*

In such an environment, education became a mere transmission of the orthodoxy and critical enquiry is eschewed. The aim of this paper is to draw the attention of our colleagues to some features of the Special Relativity theory which may enhance a critical enquiry.

1. The Larmor-Fitzgerald-Lorentz-Poincaré approach from the historical point of view

Hereinafter we will review chronologically some important dates which can be taken as mile-stones of the Pre-relativistic period.

1820:   Oersted discovered that a current flowing trough a line conductor creates a magnetic field.

1875:   Helmoltz made the hypothesis of the existence of the electron which was discovered in the following years by Lénard, J.J. Thomson and others.

1876:   Ampère and Faraday gave a strong contribution to the formulation of the e.m. theory. Maxwell, to give an explanation of the light propagation and to the experiments made by Young and Fresnel about the interference and the diffraction,pointed out that light is made up by e.m.waves travelling either through the vacuum or through a medium called "ether." In this way the cosmic ether on which the Newton physics was based founds a further support and the Luminifer theory and the e.m.theory were merged together. (1)

1879:   Lorentz wrote some papers in which he put together his idea of the ether and of the electron with the e.m. theory of Maxwell, Lorentz said that the ether is something static in which E and B propagate and therefore it is a geometrical reference system. The electron can be considered pointlike, with its inertia and radius $=1.9 \times 10^{-13}$ cm; its mass depends on the speed and it can be transformed in a magnetic field. (1)

1887:   Hertz discovered e.m waves. Michelson and Morley made an experiment to prove the ether. (2) Voigt, to explain the result of the previous experiment, introduced the" local time." (3)

$$T' = \frac{T}{1 - v^2/c^2}$$

v = speed of the moving reference system.

1889:   Fitzgerald, to explain the results of certain optical experiments, introduced the "length contraction." In a letter sent to the American journal *Science* dated May 2, 1889 he said:

> I have read with much interest Messrs Michelson and Morley's wonderfully delicate experiment attempting to decide the important questions to how far the ether is carried along by the Earth. Their result seems opposed to other experiments showing that the ether in the air can be carried along only to an inappreciable extent. I would suggest that almost the only hypothesis that could reconcile this opposition is that the lengths of the material bodies change according as they are moving through the ether or across it by an amount depending on the square of the ratio of the velocity to that of light. We know that electric forces are affected by the motion of

electrified bodies relative to the ether and it seems a not improbable
supposition that the molecular forces are affected by the motion and
that the size of the body alters consequently.

1892:   Fitzgerald (4) and Lorentz (5) in order to give an account of the
        missing detection of the first order effect of *v/c* and of the Michelson-
        Morley experiment formulated the "length contraction."

1895:   Lorentz in his article: "*Versus einer theorie der electrischen und optischen*
        Erscheinungen zwischen Körpen"(6) stated that all the bodies in a
        reference frame moving with a speed *v* with respect to the one fixed
        should have a contraction of their length in the direction of the
        motion of a factor

$$\sqrt{1 - \frac{v^2}{c^2}}$$

therefore $L = L_o \sqrt{1 - v^2/c^2}$

In the same paper he introduced the 'local time' defined as the time
measured in a reference frame moving with a speed *vT'* (local time)
was not anymore equal to *T* as it was stated by the Galilean
Relativity but T' = T – *vL/c*[2,] Lorentz conceived the local time as an
artifice to explain the results of the Michelson and Morley
experiment; it was different from the real time which was absolute as
it was for Newton.

H. Poincaré in the same year in the paper "*L'eclairage electrique*"(7)
commenting some of Larmor's remarks about the fact that the period
of a periodic dynamical systems changes when it moves trough the
ether (time dilation effect), said:

> *Experiment has provided numerous facts admitting the following*
> *generalization: it is impossible to observe absolute motion of matter,*
> *or, to be precise, the relative motion of ponderable matter and ether.*

1897:   J.J. Thomson discovered the electron

1898:   H. Poincaré in the paper "Measurement of time"(8), discussing the
        characteristics of physical time, drew the conclusion that simultaneity
        is a convention to be made and based on the constancy of the velocity
        of the light. He said:

> *This is a postulate without which it would be impossible to*
> *undertake any measurement of this velocity. The said postulate can*
> *never be verified experimentally.... The simultaneity of two events*
> *or the sequence in which they follow each other the equality of two*
> *time intervals should be determined so as to render the formulation*
> *of natural laws as simple as possible. In other words, all these rules,*
> *all these definitions, are only the front of implicit convention.*

1899:   Lénard discovered the electron.

1900:   H. Poincaré in the article: "La Théorie de Lorentz et le Principe de réaction" (9) presented to the *Lorentz Festschrift* gave an interpretation of the Lorentz local time as the time corresponding to readings of two clock synchronized by a light signal under the assumption of a constancy of the velocity of light. He also postulated that the electromagnetic energy might possess mass density equal to energy density times a constant c^2 or $E = mc^2$. He said:

> *If an instrument has a mass of 1 kg and has sent in a unique direction 3 Megajoules of mass at the speed of light, it has a recoil speed of 1 cm/sec.*

That means that the electromagnetic energy has an inertia given by $E = mc^2$.

1900:   J.J. Larmor in his paper "Aether and matter" (10) introduced the "time dilation."

1902:   H. Poincaré published "La Science et l'Hypothèse" (11) in which he stated that absolute space and time do not exist. Only relative movement should be taken into account and it is a convention to say that two intervals of time are equal. In the same year Lorentz got the Nobel Prize.

1904:   In May in the paper: "Electromagnetic phenomena in a system moving with a velocity smaller than the light"(12) Lorentz gave the transformations:

$$\begin{cases} x' = \dfrac{x - vt}{\sqrt{1 - \dfrac{v^2}{c^2}}} \\ y' = y \\ z' = z \\ t' = \dfrac{t\left(1 - v^2/c^2\right) - vx/c^2}{\sqrt{1 - v^2/c^2}} \end{cases}$$

The equation for the time contains the term $-v^2/c^2$ and the Maxwell equations are not invariant under such trasformations. Lorentz said of them:

> *But I never thought that this had anything to do with real time... there existed for me only one true time.*
> *I considered by time-transformation only a heuristic working hypothesis.*

In September in a talk delivered in Saint Louis (Missouri) to the Congress of Arts and Science, H. Poincaré extented the Principle of Relativity to electromagnetism. Poincaré stressed that:

From all these results must arise an entirely new kind of dynamics, which will be characterized above all by the rule that no velocity can exceed the velocity of light.

1905: In June in the *Comptes Rendus* of the Academy of Sciences H. Poincaré published the paper: "On the dynamics of the electron"(13) where he formulated the Lorentz transformations in the form we know today:

$$\begin{cases} x' = \dfrac{x - vt}{\sqrt{1 - v^2/c^2}} \\ y' = y \\ z' = z \\ t' = \dfrac{t - vx/c^2}{\sqrt{1 - v^2/c^2}} \end{cases}$$

They were in strict accordance with the invariance of the Maxwell equations and they form a group. The so-called Lorentz-group. Lorentz commenting this results said:

My considerations published in 1904 in the paper "Electromagnetic phenomena…" have pushed Poincaré to write this article in which he has attached my name to the transformations which I was unable to obtain…

As invariant of the Lorentz group we obtain:

$$x^2 + y^2 + z^2 - c^2 t^2 = \text{const.}$$
$$c = \text{const.} = 300{,}000 \ \text{Km} \,/\, \text{s}$$

As we can see this is the same group to which refers Minkowski in his article "Space and Time" of Sep. 21, 1905.

The September 26, Einstein published in the *Annalen der Physik* the article "On the electrodynamics of moving bodies" in which he formulated the Theory of Special Relativity.

The manuscript was submitted early the 30 of June and had as coauthor Milena Maric.

In the above we have proved that:

a) What today we call Lorentz transformations were made by H. Poincaré.

b) When Einstein wrote his paper the basis of the Special Relativity were already given by Larmor-Fitzgerald-Lorentz-Poincaré.

3. The Einstein approach

While for Lorentz the ether represented a state of real rest, for Einstein only the relative motion of two or more observers moving at constant speed was real, and the laws of the physics must be equal in any reference frame the observers may choose. This last assumption was taken as hypothesis from which, together with the constancy of the speed of light c, Einstein derived his theory in a very concise and elegant way. This was its merit. (14)

4.   Einstein and the ether

In a letter to his girlfriend Mileva Maric, dated probably July 1899 (item FK53, Mudd Library,Princeton Univ.), Einstein express some doubts about the existence of the ether. He says:

I am more and more convinced that the electrodynamics of the bodies in motion, such as it is presented today, does not correspond with reality, and that it will be possible to formulate it in a more simple way. The introduction of the word "ether" in the theories of electricity leads to the idea of a medium about the motion of which we speak without the possibility as I think, to attribute any sense to such a word.

In his paper of 1905, introducing the Special Theory of Relativity, Einstein denied the existence of the luminiferous stationary ether, which had been ruled out by the Michelson type experiment and because of that he is generally believed to be the destroyer of the ether concept. What is ignored by the hystorians of the Modern Physics is that Einstein, in 1916, modified his negative attitude and till the end of his life in 1955, he developed his own point of view about the ether. (15)

In a paper given as contribution of a special issue of *Nature* devoted to relativity and known as the "Morgan manuscript," in 1919 he wrote:

It is clear that there is no place in the special relativity theory for a notion of an ether at rest. If the reference frame K and K' are entirely equivalent for the formulation of the laws of Nature, then it is inconsequent if one attributes a fundamental role to a notion which discriminates one system in favour of the other one. If the ether is supposed to be at rest with respect to K so it is in motion with respect to K," which does not fit into the physical equivalence of the two systems.
Therefore in 1905, I was of the opinion that it was no longer allowed to speak about the ether in physics. This opinion, however, was too radical, as we will see later when we consider the general relativity theory. It is allowed much more than before to accept a medium penetrating the whole space and to regard the electromagnetic fields and the matter as well as states of it. But it is not allowed to attribute to this medium, in analogy to the ponderable matter, a state of motion in any point. This ether must not be conceived as composed of particles the identity of which can be followed in time.

In 1934 in his book *Mein Weilbild*, translated in many languages, Einstein repeated his conception of the space, the ether and the field and after that till the end of his life he used the term "total field" instead of ether. The reason for that may be was the political situation in Germany, when during the 30s, the word "ether" was used by the "German physics" against "Hebrew physics."

5. Post-relativistic Theories Based on the Ether

Lorentz in 1909 in his paper "The Theory of Electrons"(16) formulated a theory equivalent to the Einstein's one taking as assumptions the existence of the ether, rod contraction and the time dilation effect. He adopted clock synchronization based on the one-way speed of the light as suggested by Einstein.

In the thirties Ives (17), in the fifties Builder (18), in the sixties Ruderfer (19), and in the seventies Prokhovnik (20) and Mansouri and Sexl (21) developed theories based on the ether and equivalent to the Special Theory of the Relativity. In the nineties thanks to the work of Cavalleri (22) Selleri (23) and others, the topic has attracted the attention of an increasing numbers of researchers.

6. Systems Including Acceleration: Herzian Relativity

The Special Relativity is limited to inertial motions. Acceleration is not taken into account and in the real world the motion may be either inertial or not inertial (rocket launch, *etc.*) and a good Mechanics should take that into consideration.

H.R. Hertz, 15 years before the publication of the Einstein's paper elaborated an electromagnetic theory called Maxwell-Hertz Electrodynamics, valid for all motions (inertial or not) but at small velocities (24). The Hertz theory was based on the ether, but because this was not detected the theory came in conflict with observations. Such a theory, if extended at high velocities is a good basis for the formulation of a Relativistic Electrodynamics.

Mocanu (25) has formulated an Hertz Relativistic Mechanics which justifies all the modifications of classical concepts of space and time, length contraction, time dilation, variation of mass with the velocity.

7. Special Relativity and its Applications

Well known are the applications in Nuclear Physics (fission process of U-235, fusion process of He), in Particle Physics(the technique of colliding beams, based on the energy available in the center of mass system, allow to have reactions and to create particle which otherwise could be impossible to have using fixed target accelerators, which are using the energy of the laboratory system. At the I.S.R the collision between two beams of protons of 28 GeV, allowed to have in the c.m. a total energy of 1800 Gev. The time dilation effect which stretches the muon lifetime by a factor 12 when such a particle is accelerated to $v = 0.9965\,c$ from the value of 2.2×10^{-6} sec, justifies why we can detect such a particle on the earth, while it is created in cosmic rays very far away).

What is less known is that Special Relativity has applications in other fields such as Astronomy, Engineering, Geodesy and Metrology. We will give a short overview of these below.

An example of time dilation in Astronomy is given by the binary star system SS433, which is formed by a spinning neutron star (pulsar), remnant of a supernova, and a companion star. The flow of matter in the neutron star from the companion star and the spinning of the pulsar creates two symmetrical jets emerging from the neutron star and with a precessional motion with a period of 164 days because of the tidal motion of the other star. The spectra of the binary system display pairs of hydrogen emission lines, one shifted towards the red and the other toward the blue; both pairs cyclically shift over the precession period of the SS433. This can be explained with the fact that one time one jet will be moving towards the Earth while the other will be receding. If one looks at the median point of the Doppler shifted lines one will find that it is displaced from the rest wavelength observed in the laboratory. This can be explained only by the Special Relativity which predicts a slowing of 3 percent for the speed of the jets which emerge with a velocity of more than a quarter that of light. The new relativistic Doppler effect explains the anomalous redshift.

In the field of Engineering (26), Special Relativity finds an application in Navigation because of the breakdown of simultaneity, defined by the Einstein synchronization procedure and by the convention that the speed of light $c = 299,792,498$ m/sec. In fact, to locate an object within 1 foot (this is the distance traveled by a light signal in 1 nsec) of errors using the Global Positioning System the time synchronization between the clock on board the satellites and the earth stations should be better than 10^{-9}. Taking into account that the timing is made by atomic clocks having actually a stability of 10^{-13} then errors appear. They are of the order of 300m for each microsecond of deviation in the synchronization and they are compensated by corrections. If the clocks are carried on an aircraft having a speed of 1000 km/h and covering a distance of 3500 km, the corrections to be made in the synchronization with the clocks on the earth are of 108 nsec *i.e.* 24 m (108×0.23m$=24$m).

In Communicating Systems, if we consider the Communications Networks, the timing is affected by the failure of the Einstein clock synchronization in a rotating frame, due to the Sagnac effect. (27)

The synchronization around the earth equator involves an error of 208 nsec and a correction to the readings of a clock on the rotating earth should be made. To synchronize a link going from S. Francisco to New York with a second going from S. Francisco-Miami-New York the correction to be made because of the Sagnac effect is of 11 nsec. The effect we have taken into account is of the first order, there is also a time dilation due to the second order effect v^2/c^2 in this case $t' = \left[1 - \left(1/2\, v^2/c^2\right)\right]t$, due to the Doppler shift.

In Geodesy and Metrology accurate measurents of the earth's plates are made using the GPS, in fact very long base lines between fiducial points (placed on different crustal plates) placing two receivers at the end of the base line and measuring the signals from two satellites with the method of the double differential are made. The obtained values need to be corrected because of the relativistic effect due to the clock synchronization and the Sagnac effect. Even small effect of a few cm displaced by the crustal plates have a great impact on the construction codes, building restrictions and earth quake predictions.

Conclusions

We hope we have given a less than usual overview of the Special Relativity Theory.

Acknowledgments

The author wishes to express special gratitude to prof. A.A. Tyapkin and to Prof.F. Selleri for the support given to the research reported in this paper.

References

[1]   S. Bergia, La Storia della Relativitá in *La Fisica nella Scuola*,VIII, 1,1975
[2]   A.A. Michelson and E.W. Morley, *Amer Journ. of Science*, 34,333, 1887.
[3]   W. Voigt, *Nachr. KG.W. Gottingen*, 3,4 1,1987.
[4]   See O. Lodge, *London Transac*, Al84, 727, 1893
[5]   H.A. Lorentz, *Amst Verch. Akad*, 1,74,1892
[6]   H.A. Lorentz, *Versus einer...*,(E.J.Bill,London, 1895).
[7]   H. Poincaré. *L'eclairage electrique*, L5,p5, 1895.
[8]   H. Poincaré, *Revue de Metaphysique et de Morale*, 6,1,1898
[9]   H. Poincaré *Archives Neerland*, 5,252,1900.
[10] J.J. Larmor, *Aether and Matter*, (Univ.Press.Cambridge, 1900)
[11] H. Poincaré, *La Science et l'Hypothese*, Paris,Flammarion,1902,2
[12] H.A. Lorentz, *Proc.Acad.Scientific*, Amsterdam, 6,809, 1904
[13] H. Poincaré, *Comptes rendus*, 140,1504,1905.
[14] John S. Bell, How to teach special relativity in: *Speakable and Unspeakable in Quantum Mechanics*, Cambridge University Press (1987).
[15] L. Kostro, The physical meaning of Albert Einstein's relativistic ether concept in: *Frontiers of Fundamental Phyisics*, Editors M. Barone and F. Selleri, Plenum Press.
[16] H.A. Lorentz, *The Theory of Electrons and its Application to the Phenomena of Light and Radiation Heat*. (Columbia U.P.,New York, 1909,1915.)
[17] Ives, H.E., *J.Opt.Soc.Amer.*29,472, 1939.
[18] Builder, Ether and Relativity, *Australian J.Phys*, 1 1,294,1958.
[19] Ruderfer, *Phys.Rev.Letters* 5,192,1960.
[20] J.J. Prokhovnik, An Introduction to the Neo-Lorentzian Relativity of Builder, *Speculation in Science and Technology* Voll2, N3,225,1979.
[21] Mansouri and Sexl, *General Relativity and Gravitation*,Vol 18,N7,p.497,1977.
[22] Cavalleri, C. & Bernasconi *Il Nuovo Cimento*,Vol 104,13,N5,pag.545,l989.
[23] Selleri, F., *Found.Phys* 41,369,1996.
[24] H. Hertz, *Ann. der Phys* 41,369,1996.

[25] Mocanu, Herztian extension of Einstein's special relativity to non uniform motions in: *Advances in Fundamental Physics*, Editors M. Barone and F. Selleri; Hadronic Press, Palm Harbor, 1995.

[26] N. Ashby, Relativity in the Future of Engineering, IEEE 1993, *International Frequency Control Symposium*.

[27] Sagnac, G., *Comptes Rendus*, 157,708.1410,1913.

Further Reading

[1] A. Tyapkin, *Relativitá speciale*, Editrice Jaca Book, Milano,1992.

[2] A. Tyapkin, On the History of Special Relativity Concept, in: *New Frontiers in Physics*,Vol. 1, Hadronic Press, Editor T.L. Gill, 1996.

[3] F. Goy and F. Selleri, Time on a rotating platform. Pre-print-1996 Bari University, Italy.

Correspondence and Commensurability in Modern Physics (a Study of the Compton Effect)

Jenner Barretto Bastos Filho
Departamento de Física da Universidade Federal de Alagoas
Campus da Cidade Universitária, Cep 57072-970
Maceió – Alagoas – Brasil

The concepts of correspondence and commensurability are studied in the light of the Compton effect. We compare the Compton formula for the photon-electron scattering with another derivation in which the electron is considered as a non-relativistic particle. Although based on a less general theory, the second derivation leads to a formula which seems to be more general than the original Compton scattering formula. We show that this generality is only apparent. This circumstance means that the inclusion of additional terms does not necessarily imply more generality. We discuss some difficulties with the concepts of commensurable theories and correspondence, in particular those involving the passage from one theory into another through the concept of a limit of mathematics. We conclude that the problem goes beyond the mere mathematical limit.

1. Correspondence and commensurable theories

It is widely thought that a more general theory ought to pass into a particular one which is a special case of the former. With respect to mathematically formulated theories this procedure constitutes, in the overwhelming majority of cases, the passage from the general theory to the particular one by using the concept of limit in mathematics[1].

The above point of view would be compatible with the inference following which the theories must be *commensurable* with one another. Perhaps the main argument in favour of this point of view is the following: if a given particular theory is proved to be correct in a given domain of validity, then a more general theory must reproduce all the results of the former. In other words, the general theory passes into the particular one in this limit. This constitutes a possible formulation of the *correspondence principle*.

K.R. Popper,[2] for example defines the *correspondence principle* as "*a demand that a new theory should contain the old one approximately for appropriate values of the parameters of the new theory.*" In this formulation of the correspondence principle, the word *new* means *general* and the word *old* means *particular*. At first sight, this terminology could imply that *commensurability* between two given theories and *induction* are inherently connected concepts. However, Popper is a bitter adversary of *induction* and thus, to him, the *inductive* process, understood as a transition from the *particular (old)* to the *general (new)* theory, is impossible.

According to Popper,[3–4] the passage from a particular to a general theory cannot be reached by means of *inductive inference*. The inductive process is not possible because it contradicts the starting premises of the particular theory. However, the inverse process—from the general to the particular theory—can be perfectly understood as a limiting case. For instance, it is absolutely impossible to pass inductively from Galileo's law of falling bodies to the Newtonian gravitational law because Galileo's law asserts that all the bodies fall with constant acceleration g, whereas Newton's law asserts that this acceleration increases during the fall. Following Popper, the result where *acceleration increases during the fall* is contradicted by the starting premise whereby *acceleration is constant during the fall*; consequently, the inductive process is impossible. However, the inverse way is perfectly understandable because the increase of acceleration is small, and thus constant acceleration constitutes an excellent and mathematically justified approximation. According to this view, *induction* is denied but *commensurability* is kept.

In this context, the concepts of *commensurability* and *correspondence* seem to be inherently correlated.

The rational and logical *desideratum* of *commensurability* between two given theories connected by a *correspondence principle* is different from their historical development. Indeed, theories are built on different conceptual bases and in different historical contexts.

Nevertheless, some major thinkers vehemently oppose the *correspondence principle*. Thomas Kuhn [5] argues against correspondence among theories divided by a scientific revolution. Paul Feyerabend [6] recognises the important role played by the *correspondence principle* in the context of the quantum mechanics, but feels that the idea of correspondence *(Korrespondenzdenken)* hinders revolution in science.

In this paper, we intend to present a discussion of some aspects of the above concepts in the context of a single example: the Compton scattering theory. As we know the Compton scattering formula is obtained from the assumptions of validity of Planck-Einstein and de Broglie dualistic formulas for the photon, and the validity of the relativistic formulas expressing the conservation laws (energy and linear momentum) for photon-electron scattering. There is an alternative derivation of the Compton scattering formula in which, with respect to the electron, relativistic formulas are not used. This second approach [7] leads to a new scattering formula which "contains" the original one due to Compton, plus an additional term. When this "additional term" is very small, then the new formula "passes into" the old Compton scattering formula. This situation seems strange because non relativistic formulas are less general than the corresponding relativistic ones.

We propose to discuss this problem in the light of the *correspondence principle*.

2. Derivation of the Compton scattering formula

The Compton scattering consists of a collision involving an incident photon of energy hv with an electron at rest with rest energy $m_0 c^2$. Planck's constant is denoted by h; v refers to the frequency of the incident photon, m_0 denotes the rest mass of the electron and c denotes the velocity of light in vacuum.

After the collision the outgoing photon and electron acquire the energies respectively hv' and mc^2, with

$$m = m_0 \left(1 - \frac{v^2}{c^2}\right)^{-\frac{1}{2}} \tag{1}$$

where v denotes the final velocity of the electron and v' refers to the frequency of the outgoing photon. The symbol ϕ denotes the angle formed by the directions respectively of the incident and outgoing photon, and β denotes the angle formed by the directions respectively of the incident photon and outgoing electron.

The ingredients appearing in the Compton scattering are:

(*i*)   The Planck-Einstein and de Broglie dualistic relations for the photon, respectively

$$E = hv \tag{2}$$

and

$$p = h/\lambda = hv/c \tag{3}$$

(*ii*) The conservation of energy and the conservation of linear momentum in the context of the relativistic theory.

Mathematically, the conservation of energy and the conservation of linear momentum are expressed by

$$hv + m_0 c^2 = hv' + mc^2 \tag{4}$$

$$\frac{hv}{c} = \frac{hv'}{c}\cos\phi + mv\cos\beta \tag{5}$$

$$0 = \frac{hv'}{c}\sin\phi - mv\sin\beta \tag{6}$$

Straightforward and well-known calculations lead to the result

$$(\lambda' - \lambda) = \Delta\lambda = 2\lambda_o \sin^2\left(\frac{\phi}{2}\right) \qquad (7)$$

where,

$$\lambda_o = \frac{h}{m_o c} \qquad (8)$$

The quantity λ_o is the so-called Compton wavelength, and

$$\lambda = \frac{c}{v}; \quad \lambda' = \frac{c}{v'} \qquad (9)$$

are the wavelengths of the incident and outgoing photons respectively.

3. A derivation of the scattering formula assuming the electron to be a non-relativistic particle

The ingredients appearing here are (i) and
(ii') The conservation of energy and the conservation of linear momentum. The electron is supposed to be a non relativistic particle.

According to the assumptions (i) and (ii'), the above conservation laws are expressed by,

$$hv = hv' + \frac{p_e^2}{2m_o} \qquad (10)$$

$$\frac{hv}{c} = \frac{hv'}{c}\cos\phi + (m_o v)\cos\beta \qquad (11)$$

$$0 = \frac{hv'}{c}\sin\phi - (m_o v)\sin\beta \qquad (12)$$

In the context of the non relativistic approximation [$1 >> (v/c)^2$ i.e. $c^2 >> v^2$], the electron mass is

$$m = m_o$$

Straightforward calculations (see appendix) lead to the result

$$(\lambda' - \lambda) = \frac{\lambda_o}{2}\left[\frac{(\lambda' - \lambda)^2}{\lambda\lambda'}\right] + 2\lambda_o \sin^2\left(\frac{\phi}{2}\right) \qquad (13)$$

If

$$\frac{(\lambda' - \lambda)^2}{\lambda\lambda'} \approx 0 \qquad (14)$$

then, (13) passes into (7). We arrived at a strange situation in which a less general theory "leads to" a more general formula.

4. Some conceptual problems involving the correspondence principle

Can a less general theory lead to a more general formula?

At first sight, the answer to this question is an emphatic *no*. It is widely accepted that a more general theory leads to a more general formula. The less general formula of the less general theory is obtained from the more general theory when the later passes into the former (limit case). However, we found a case that seems to be a counter-example of the above statement. We will see that this generality is only apparent.

Is this an inverse correspondence principle?

In order to answer this question we must analyse the passage from the relativistic theory to the classical one. It is not enough to say simply that Lorentz transformations pass into Galileo's transformations when $c \rightarrow \infty$. This limit is not quite correct. The more correct situation is $1 >> (v/c)^2$, *i.e.* $c^2 >> v^2$. We ought to consider a transition in which some kind of difficulty takes place.

Let us consider, for instance, the passage from the relativistic theory to the classical one in the case of the Lagrangean expression of the free particle. By using the principle of least action and the concept of interval, ds, of the Minkowski space, Landau and Lifshitz [8] conclude that the Lagrangean for the free particle in the relativistic domain must assume the expression

$$L = -\alpha c \left(1 - \frac{v^2}{c^2} \right)^{\frac{1}{2}}$$

where α is a constant which will be identified by means of the correspondence process from the relativistic to the classical theory. This transition is carried out when L is expanded in powers of v/c. Neglecting terms of higher order, we obtain

$$L = -\alpha c \left(1 - \frac{v^2}{c^2} \right)^{\frac{1}{2}} \approx -\alpha c + \frac{\alpha v^2}{2c}$$

Comparison with the classical expression $L = m_o v^2/2$ leads to the value

$$\alpha = m_o c$$

Thus, the Lagrangean expression of the free particle in the classical domain will be

$$L \approx -m_o c^2 + \frac{m_o v^2}{2}$$

The correspondence process is logical and intelligible, but does not reproduce exactly the classical result $L = m_o v^2/2$.

How can the "spurious" term $(-m_o c^2)$ be interpreted in the light of the correspondence principle? Landau and Lifshitz write[9]:

"Constant terms in the Lagrangean do not affect the equation of motion and can be omitted."

However this procedure does not solve the conceptual problem because the term $(-m_o c^2)$ is not merely a simple constant term. It is a term out of the conceptual domain of Newtonian physics, and thus Newtonian mechanics is not genuinely derived.

At first sight, correspondence is regarded as a process involving a limit when the general theory passes into a particular one. But this process (logical and intelligible) cannot be used to hide conceptual problems.

Let us consider another interesting example. It is widely accepted that the General Theory of Relativity (GTR) passes into the Newtonian Gravitational Theory (NGT) when the famous Newtonian limit takes place. In spite of this the above two theories are based on radically different concepts of space. With respect to this important fact, Einstein [10] pointed out that:

> "These two concepts of space may be contrasted as follows: (a) space as positional quality of the world of material objects; (b) space as container of all material objects. In case (a), space without a material object is inconceivable. In case (b), a material object can only be conceived as existing in space; space then appears as a reality which in a certain sense is superior to the material world. Both space concepts are free creations of human imagination, mean devised for easier comprehension of our sense experience."

According to conception (a), space does not constitute an independent category. The space-time curvature of the general theory of relativity is inherently connected with the distribution of matter in the universe.

According to conception (b), space constitutes a reality that, in a certain sense is superior to the material world. If the material objects are removed from a given room, the corresponding space remains there as before.

In fact, NGT is based on the concept (b) whereas GTR is based on the concept (a). In spite of the radical difference between the concepts (a) and (b), there is a mathematical limit (the so-called Newtonian limit) from GTR into NGT. We arrive at the conclusion that the conceptual problem goes beyond a mere mathematical limit.

With respect to the passage from quantum to classical mechanics, it is widely considered that quantum mechanics passes into the classical mechanics when Planck's constant h "tends" to zero $(h \to 0)$. In order to justify this point of view, Landau and Lifshitz [11] argue that,

> "There is an interrelation, somewhat similar to that between quantum and classical mechanics, in electrodynamics between wave optics and geometric optics.... The transition from quantum mechanics to classical mechanics, corresponding to large phase, can be formally described as a passage to the limit $h \to 0$ (just as the transition from wave optics to the geometric optics corresponds to a passage to the limit of zero wavelength $\lambda \to 0$) "

However, in another context, Landau and Lifshitz [12] point out certain difficulties. They write,

"A more general theory can usually be formulated in a logically complete manner, independently of a less general theory which forms a limiting case of it. Thus, relativistic mechanics can be constructed on the basis of its own fundamental principles, without any reference to Newtonian mechanics. It is in principle impossible, however, to formulate the basic concepts of quantum mechanics without using classical mechanics.... Hence it is clear that, for a system composed only of quantum objects it would be entirely impossible to construct any logically independent mechanics."

Thus, the above correspondence contains a kind of vicious circle due to the alleged lack of logical independence of the more general theory. Consequently, the argument following which there is more than a mere mathematical limit in the idea of correspondence is reinforced.

In spite of the above alleged problems, the *correspondence principle* constitutes an important cognitive expedient in physics. Broadly speaking, we can say that the correspondence principle was used in the 19-th century: Van der Waal's law passes into the Boyle-Mariotte's law when we may neglect the intermolecular forces ($a \to 0$) and the volume of molecules ($b \to 0$). Another example is the passage from wave optics to geometric optics: wave optics passes into geometric optics when the wavelength λ "tends" to zero ($\lambda \to 0$). The *correspondence principle* was used by Einstein [13] in his theory of light in 1917 when, in order to derive Planck's radiation formula, he introduced the important concept of *stimulated emission*. By means of the *correspondence principle* Bohr [14] was able to pass from his atomic theory to the classical theory in the limit of *large quantum numbers*. With this procedure he was able to express Rydberg's constant as a combination of three basic constant of nature: the electronic charge and mass and Planck's constant.

The above eloquent examples are enough to suggest that we conclude that *correspondence principle*, in spite of some conceptual problems, constitutes an important achievement of science.

5. Correspondence and complementarity

Let us return to the last quotation [12] of Landau and Lifshitz. When they write that *"It is in principle, impossible, however, to formulate the basic concepts of quantum mechanics without classical mechanics"* the underlying Copenhagen conception is clearly present. This circumstance is broadly confirmed when Landau and Lifshitz write: *"Hence it is clear that, for a system composed only of quantum objects it would be entirely impossible to construct any logically independent mechanics."*

It is important to emphasise that the word *"impossible"* appears two times. As we know, Landau was adherent of Copenhagen view, and thus we can perfectly infer that his term *"impossible"* is clearly connected with "impossibility proofs" à la von Neumann, and with Bohrian claim of *mutual exclusion* of the *space-time* and *causal* descriptions of the microscopic world.

A similar attitude was adopted by Julius Robert Oppenheimer [15] in his book *Science and the Common Understanding*. In it, Oppenheimer argues that the word *correspondence* expresses the conservative features of physics due to its need to refer to the scientific tradition, whereas the word *complementarity* contains entirely new features which provide a larger and more human comprehension of the natural world.

We would like to make some comments on both, Landau/Lifshitz's and Oppenheimer's opinions.

The part of the argumentation where we can agree with Landau/Lifshitz and Oppenheimer concerns the great importance of tradition. From past knowledge, we go to present knowledge. However, the later is not a *deterministic* consequence of the former. It is a consequence that originates from many complex tendencies. Otherwise, the concept of scientific creativity would be an illusion. Although the reference to the tradition is necessary, modification of the past knowledge in order to create genuinely new knowledge does not take place as an *accumulation*. Consequently, the need to refer to tradition as a requirement of *correspondence* must be understood neither as a mere *accumulation* process nor as a *deterministic* one.

Concerning the term *impossibility* quoted by Landau and Lifshitz, we can easily ascribe this attitude to Bohr's idea of *complementarity*. We note the following quotation from Bohr [16]:

> "I advocated a point of view conveniently termed complementarity, suited to embrace the characteristic features of individuality of quantum phenomena, and at the same time to clarify the peculiar aspects of the observational problem in the field of experience. For this purpose, it is decisive to recognise that, however far the phenomena transcend the scope of classical physical explanation, the account of all evidence must be expressed in classical terms."

It is important to stress that, following Bohr, *the quantum phenomena transcend the scope of physical classical explanation*, but, *must be expressed in classical terms*. Clearly, this methodological attitude centred on the term *must be* does not constitute a necessity. Consequently, it is an arbitrary and conservative methodological choice. It is the same methodological choice of von Neumann's *impossibility proof*, forbidding the causal completion of quantum mechanics, which has been rebutted by Bohm, Bell, Selleri among others (see for example the reference [17]).

Consequently, Oppenheimer's arguments in favour of Bohr's *complementarity*, consisting of the defence that this principle introduces new features in science, are not correct. The "necessity" of the adoption of classical mutually exclusive concepts as a requirement of *correspondence* with the macroscopic world forbids other legitimate choices.

Einstein's and de Broglie's choice based on the idea of duality as an objective property of the quantum object constituted a genuinely new scientific contribution whose development was absurdly discarded by Bohr's conservative complementarity idea (see reference[18]). This point is very relevant and constitutes an important part of the competition between Einstein's and Bohr's scientific research programmes, whose underlying concepts are *space-time* and *cause*[19].

With respect to Bohr's instrumentalist description, Stapp[20] has pointed out that *"according to Bohr quantum theory must be interpreted, not as a description of nature itself, but merely as a tool for making predictions appearing under conditions described by classical physics."* However, recognition of this point is not enough to characterise the arbitrariness and negative aspects of the Copenhagen view. It is necessary to say something else. Bitsakis[21] has argued more emphatically, writing:

> *"The conclusions of this analysis are that complementarity is not a scientific concept; that it is an epistemological and,* par *extension, an ontological postulate, incompatible with realism. Finally, its function was that of an obstacle for the investigation of the physical foundations of physics."*

We conclude this section expressing our disagreement with Oppenheimer. In fact there are two different aspects of the same Bohr. *Bohr's correspondence* principle is a rational one and it does not imply necessarily a conservative attitude. On the other hand, *Bohr's complementarity*, different from what Oppenheimer proposes, contains many conservative and even irrational aspects imported from the Danish existentialism[22].

6. Discussion and concluding remarks

Let us now consider the derivation of the scattering formula from Section 3. When we say "without relativistic" formulas this circumstance refers to the electron motion, which is treated by means of non-relativistic formulas. In (10) the kinetic energy acquired by the electron is written in the non-relativistic domain *i.e.*, the Newtonian mass of the electron is equal to the corresponding rest mass of the relativistic theory. However, with respect to the photon the situation is different. The photon is an ultra-relativistic particle and thus it cannot be considered as a particle obeying non-relativistic laws because such a procedure would be self contradictory.

The above circumstance is easily understood when we write the relativistic formula for the photon

$$(E_{\text{photon}})^2 = (p_{\text{photon}})^2 c^2 + [(m_{\text{o}})_{\text{photon}}]^2 c^4 \qquad (15)$$

According to the relativistic theory the photon rest mass is equal to zero. Consequently,

$$E_{\text{photon}} = p_{\text{photon}} \, c \qquad (16)$$

which is the so-called ultra-relativistic formula. By combining (16) with (2) we obtain (3) This means that (3) contains the ultra-relativistic regime and thus the derivation carried out in the section 3 is not entirely non-relativistic.

The arguments articulated above show that the comparison between the approaches of the sections 2 and 3 does not constitute a comparison of a relativistic theory and a genuine non relativistic theory.

The analysis of this circumstance is very important in virtue of the important difference between the formulas (7) and (13). The formula (7) [Compton scattering formula] expresses the result following which $\Delta \lambda$ does not depend on the wavelength of the incident photon. It depends only on the scattering angle ϕ.

On the other hand, the formula (13) contains an "additional" term which expresses a possible dependence of $\Delta \lambda$ on the product of the percent variation of λ and λ'. This term is very small and does not correspond to the experimental situation.

Our results can be summarised as follows:

[1] *If a given formula is derived in the context of the theory B (for instance the approach of Section 3) and this formula contains more terms than the corresponding formula of the theory A (for example the approach of Section 2), then this fact does not necessarily imply that the theory A is less general than the theory B.*

[2] *By using the de Broglie formula $p=(h/\lambda) = h\nu/c$ for the photon, we also apply the ultra-relativistic formula (16). Consequently, we do not have a clear comparison between a relativistic theory and a genuine non relativistic one.*

[3] *Consequently the formula (13) cannot be considered more general than the Compton scattering formula (7). The additional term in (13) is not a genuine "additional" term.*

Appendix: Santarine and Stein-Barana's Scattering Formula

In order to present Santarine and Stein-Barana's scattering formula, let us re-write the formulas (10–12) in the following form,

$$E = E' + E_e \tag{I}$$

$$p = p' \cos\phi + p_e \cos\beta \tag{II}$$

$$0 = p' \sin \phi - p_e \sin \beta \tag{III}$$

where,

$$E = h\nu = hc/\lambda; \; E' = h\nu' = hc/\lambda' \tag{IV}$$

$$p = h/\lambda = h\nu/c; \; p' = h/\lambda' = h\nu'/c \tag{V}$$

$$E_e = p_e^2 / 2 m_o; \; p_e = m_o v \tag{VI}$$

From the expression (II), we have,

$$\cos \beta = (p - p' \cos \phi)/ p_e \tag{VII}$$

From the expression (III), we have,

$$\sin \beta = p' \sin \phi /p_e \tag{VIII}$$

Applying the trigonometric formula $sin^2\beta + cos^2 \beta = 1$ to the expressions (VII) and (VIII), we have

$$p_e^2 = p^2 + (p')^2 - 2\,pp'\cos\phi \tag{IX}$$

Putting (IX) and (V) into the first expression of (VI) we obtain,

$$E_e = \frac{h^2}{2m_o}\left\{\frac{(\lambda'-\lambda)^2}{(\lambda\lambda')^2} + \frac{2}{\lambda\lambda'}(1-\cos\phi)\right\} \tag{X}$$

From (I) and (IV) we have,

$$E_e = E - E' = hc\frac{\lambda-\lambda'}{\lambda\lambda'} \tag{XI}$$

Putting (X) = (XI) and considering the trigonometric formula

$$2\sin^2(\phi/2) = (1-\cos\phi),$$

we obtain,

$$\lambda'-\lambda = \Delta\lambda = \frac{\lambda_o}{2}\frac{(\lambda'-\lambda)^2}{\lambda\lambda'} + 2\lambda_o\sin^2\left(\frac{\phi}{2}\right) \tag{XII}$$

where $(\lambda_o = h/m_o c)$ is the Compton wavelength defined in (8).

Acknowledgements

I would like to express my sincere gratitude to Prof. Franco Selleri for the invitation to the International Conference *Relativistic Physics and Some of its Applications*. I'm very grateful to Prof. E.Bitsakis of the Universities of Athens and Ioannina for discussions and to Prof. Roberto Jorge Vasconcelos dos Santos for his suggestions.

References

[1] W. Krajewski, *Correspondence Principle and Growth of Science*, D. Reidel Publishing Company, Dordrecht – Holland / Boston USA (1977)

[2] K.R. Popper, *Objective Knowledge*, Oxford (1972) p.202

[3] K.R. Popper, *ibid.* Ch. 5

[4] J.B. Bastos Filho, *Rev. Bras. Ens. Fis.*, **17**, n°3 (1995) p. 233

[5] T.S. Kuhn, *The Structure of Scientific Revolution*, Chicago (1970)

[6] P. Feyerabend, *Problems of Microphysics*, In: *Frontiers of Science and Philosophy*, Colodny (Ed.), Pittsburgh, (1962)

[7] G.A. Santarine and A.C.M. Stein-Barana, *Rev.Bras. Ens. Fis.*, **19**, n°1 (1997) p. 64

[8] L.D. Landau and E. Lifshitz, The *Classical Theory of Fields*, Pergamon Press, Oxford, New York, Toronto, Sydney, Paris, Frankfurt,Ch.2, (1975)

[9] Reference [8] p.25

[10] A. Einstein, foreword of the book *Concepts of Space of Max Jammer*, Harvard University Press, Cambridge, Massachusetts (1970)

[11] L.D. Landau and E. Lifshitz, *Quantum Mechanics (Non Relativistic Theory)*,Pergamon Press, Oxford, New York, Toronto, Sydney, Paris, Frankfurt Ch. 1 section 6, (1977)

[12] Reference [11] Ch 1 section 1

[13] A. Einstein, *Physik Z.*, **18**, 121 (1917)

[14] N. Bohr, *Drei Aufsätze über Spektren und Atombau*, Brausnsweig (1922)

[15] J.R. Oppenheimer, *Science and the Common Understanding*, Princeton University Press, Princeton (1954); Franc. Translation: *La Science et le Bom Sens*, Gallimard, France (1972) p.53

[16] N. Bohr, Discussions with Einstein on Epistemological Problems in Atomic Physics 1949, In: *Atomic Physics and Human Knowledge*, Science Edition INC. NY (1961) p. 39

[17] F. Selleri, *Quantum Paradoxes and Physical Reality*, Kluwer Academic Publishers, Dordrecht, Boston, London (1990).

[18] J.B. Bastos Filho, Dangerous Effects of the Incomprehensibility in Microphysics, In: *Frontiers of Fundamental Physics*, Edited by M. Barone and F. Selleri, Plenum Press, New York, (1994) p. 485

[19] J.B. Bastos Filho, Descartes, Leibniz, Newton and Modern Physics: Plenum, Action at a Distance and Locality, paper presented at the *International Conference Descartes and Scientific Thought* (Perugia – Italy, 4-7 September, 1996)

[20] H.P. Stapp, Light as foundation of Being, In: *Quantum Implications (Essays in Honour of David Bohm)*, Ed. by B.J. Hiley and F.D. Peat, Routledge / Kegan Paul, London and New York (1987)

[21] E.I. Bitsakis, Complementarity: Its Status and Function, In: *Bell's Theorem and the Foundations of Modern Physics*, Edited by van der Merwe, F. Selleri and G. Tarozzi, World Scientific, Singapore, New Jersey, London, Hong Kong (1992) pp. 73-91

[22] F. Selleri, *Rivista d. Storia d. Scienza* 2 (1985) p. 545.

Space and Time: Who was Right, Einstein or Kant?

Eftichios Bitsakis
Department of Physics, University of Athens
Department of Philosophy, University of Ioannina

The main answers given to the problem of space and time are: 1. The realist, according to which space and time are forms of existence of matter. 2. For positivism, space and time are concepts convenient for the description of phenomena. For Poincaré, to take a notorious case, "the characteristic property of space, that of having three dimensions, is only a property of our distribution board, a property residing, so to speak, in the human intelligence. The destruction of some of these connections, that is to say, of these associations of ideas, would be sufficient to give us a different distribution board, and that might be enough to endow space with a fourth dimension"[1]. According to E. Borel, on the other hand, "it is *convenient* to use intervals, just as it is *convenient* to assume that the earth is rotating and the sun is standing still (to a first approximation). Moreover we must not forget that for Poincaré *convenience* is identical to scientific truth"[2]. Hume, Mach, Poincaré, Wittgenstein, the multiform schools of classical and modern positivism, deprived the concepts of space and time of their ontic counterpart. 3. Kant, on the contrary, was, from a certain point of view, a realist. He accepted the existence of the *things in themselves*, independent of the subject. For him, however, space and time are the *a priori* forms of the intuition. They are not forms of existence of matter.

I will not try to refute the conception of positivism[3]. The case of Kant is more delicate and interesting and I will try to confront it with that of Einstein. For contemporary realism and materialism, the realist Einstein is right. Not Kant. However the problem is not so simple. So, in order to discuss it, let us at first summarize the Kantian conception.

1. Space and Time: *a priori* forms of intuition

For Kant also, knowledge begins with experience. From this fact, however, does not follow that all arises out of experience. Beyond the sphere of experience, Kant accepted the existence of a "transcendental or supersensible sphere," where experience affords neither instruction nor guidance. In this sphere lies the investigation of Reason.

Kant accepted the existence of knowledge independent of experience. "Knowledge of this kind is called *a priori*, in contradiction to empirical, which has its sources *a posteriori*, that is, in experience." Knowledge *a priori*, in its turn, is pure or impure. Pure knowledge *a priori* is that with which no empirical element is mixed up. Example: The science of mathematics. *A priori* impure knowledge contains elements derived from experience. Example: "Every change has a cause." The *a priori* knowledge is characterized, according to Kant, by *universality* and *necessity*. A proposition which contains the idea of necessity in its very conception, is a judgment *a priori*.

Kant insisted on the division of propositions in analytical and synthetic. A proposition is analytical, if the predicate B belongs to the subject A. Example: All bodies are extended. If the predicate B does not belong to A, then the proposition is synthetic. Example: All bodies are heavy. Judgments of experience are always synthetic.

Synthetic propositions are *a posteriori*, or *a priori*. Example of an *a priori* synthetic proposition: "Every thing that happens has a cause." Mathematical science, Kant maintains, "afford us a brilliant example how far, independently of experience, we may carry our *a priori* knowledge." Pure mathematics consists, according to him, of knowledge all together non-empirical and *a priori*. Arithmetical propositions are always synthetic. Example: $7 + 5 = 12$. Geometrical propositions are also synthetic. Example: "A straight line between two points is the shortest." Yet, it is true that for Kant some mathematical propositions are analytical, but they are not principles. They are links in the chain of method. Example: $a = a$, or $a + b > a$.

On the basis of the preceding definitions of Kant, it is possible to expose his conception about the *things in themselves* and the *phenomena*. In particular, about space and time.

Our knowledge relates to objects. However, it relates to them by means of intuitions. The undetermined object of the empirical intuition is called by Kant, *phenomenon*. "That which in the phenomenon corresponds to the sensation, I term it *matter*." That which affect that the content of the phenomenon can be arranged under certain relations, is called by Kant, *form*. The form must be ready *a priori* for them in mind. Consequently form can be regarded separately from sensation: we find in the mind the *a priori* forms of intuition. Correspondingly, a *pure representation* contains nothing belonging to sensation. *The pure forms of intuition, are space and time.*

Space, Kant maintains, is a pure, *a priori* form of intuition. It contains principles of the relations of objects prior to experience. Consequently, space is not a form which belongs to a property of things. It is not a conception derived from outward experiences. It is a necessary, *a priori* representation, which serves for the foundation of all external intuitions. Space does not represent any property of objects as things in themselves. It is impossible, therefore, to know things in themselves—we know only phenomena. In fact, according to Kant, space contains all which can appear to us externally, but not all things, considered things in themselves. Thus, properties belonging to objects as things in themselves never can be presented to us through the medium of senses. Space is represented as an infinite given quantity.

Correspondingly, time is not an empirical conception. It is given *a priori* and constitutes the universal condition of phenomena. Different times are parts of one and the same time. This form of *inward* intuition can be represented before the objects and, consequently, *a priori*. Time is the subjective condition of our intuition. All phenomena are in time. However, independently of the mind, time is nothing.

Space and time have a certain absolute character for Kant. This fact recalls the absolute space and time of on the newtonian physics. However, space and time for Newton are objective "realities." The first is *"sensorium Dei."* The second corresponds to the omnipresence of the Creator.

Space and time are, according to Kant, two sources of knowledge, from which various *a priori* synthetic propositions can be drawn. It follows that "geometry is a science which determines the properties of space synthetically and yet, *a priori*." Our representation of space, Kant maintains, must be an intuition, not a mere conception, an intuition found in the mind *a priori*. *It is, a pure, non empirical intuition*. Geometrical principles are always apodictic, that is, united with the consciousness of their necessity, as: "Space has only three dimensions." Propositions of this kind cannot be empirical judgments, nor conclusions from them.

Yet, the existence of *a priori* knowledge is not self-evident. Thus, Kant puts the question: "How can an external intuition anterior to objects themselves, and in which our conception of objects can be determined *a priori*, exist in human mind?" His answer is as follows:" Obviously not otherwise than in so far as, it has its seat in the subject only, as the formal capacity of the subjects being affected by objects, and thereby of obtaining immediate representation, that is, intuition: Consequently, as the *for*m of the external sense in general." Evidently, this is not an answer.

Geometry, Kant affirms, is based on *a priori* principles. In a similar way, the science of Natural Philosophy (Physics)contains synthetic judgments *a priori* as principles. According to Kant the proposition: "In all changes of material world the quantity of matter remains unchanged," as well as the proposition: "in all communication of motion, action and reaction must be always equal," are *a priori*, synthetic propositions[4].

Some preliminary remarks are needed here.

All knowledge, Kant accepts, begins with experience. Reason without experience is void. Yet, at the same time he maintains that synthetic *a priori knowledge* is possible, that is to say, knowledge independent and prior to experience. However, in what way we acquire such a knowledge? And how is it possible to justify the assertion that space and time are *a priori* pure forms of intuition? Kant considers these assertions as eternal truths. Yet, as I will try to show, the alleged *a priori* knowledge is historically and socially acquired. Its incontestable verity gives the impression that it is independent of experience and history. In reality, as knowledge socially and historically acquired and transmitted, it is prior and independent of the individual experience and is imposed to individuals as absolute and a-historical truth. Space and time, in their turn, are forms of intuition genetically-historically determined. Consequently, it would be in principle possible to explain our possibility to see, to hear, etc, in a three dimensional space and to arrange phenomena in a unique temporal succession. There is a certain *a priori* in our intuition. Yet this is "simply" a *possibility* for a certain form of representation, not a knowledge. Mathematics, on the other side, do not represent pure *a priori* knowledge. Their propositions are *a posteriori*, not *a priori* synthetic. They are acquired through experience, abstraction, generalization and transcendence of the experience. The geometrical triangle for example, has its prototype in the real and imperfect triangles, not in the ideal triangle of the platonic world of ideas. The equation:a+b+c=180 is not an *a priori* synthetic proposition. It was formulated after a long historical period of practical experience through generalization and abstraction of the real properties of the real triangles. The ideal, mathematical property has its basis on reality, contrary to the Pythagorean, Platonic and Kantian epistemology.

In a analogous way, the laws of theoretical physics are *a posteriori* and not *a priori* synthetic propositions. Necessity and universality are not a sufficient reason and proof of an *a priori* status. The necessity and the universality of the laws of physics, is due to the fact that they are not deduced by induction, but *via* a process of abstraction, generalization and transcendence of their experimental or observational basis. By this way they express the ideal limit of relations existing in nature itself and they acquire thereby their phenomenal independence from experience.

The alleged absolute and necessary character, on the other hand, has been proved relative and not necessary by new observations and the theoretical generalization of the experience (not to speak of the ideological presuppositions of the "paradigms" of the theoretical physics). But I will discuss these problems in the last part of this paper.

2.   Relativity: Einstein against Kant

We must dispense justice to Kant. In his time only Euclidean geometry existed. Euclidean space on the other hand, was the indisputable frame of physics. Newton, before Kant, professed the existence of a space, "absolute in its nature, without relation to anything external, always similar and immovable." This objective, three dimensional space with its absolute, positive metric, was transformed by Kant to a subjective *a priori* pure form of the intuition. In the same spirit Newton hand admitted the existence of an absolute mathematical time which, "from its own nature flows equably, without relation to anything external"[5]. Kant, following Newton, considered the different, local times, as part of one and the same, universal time.

However, and contrary to Newton, he conceived time as a subjective, *a priori* form of the inward intuition. The categories of the mechanistic-realist ontology of Newton, were transformed by Kant into the incomprehensible *a priori* forms of the so-called pure intuition.

It is well known that the conceptual nucleus of classical mechanics is that of the action-at-a-distance, and that its ontological premises are formally expressed by the Galilean group of transformations. Therefore, the incompatibility of this group with the Maxwellian electromagnetism was inevitable. Einstein and Minkowski demonstrated that the natural spatio-temporal frame of electromagnetism was a quadri-dimensional pseudo-Euclidean space. The Lorentz group ensured, as is well known, the invariance of the equations of Maxwell.

The special theory of relativity brought to light the relativity of space and time. At the same time revealed their intrinsic unity. The absolute character of the four-dimensional space-time interval and the equally absolute character of the new relativistic physical quantities (velocity, force, acceleration, current, etc) expressed in a tensorial form, are incompatible with the modern gnoseological relativism and concord with the realist, causal and local interpretation of relativity. General theory of relativity in its turn, whose physical content is the theory of gravitation of Einstein, expresses, as is well known, the laws of nature in a form invariant for all systems of coordinates, Galilean or accelerated. This is a generalized, stronger objectivity. At the same time general relativity revealed the intrinsic unity of space, time, matter and motion, "since the ten functions representing the gravitational field, at the same time define the metrical properties of space"[6].

Historicity of the concepts of space and time? In fact, from the Euclidean-Newtonian absolute space and time, we passed to the pseudo-Euclidean space of Minkowski and further to the Riemannian space of general relativity. However the historicity of the mathematical spaces does not constitute an argument in favour of the gnoseologiacal relativism, because their epistemic difference does not exclude their dialectical compatibility, in opposition to the alleged incommensurability, which is professed by contemporary formalistic epistemologies. In fact, as is well known, we pass from the space of Riemann to that of Minkowski, if we consider space practically void of matter. And we can dissociate the space of Minkowski into two subspaces, if we consider very slow velocities. As A. Papapetrou puts is, "the Minkowski space is the simplest form of a Riemannian space—it is a flat space"[7]. Historicity and commensurability are incompatible with the gnoseological relativism related to relativity.

Till now we have to do with mathematical spaces. Consequently, the historicity of the concepts of space and time concerns the epistemological aspect of our problem. Yet, what is the ontic status of space and time? Is there a relation between mathematical spaces and physical space? Is there a kind of correspondence, of *morphism,* between them? And if yes, then how and when mathematics can represent physical reality? I will try an answer to these questions in the last part of this paper.

For the moment, let us note the objections of Einstein to the Kantian *a priorism*. From a certain point of view Einstein was an empiricist. As he admits in his *Autobiographical Notes:* "It was Ernst Mach who, in his *Science of Mechanics,* shook this dogmatic faith. This book exerted a profound influence upon me in this regard, while I was a student. I see Mach's greatness in his incorruptible skepticism and independence." However the philosophy of Mach appeared latter to Einstein, "essentially untenable"[8].

Einstein was a realist. Bodily objects for him are independent of the sense impressions they provoke. They have "a real, objective existence." Subjective time, Einstein writes, leads through the concept of bodily object and space, to objective time. "Ahead the notions of objective time there is, however, the concept of space, and ahead the latter we find the concept of bodily objects"[9].

Einstein founded his realist conception of space to the existence of bodies, independently of the fact that they are perceived. "In my opinion," he writes, "the fact that every bodily object situated in any arbitrary manner can be put into contact with the Bo (body of relation) this fact is the empirical basis of our conception of space"[10]. Space and time have an empirical basis. And in spite of this, one may led into the error of believing that the concepts of space and time are *a priori,* that they preceed all experience, and they constitute the basis of the Euclidean geometry and of the concept of space belonging to it. "This fatal error, Einstein notes, arose from the fact that the empirical basis on which the axiomatic construction of Euclidean geometry rests has fallen into oblivion"[11].

Einstein insisted on the need that the axiomatic form of the Euclidean geometry must not conceal its empirical origin. The three dimensions of space, in particular, and its Euclidean character are, for him, of empirical origin[12]. In a higher, abstract level, the general theory of relativity demonstrated the intrinsic unity of physics and geometry. As Paul Langevin puts it, "our physics became a geometry of the universe"[13]. In an inverse sense, one could say that our geometry became a branch of Physics.

From practical experience to Euclidean absolute space and time. The laws of electromagnetism demonstrated the intrinsic unity of space and time. The relativistic theory of gravitation is formulated in the frame of a Riemannian geometry. These abstract geometries contradict the Kantian dogma. At the same time they are not abstract forms, but forms full of concrete physical content. Our intellect transcended the limits of the immediate intuition.

Einstein opposed his realist conception to that of Kant. One cannot take seriously, according to him, the tentative of Kant to deny the objectivity of space[14]. The non-Euclidean geometries have been, Einstein maintains, a fatal blow to the conception of Kant. For him the physical notion of space, as originally used by Physics, is tied to the existence of rigid bodies. However, for Einstein, as we have noted, one may easily led to the error, that the notions of space and time, the origin of which has been forgotten, are necessary and unalterable accompaniments to our thinking. "This error may constitute a serious danger of science[15]." Kant, according to Einstein, "was misled by the erroneous opinion –difficult to avoid in his time– that Euclidean geometry is necessary to thinking and offers assured (*i.e.* not dependent upon sensory experience) knowledge, concerning the objects of the "external" perception." From this error, Einstein notes, Kant concluded "the existence of synthetic judgments *a priori*, which are produced by the reason alone, and which, consequently, can lay claim to absolute validity"[16]. Einstein, finally, notes, that his attitude is distinct from that of Kant, by the fact that he does not conceive the categories "as unalterable (conditioned by the nature of understanding) but as (in the logical sense) free contentions"[17].

Einstein rejected the *a priori* character of space, time and categories. He also rejected the existence of *a priori* synthetic judgments. Yet, Kant was completely wrong?

3.   Beyond Kantian *a priorism* and beyond empiricism

Empiricism rejects the existence of the *a priori* forms of intuition, the *a priori* nature of categories, and the existence of *a priori* synthetic knowledge. However, the problem is not to reject Kant, but to go beyond his philosophy and at the same time beyond empiricism.

Let us recall the question of Kant: "How can an external intuition, anterior to objects themselves, and in which our capacities of objects can be determined *a priori*, exist in human mind? Obviously not otherwise than in so far as it has its seat in the subject only, as the *formal* capacity of the subjects being affected by objects, and thereby of obtaining immediate representation, that is, intuition"[18].

It is evident that Kant does not give an answer to his question.

However Kant admits "the formal capacity of the subject's being affected by objects." How this is possible? And why we see objects in a three-dimensional Euclidean space? And this possibility is compatible with the a-historical and incomprehensible *a priori* of Kant?

According to Kant, space and time are pure forms ready *a priori* in mind. Is it possible to reject this *a priori* and at the same time to understand our *possibility* to see, to hear, *etc.*, in a three dimensional Euclidean frame? Biology and physics give us the necessary elements in order to try to give an answer. Let us take the case of the eye-of vision. In the inferior animals, the cells sensible to light were distributed on the surface of the organism. The photosensivity of these animals was therefore very diffused. The first animals having photosensitive cells concentrated on the cephalic extremity were the worms. With the evolution of the species, these cells took the form of a plate. This permitted already the orientation of the animal to the light. In a more developed phase, these plates constituted an internal photosensitive cavity of spherical form, permitting the perception of the movement of the objects[19]. More generally, our sense organs were developed during the long period of phylogenesis, *via* the interaction of the organisms with their environment and, in particular, *via* the reception of physical signals: light, sons, chemical molecules.

Let us take again the case of light. Electromagnetic waves are propagated in space. If, then, we accept the relativistic conception, according to which out space is locally Euclidean, then it is in principle possible to understand why we see objects as existing in a three dimensional Euclidean space. Our sense organs were constituted, developed and adapted to the local Euclidean space.

The structure and the function of our sense organs and of our brain we developed in an a locally Euclidean space in interaction with our environment. The evolution of our brain was determined latter by the practical relations with nature, the work and the whole of the social life. From the simple excitation, the stimulus, the sensory-motor activity and the representation, and by generalization of the empirical knowledge, we acquired the possibility to use rudimentary symbolic languages (gestures, cries, *etc.*) and finally the use of concepts and of conceptual thinking. Our scientific concepts have an empirical origin, and at the same time they transcend immediate intuition.

There is an *a priori* concerning our intuition. This *a priori*, however, is radically different from the a-historic and incomprehensible *a priori* of the Kantian theory of knowledge. This *"a priori"* does not presuppose the existence of knowledge anterior to experience. It simply means that we have *an a priori possibility* to see, to hear, *etc.*, in a three dimensional space. Consequently, space and time are not the subjective, *a priori* forms of our intuition. Intuition, on the contrary, presupposes the objective existence of space and time. On this ontological premise is possible to explain our *a priori* possibility, presupposition of the intuition. This conception is incompatible with the Kantian one, as well as with the conventionalism of Mach, Poincaré, *etc.* (Mach, for example maintained that space and time as regards to Physics they stand for functional dependencies upon one another of the elements characterized by the sensation)[20].

Accordingly, Euclidean geometry is not based on *a priori* principles. It is a science of empirical origin and its axioms are the outcome of a long process of abstraction and generalization. The evident truth of its axioms and their universality and necessity are not proof of an *a priori* origin. More than that: these axioms are necessary and universal only in the frame of this geometry. They are not compatible with other, non Euclidean geometries. Finally the distinction between pure geometry and the geometry of physical space is not absolute. The so-called pure geometry is not an expression of *a priori* forms. It is an abstraction from the real properties of matter.

However, let us try to imagine the four-dimensional space of Minkowski. This is impossible. Why? Because intuition cannot go beyond its *a priori* restrictions. From this point of view, it seems that Kant is right. This *a priori*, however, is radically different from the Kantian one.

Our intuition has this limited possibility. However, we can think about things in the absence of things. Our reason is liberated from the restrictions of the senses, and can create abstract theories having not a visual counterpart, and non-Euclidean geometries in particular.

The laws of Physics are, according to Kant, *a priori* synthetic propositions. "The Science of Natural Philosophy (Physics)," he writes, "contains in itself synthetic judgments *a priori*, as principles. I shall adduce two propositions. For instance, the proposition: "In all changes of natural world the quantity of matter remains unchanged, or that "in all communications of motion, action and reaction must always be equal." In both of these, not only the necessity, and therefore their origin *a priori* is clear, but also that are synthetic propositions"[21].

Three remarks on this subject: 1. The principle of the conservation of matter is not a law of Physics. It is an ontological postulate. Matter, on the other hand, is not a concept. It is an ontological category. There exists not a *measure* of matter. Consequently, its alleged conservation is impossible to be proved or refuted. This finally, is not a synthetic proposition. It is an ontological postulate and its epistemological status is to be discussed. 2. The necessary character of a proposition, as that of the equality of action and reaction, is not a proof of its *a priori* synthetic character. This proposition is *a posteriori* synthetic, formulated *via* the generalization of empirical data. 3. More generally, the laws of theoretical physics are not *a priori* synthetic propositions. They are theoretical propositions *a posteriori* synthetic, generalizing and transcending their empirical basis and, because of that, formulated axiomatically and not by induction. The historicity of these laws, on the other hand, is an argument against the a-historical necessity and universality attributed to them by Kant.

Now, concerning the analytical or synthetic status of propositions. "Every dog is an animal." From the formal, logical point of view, this is an analytic proposition. But why is it *a priori*? Its formulation presupposes the social life, the distinction of animal species, *etc.* Consequently, its is not an *a priori* proposition. Evidently, it is not necessary to relate it to experience, in order to certify its status. However, its evident truth is of social character. On the other hand, the proposition: "The moon turns around the earth" is considered as an *a priori* synthetic proposition. In fact this propositions presupposes the relevant astronomical observations, and on the theoretical level, the theory of gravitation. It is, for these reasons, an *a posteriori* synthetic proposition. His evident verity is not an argument in favor of the alleged its *a priori* status. *Conclusion:* The antithesis between analytic and synthetic propositions is not a formal one, as it seems to be in the Kantian philosophy. Both of them presuppose the social life, experience, observation, abstraction and generalization. Finally, theoretical laws are not, in general, formulated by induction. As Hume, Kant, and Popper in our days explained, universal statements cannot be justified by induction. This is true, to a higher degree, for theoretical physics.

Finally, concerning the categories—causality *etc.* For Kant they are *a priori* categories of Reason. However, categories are also historical from the gnoseological point of view, because they represent the generalization and transcendence of human experience. The idea of causality, for example, was for the first time formulated in the frame of an animistic conception of nature. It was embodied latter in the religious conception of the world. And after Galileo and Newton we know at least four forms of causal determination: the mechanistic, the dynamic, the classical statistical and the quantum-statistical form. This social-politismic category was transformed by Kant into an a-historical and unexplained *a priori* attribute of Reason.

Now, concerning space and time. These concepts also are historical. They do not correspond to *a priori* forms of intuition. As Paul Langevin puts it, there is neither space nor time *a priori*. To every phase of our theories corresponds a different conception of space and time. Mechanics implied the ancient conception. Electromagnetism requires a new one, and nothing permits to maintain that this new conception will be the final one[22].

The argument of Langevin concerns the concepts of space and time, that is to say, the gnoseological aspect of the problem. These concepts are related to different mathematical spaces and their historicity and *a posteriori* character is evident. However Kant was a philosopher and he speaks about categories. The status of categories is different from that of concepts. How then is it possible to pass from the level of concepts to that of categories? Let us accept that between the concept of space and time and the physical space there is a certain correspondence, a certain kind of *morphism*. That concepts tell us something about the real properties of space and time. How then is it possible to pass from the domain of the scientific knowledge to the level of the categories? Scientists use the *words* of space and time as well defined scientific concepts. At the same time they refer to them as the general frame of their experiences and theories. The same words are used by philosophers, as categories. One could say that in the general case, space and time are used by scientists as *quasi-philosophical* concepts. Because of that they assure a kind of junction between the scientific and the ontological level[23]. The historical character of the concepts of space and time, the deepening of our knowledge concerning these concepts, determine the historical character of the categories of space and time from the gnoseological point of view. Matter is also an historical category even from the ontological point of view. In fact, we know to day that different forms of matter correspond to the different phases and regions of the Universe. If, therefore, we accept a realist epistemology, namely that space and time are forms of existence of matter, then it would be possible to maintain that space and time are historical categories from the ontological point of view also, because of the fact their ontic conterpart changes, as a consequence of the evolution of the Universe.

It is possible to maintain that the Universe is infinite in space and time. Infinite, however, is always different and different (Aristotle). In that case matter and its forms do not correspond to an eternal and immovable being, but to a changing totality. Any attempt to construct a metaphysical ontology would be, therefore obsolete.

Final question: How to explain the efficacy of natural sciences, if we can know only phenomena? How phenomena are in a certain correspondence with the inaccesible things in themselves? Kant does not explain this. However, phenomena are the manifestation of hidden relations and properties of things. They reveal hidden essentialities and at the same time they cover them. Things in themselves are accessible to reason. But an exhaustive, absolute knowledge of them is impossible. Concerning space and time: Our *aporia* will not find a final answer, because it concerns the *απειρον.*"

References

[1]   H. Poincaré, *Science and Method*, Dover, N.Y, pp. 112-113.

[2]   E. Borel, *Space and Time*, Dover, N.Y., 1960, p. 163.

[3]   See, *e,g.*, E. Bitsakis, *Physique et Matérialisme*, Editions Sociales, Paris, 1983.

[4]   The above is a summary from the *Critique of Pure Reason*, J.M.D. Meiklejohn (transl.) London MDCCCLV.

[5]   I. Newton, *Principia*, Univ. of California Press, 1947, p. 6.

[6]   A. Einstein, in *The Principle of Relativity*, Dover, N.Y. 1923, pp. 117-120.

[7]   A. Papapetrou, *Lectures in General Relativity*, Reider, Dordrecht, 1974.

[8]   A. Einstein, "Autobiographical Notes," in *Albert Einstein, Philosopher-Scientist*, P.A. Schilpp (Ed.), The Library of Living Philosophers, Evanston, Illinois, 1949.

[9]   A. Einstein, *Journal of the Franklin Institute*, 221, 349 (1936).

[10] A. Einstein, *Ibid.*

[11] A. Einstein, *Ibid.*

[12] A. Einstein, *La Relativité Restreinte et la Relativité Généralle*, Gauthier-Villars, Paris, 1978, p. 157.

[13] P. Langevin, *La Relativité*, Hermann, Paris, 1932, p. 86.

[14] A. Einstein, *La Relativité Restreinte et la Relativité Générale, op.cit.*

[15] A. Einstein, *Journal of the Franklin Institute, op.cit.*

[16] A. Einstein, "Reply to Criticisms," in *Albert Einstein, Philosopher-Scientist, op.cit.*, p. 679.

[17] A. Einstein, *Ibid*, p. 674.

[18] E. Kant, *Critique of Pure Reason, op.cit*, p. 25.

[19] A. Leontiev, *Le Développement du Psychisme*, Ed. Sociales, Paris, 1976, p. 18.

[20] E. Mach, *The Analysis of Sensations*, Dover, 1959.

[21] E. Kant, *Critique of Pure Reason, op.cit.*, p. 11.

[22] P. Langevin, *La Pensée et l'Action*, Ed. Sociales, Paris, 1964, p. 70.

[23] E. Bitsakis, *Science et Philosophie*, Dodoni (Univ. of Ioannina) KA, 1992, pp. 9-44.

Einstein and the Development of Physics

A. Jannussis
Department of Physics
University of Patras,
26110 Patras, Greece

The two theories where Einstein has made his greatest contribution are the Theory of Relativity and the Quantum Theory. A remarkable year in the life and scientific development of Einstein is undoubtedly the year 1905, when Einstein in Bern formulated at the same time his theories of the photoelectric effect, Brownian motion and the theory of special relativity (as it came to be known later).

The decisive points in the scientific life of Einstein were the Boltzmann statistical significance of entropy, the Planck theory of thermal radiation and the works of Lorentz on electrodynamics.

Nowadays we correctly call the transformations of the theory of relativity "the Lorentz group," but the truth is that at the time (1895-1905) Lorentz himself did not know the group character of his transformations. This was derived independently by Poincaré and Einstein. It is interesting to note that at one time there was an argument over priority, but the important thing is to distinguish the manipulation of the same problem in different ways by the mathematician Poincaré and the physicist Einstein. Poincaré begins with the Maxwell equations and shows that these accept some known transformations. Einstein's motivation is revealed by a letter from him to Dr. Seelig, which has been communicated by Born. From his own works about the photelectric effect, Einstein had noted that the theory of Maxwell could accommodate some improvements. Therefore, he formulated the invariance of the laws of nature under Lorentz transformations as a general postulate, which is more elegant than the Maxwell equations. Independently of the Maxwell equations he founded this postulate on kinematic relations and gedankenexperiments relating the unity of the principle of relativity under translations with the principle of the constancy of the velocity of light in the relativity of time.

In the works of Lorentz, we already find the electrodynamic theory of Maxwell freed from the older notion of the aether. All the models with an aether were not accepted by Einstein. He said: "The emancipation of the concept of the field from the installation of a material bearer belongs to the psychologically interesting examples of the development of physical thought."

We can see the degree to which the theories of Einstein in relativity and quantum mechanics have been developed together. There is the prologue of a collection of works published in 1908 in Salzburg, where, after his name as the author, "Zurich" appears, and the title: *Über die Entwicklung unserer Anschauungen über das Wesen und die Konstitution der Strahlung."* (On the development of our theories about the nature and constitution of radiation). Then, after once more putting forward his famous conclusion concerning the equivalence of mass and energy, he proceeds to the quantum structure of radiation. He was expecting a theory which should be the mixing of wave theory and emission.

Einstein's idea of a quantum is pictured in an answer he had once given to Planck in the course of a discussion.

> *I think of a quantum as a singularity surrounded by a large vector field. A large number of quanta span a vector field similar to the one we encounter in radiation. The aggregation of rays on a plane results in interaction and separation of quanta. The equations for the resulting field will not be very different from the ones we already have in our theories... I do not see any difficulty in principle in the radiation of interference.*

But things were not so simple. The interference radiation of the intensity of light, even when we have only a few light quanta. Einstein knew that, but his insight had led him to search for the solution of the quantum riddle, which he did later.

Another decisive point in the work of Einstein was in 1917. Then he completed his theory of general relativity, different from the special, without contemporary research from other authors. This theory is exclusively from Einstein. It is characterized by an elegant mathematical structure and has influenced and helped research on problems concerning the structure of the Universe. There have been projects in performing experiments verifying the universality of the acceleration of gravity, by Dicke in Princeton. Other experiments have been performed by Zacharias in Cambridge, Massachusetts, concerning the accuracy of a clock in the field of gravity.

In 1917, Einstein not only developed general relativity, but also made a contribution to quantum theory. He formulated the theory that the emission of light is in straight rays, by the use of the statistical quantum laws of emission and absorption. At the end of this paper, Einstein wrote a famous paragraph on the meaning of chance, which has been quoted many times: "The weakness of the theory presented is that it does not bring us nearer to the task of competing wave theory, and that it leaves the time and direction of elementary processes to chance: but I believe that the chosen way is incidental."

Einstein could not accept *a priori* probabilities. He said: "For the rest of my life I will be thinking about what light is!" Perhaps his continuing absorption in the theory of relativity, created in him the faith that there is no other way for the development of science except determinism.

In the further elaboration of general relativity, he was involved with a problem that he could not finally resolve. Ernst Mach had proposed that inertia is due completely to the influence of remote masses. If this Mach principle was right, then Einstein's G-field should vanish when all masses are removed. Einstein, because of his own theory, favoured this principle and considered it to be correct. But the equations of the theory did not lead him anywhere.

When Einstein was working on the unification of electromagnetism and the theory of gravitation, he stated that the quantum of interaction is encountered not only in light, but also in material substances. He was very much interested in the work of de Broglie on matter waves, and was one of the first scientists who shared these ideas. In a paper with Bose as co-author, he stated in brief the statistics of a system composed of identical particles, and this is what is called "Bose-Einstein statistics" today.

Another decisive point in Einstein's scientific life started in 1927, when the new "wave-mechanics" was formulated and announced. The major discussions between Einstein, Bohr and others about wave mechanics in the Solvay congress in Brussels are well known. Einstein admitted that the new theory had no contradictions, but he regarded the statistical character of the theory as incomplete. "By saying 'perhaps' one cannot make a theory," he often said, and also "It is deeply wrong, even if it is empirically and logically right.." Einstein could not accept *a priori* probabilistic results.

As the Copenhagen interpretation of Bohr was more and more accepted by the majority of his contemporary theoretical physicists, Einstein remained fixed in his beliefs. His opposition is reflected in his works. In his famous paper with Podolsky and Rosen, he criticizes the notion of interaction at a distance posed by the quantum theory. He published to similar papers later without co-authors, expounding his objections openly. He once said sarcastically to Pauli: "Physics is the description of reality, or should I say that Physics is the description of what one pictures only in one's own mind?" Einstein was concerned that the new theory, by refuting reality, would lead to the rejection of objectivity, so that a physical theory could fail to distinguish what is real from what is a dream or hallucination. Einstein always supported an objective description of physical reality, without interference of the observer. It is characteristic that till the end of his life he expected clarification of the atomic constitution of matter through the classical theory of the field.

From 1927 on, Einstein was disappointed by the further development of physics. Slowly he retired in to an intellectual solitude. His subsequent papers on field theory are almost of the same mathematical artfulness as the earlier ones, but they lack the close contact with nature and contemporary science. It is doubtful whether his later works have found any immediate application in current research. He chose separation from the conventional quantum mechanical trade, and perhaps consequently this also affected his field theory, by making it not very popular. The life of Einstein ended with a question about the science of Physics and the requirement for us to search for synthesis.

The Physical and Philosophical Reasons for A. Einstein's Denial of the Ether in 1905 and its Reintroduction in 1916

Ludwik Kostro
Department for Logic, Methodology and Philosophy of Science,
University of Gdańsk, Bielańska 5
80-851 Gdańsk, Poland, e-mail: fizlk@univ.gda.pl

In the first part of this paper the author presents physical and philosophical reasons which, in 1905, finally impelled A. Einstein to deny the existence of the ether of 19th century physics, especially H.A. Lorentz's ether. Since Einstein, under the influence of P. Drude, identified Lorentz's ether with absolute space, the denial of Lorentz's ether meant, for Einstein, the denial of the existence of absolute space. The author also presents reasons of the same kind which impelled Einstein, when he was formulating the general relativity, to entirely deny the existence of physical space and time. This denial lasted from 1913 to 1916.

In the second part the author presents physical and philosophical reasons that impelled Einstein in 1916, to again recognise the real existence of space and time and to call them (connected in his theory into one space-time) "the new ether."

I. Physical and philosophical reasons for denying the ether, in 1905, and physical space and time in the period from 1913 to 1916

1. The physical reasons for Einstein's denial of the ether have their roots in 19th century physics and in the results of his special relativity, formulated finally, as is well known, in 1905. Einstein did possess a sufficiently good knowledge of the conceptual system of 19th century physics. In this system, the notions of space and time were closely connected with the notion of reference system. Therefore, in Einstein's mind, at the beginning of his scientific activity, the notions of space and time were also closely connected with the notion of reference frame, and, therefore, Einstein, in his papers, called the reference systems "reference spaces" (*Bezugsraume*) and understood

absolute space and time as a privileged reference system. Later, however, after the definitive formulation of the General Relativity Theory, Einstein began to make a clear and explicit distinction between physical space and reference systems. According to him, as we will see later, physical space as such can never be considered a reference system.

When Einstein began his scientific activity, he was intellectually dissatisfied with the existence of an asymmetry between mechanics and electrodynamics. On the one hand, in mechanics, all the so-called inertial reference systems were equivalent for the formulation of the laws of nature but on the other hand, in electrodynamics and optics one reference system was privileged. It was the reference system of the ether, especially Lorentz's ether, that Einstein, under the influence of the German physicist Paul Drude, identified with absolute space. The unnaturalness of this asymmetry, in Einstein's opinion, was the first and the principal physical reason that impelled him to remove absolute space and time, and at the same time the ether, by introducing the Special Principle of Relativity. As is well known, according to this principle, not only in mechanics but also in electrodynamics and optics, all inertial reference systems are equivalent for the formulation of the laws of nature.

Therefore, in Einstein's opinion, any privileged reference system became superfluous. And that is why we find in his Special Relativity paper the following sentence:

> *The introduction of a "luminiferous ether" will prove to be superfluous inasmuch as the view here to be developed does no longer need an absolute space, at absolute rest, with physical properties* [1]

This quotation shows, in a clear way, that Einstein identified the luminiferous ether with absolute space at absolute rest. He was convinced that in Lorentz's conception of the ether such an identification existed. It must be noted, however, that Einstein did not know exactly what Lorentz's ether was [2]. Lorentz did not consider his ether to be at absolute rest; he only maintained that the parts of his ether were at rest with respect to each other, and he attributed a certain substantiality to his ether. According to him, the vacuum was "too empty" to qualify as an ether [3]. As it has already been mentioned, Einstein identified Lorentz's ether with absolute space at absolute rest under the influence of Paul Drude's textbooks and papers [4,5], which he read during his studies at the ETH (*Eidgenossische Technische Hochschule*) in Zurich. It was Drude who presented Lorentz's results in electromagnetism in an ether that was identified with physical space at absolute rest, endowed with physical properties.

The existence of the above-mentioned asymmetry was already indicated by Einstein in the first sentence of his Special Relativity paper:

It is well known that Maxwell's electrodynamics—in its modern understanding—leads to an asymmetry when it is applied to bodies in motion. [1]

As we can see, this asymmetry—we repeat once again—was the first and the principal physical reason which disposed Einstein to introduce his Special Relativity Principle and inclined him to deny the existence of the ether, especially the Lorentz ether.

2. The negative result of Michelson-Morley experiment was the second physical reason which disposed Einstein to introduce his Special Principle of Relativity and inclined him to deny the ether. Einstein himself indicated this reason in his speech at Kyoto University in 1922:

While I was thinking of this problem in my student years, I came to known the strange result of Michelson's experiment. Soon I came to the conclusion that our idea about the motion of the earth with respect to the ether is incorrect, if we admit Michelson's null result as a fact [6].

Einstein came to know of Michelson's experiment in his student years when he read Lorentz's book of 1895 [7] in which this experiment was presented and interpreted in the paragraphs 89-92 entitled *Der Interferenzversuch Michelsons*. In the above-mentioned speech at Kyoto University he said: "*I had a chance to read Lorentz's monograph of 1895*" [6]. However, in his Special Relativity paper, when he presented the results of this experiment, he did not mention Michelson's name.

3. The physical reasons presented above were accompanied by philosophical ones. Einstein was under great influence by the so-called second positivism represented by E. Mach [8], W. Ostwald [9] and R. Avenarius [10]. He studied the books of these authors and discussed them with his colleagues during sessions of their Olymp Academy [11]. According to the positivistic philosophy of these authors, in science the principle of pure experience must be respected, and therefore science must be purified of all metaphysical intercalations or interpolations and of all foreign matter which have nothing to do with experience. According to Mach, space and time (especially absolute space and time) were such metaphysical interpolations that have to be removed from physics. They have to be removed because they have no experimentally accessible properties.

According to Mach the inertial properties of physical bodies did not have to be considered with respect to the absolute space but they have to be conceived as a result of the influence of the distant masses. Mach, therefore, admitted the existence of the ether. He needed the ether hypothesis to explain the action of the distant masses. W. Ostwald made the next step and also considered the ether to be a metaphysical interpolation. When Einstein published his Special Relativity paper with the denial of absolute space and the ether, he was convinced that he was realising the programme of Mach's, Ostwald's and Avenarius' philosophy. We quote a passage from Einstein's

1905 paper which shows, in a clear way, the influence of the positivistic philosophy on Einstein's opinions concerning the ether.

> *The unsuccessful investigations, the purpose of which was the ascertainment of the Earth's motion with respect to the "luminiferous medium," lead to the supposition that not only in mechanics but also in electromagnetism, to the notion of absolute rest correspond no properties typical of physical phenomena.* [1].

As we can see, according to Einstein, as in to the positivistic philosophy of Mach, Ostwald and Avenarius, to the notion of absolute rest there correspond no properties typical of physical phenomena. In other words, absolute rest cannot be experimentally proved. It is not accessible to any experience.

In this way we arrive at the main philosophical reason that inclined Einstein to deny the existence of absolute space and time and the existence of the ether in 1905. Concluding, we can say that he, like the positivist philosophers, considered them to be metaphysical interpolations for which there is no place in physics.

4. Einstein was not satisfied with the denial of absolute space in Special Relativity. According to him this denial was only the first step in carrying out Mach's program. His Special Relativity recognised still a privileged class of reference spaces, the so-called class of inertial reference systems. Subsequently, when he arrived at the conclusion that in his General relativity he succeeded in removing the privileged set of inertial reference frames and when he became aware that, in this new theory of gravitation, the co-ordinate systems lost physical meaning, he begin to be convinced that General Relativity had achieved Mach's goal and removed space and time from physics as metaphysical interpolations. According to him, the only real things that remained in physics were the space-time coincidences of events.

In 1916, in his General Relativity paper he wrote:

> *That this requirement of general covariance, which takes away from space and time the last remnant of physical objectivity, is a natural one, will be seen from the following reflection. All our space-time verifications invariably amount to a determination of space-time coincidences* [12]

In a letter to Moritz Schlick, on December 14th, 1915, he stated:

> *Thereby (through the general covariance of the field equations) space and time lose the last remnant of physical reality* [13]

In a letter to P. Ehrenfest, on December 26th, 1915, he emphasised:

> *The physical real that happens in the world (as opposed to what depends on the choice of the reference system) consists of spatio-temporal coincidences. (In a footnote, Einstein added: "And nothing else!")* [14]

It can be proved that in the period from 1913 to 1916 Einstein did not believe in the existence of physical space endowed with real physical properties. He expressed his disbelief, *expressis verbis*, in a letter to E. Mach, in late 1913 or early 1914:

> For me it is absurd to attribute physical properties to "space" [15].

And in a paper of 1914 he stated:

> As much I am not disposed to believe in ghosts so I do not believe in the enormous thing about which you are me talking and which you call space [16].

Since for Einstein ether meant "physical space with real properties" therefore, at that time, he solidified his disbelief in the existence of the ether as well.

Concluding, we can say that the denial of the existence of space and time had in Einstein's mind, on the one hand scientific reasons—the removal of the privileged class of inertial reference frames and the loss of physical meaning by the co-ordinate systems—and on the other hand, philosophical reasons—the program of the second positivism.

II. The physical and philosophical reasons that impelled Einstein to again recognize the existence of space and time and interpret real space-time as a new ether

1. In June 1916, Einstein changed his mind under the influence of a letter exchange with A.H. Lorentz and because of a polemics with Ph. Lénard. Lorentz, after reading Einstein's General relativity paper, became convinced that Einstein's new theory of gravitätion allowed the hypothesis of a stationary ether. On June 6, 1916, Lorentz wrote a long letter to Einstein in which he tried to convince him of his views. Einstein did not agree with Lorentz because a stationary ether violated his Principle of Relativity, but he did come to the conclusion that in General Relativity space-time has real physical properties, and therefore a new ether can be introduced.

He wrote to Lorentz:

> I agree with you that the general theory of relativity is closer to the ether hypothesis than the special theory. This new ether hypothesis, however, would not violate the principle of relativity, because the state of this $g_{\mu\nu}$ = ether would not be that of a rigid body in an independent state of motion, but every state of motion would be a function of position, determined through the material processes [17].

As we can see, physical space (intimately connected with time) the local state of which is described by the components $g\mu\nu$ of the metrical tensor g, was regarded by Einstein as a new ether. In this new conception the ether is no longer considered as a rigid quasi-object to which we can apply the notion of rest and motion, and that is independent of material processes, but it is conceived as a field-medium the structure of which depends upon the presence and motion of material bodies, and which determines the inertio-gravitätional behaviour of these bodies.

Einstein did not publish his new ether conception in either 1916 or 1917. A controversy between Einstein and Lénard provoked the first appearance of the new view in print. In a lecture in 1917, and the next year in a published paper [18], Lénard raised the objection against Einstein's general theory of relativity that in this theory the disqualified ether came back under the new name "space." In reply Einstein wrote an essay [19] in which he recognised that the physical space, in general relativity, has real physical properties and therefore it can be called "new ether." He emphasised, however, that the notions of rest and motion cannot be applied to the new ether and that, therefore, it can never be considered as a reference frame.

As we can see, after June 1916, in Einstein's mind the notion of physical space broke off from the notion of reference frame. In his opinion, the real physical space, that is one and unique, can never be a reference frame because it is not composed either of points or parts the motion of which could be followed in time.

> *This ether, however, should not be thought of as endowed with the properties characteristic of ponderable media, as consisting of parts that may be traced through time. The idea of motion may not be applied to it* [20].

In identifying the new ether with physical space, Einstein made a very clear distinction between space as such (*Der Raum als solche*), conceived as it indicated above and the reference spaces (*Die Bezugsraume*). According to Einstein there is only one single physical space as such which physically manifests itself through field properties which are mathematically represented by the components $g_{\mu\nu}$ of the metrical tensor g interpreted as gravitätional potentials.

There is also an infinite number of reference spaces which are artificial extensions of reference bodies. We introduce a reference space through an infinite number of points that we connect with a reference body. Therefore, every reference space is composed of points like every reference body is composed of particles. Every reference space like every material medium can serve as a reference frame.

When we move with respect to a material medium we feel either a wind or a change of temperature. When we move with respect to a reference space we "feel" a wind of points. Either the particles of a material medium or the points of a reference space can be followed in time. Therefore, the notions of rest and motion are applicable to material media and reference spaces. References spaces are able to move with respect to one another.

The notions of rest, motion, and velocity, however, are absolutely inapplicable to the new ether identified with physical space as such.

According to Einstein, physical space as such constitutes an ultra-referential fundamental reality that makes possible the existence and the motion of references spaces although it is neither at rest or in motion itself. It manifests its reality through its field properties that determine the behaviour of free moving particles and bodies.

As we can see, the letter exchange with Lorentz and the controversy with Lénard provided the physical reasons for the reintroduction of the ether under a new relativistic form. Einstein became aware that the space-time of his relativity theory has real physical properties that are mathematically describable. Therefore, he wrote, in 1919, in the so-called "Morgan Manuscript":

> *Thus, once again "empty" space appears as endowed with physical properties, i.e. no longer as physical empty, as seemed to be the case according to special relativity. One can thus say that the ether is resurrected in the general theory of relativity, though in a more sublimated form* [21].

There is also another physical reason that impelled Einstein to consider the physical space as a new ether. When Einstein became aware that the space-time of his relativity theory has real field properties he became aware as well that it possesses a certain energy density and therefore constitutes a material medium *sui generis*. And therefore, according to him, the name "ether" is very appropriate for it.

2. The philosophical reasons were provided by Moritz Schlick. After having read Einstein's letter in which he informed him that in General Relativity *space and time lose the last remnant of physical reality* [13], Moritz Schlick wrote a book that constituted his *Habilitationsschrift* [22]. In this book, among other things, he makes a criticism of Mach's philosophy. He criticises Mach's opinion, according to which the physical world investigated by science is composed only of sensations of our senses called in Mach's philosophy "elements." Schlick shows that:

> *."..all physical quantities with which we are dealing in the laws of physics do not constitute "elements" in Mach's sense. The coincidences, that are expressed by the differential equations, are not immediate experiences of our senses. They do not mean immediate convergence of the data of our senses but they constitute unimaginable quantities, like for example forces of the electric and magnetic fields"* [22, p. 85.]

Einstein, after reading Schlick's criticism, arrived at the conclusion that there are no reasons to maintain that only coincidences of events are real, because these coincidences are also not immediate data of our senses when considered by theoretical physics. He abandoned the epistemology of the second positivism and recognised that the criterion of exclusive pure experience in physics is unacceptable.

Thus, Schlick helped Einstein to recognise the real existence of space-time endowed with physical properties expressed mathematically by the components $g_{\mu\nu}$ of the metrical tensor g. These components, of course, are not immediate data of our senses.

From 1916 until 1938 Einstein used the words "new ether" to refer to space-time with physical properties. The last time Einstein mentioned the new ether was in his book *The Evolution of Physics* written with Leopold Infeld in 1938. We quote their words.

> *We may still use the word ether, but only to express the physical properties of space. This word ether has changed its meaning many times in the development of science. At the moment it no longer stands for a medium built up of particles. Its story, by no means finished, is continued by the relativity theory* [23]

Nowadays we are dealing with a renewal of ether theories because we have become aware, like Einstein, that the vacuum has a real structure with physical properties. In this domain the papers published by F. Selleri are very interesting. Here we may mention one of them entitled *Space-time Transformation in Ether Theories* [24]. Also papers and books by S.J. Prokhovnik are worth reading *e.g.* [25].

Acknowledgements

The author would like to express his sincere thanks to the Volkswagen-stiftung for a scholarship that made also this research possible, and to the Hebrew University of Jerusalem for the kind permission to quote from Einstein's papers and letters.

References

The abbreviation EA and the numbers in brackets refer to the control index of the Einstein Archive at the Hebrew University of Jerusalem.

[1]   A. Einstein, Zur elektrodynamik bewegter Körper, *Annalen der Physik*, **17**, (1905) p. 891.

[2]   J. Illy, Einstein Teaches Lorentz, Lorentz Teaches Einstein. Their Collaboration in General Relativity, *1913-1920, Archive for History of Exact Sciences* **39**, (1988), pp.247-289.

[3]   A.J. Kox, Hendrik Antoon Lorentz, the Ether, and the General Theory of Relativity, *Archive for History of Exact Sciences*, **38**, (1988), pp.67-78.

[4]   P. Drude, *Physik des Äthers auf elektromagnetischer Grundlage*, Ferdinand Enke, Stuttgart 1894.

[5]   P. Drude, *Lehrbuch der Optik*, S. Hirzel, Stuttgart 1900.

[6]   A. Einstein, Speech at Kyoto University, December 14, 1922, *NTM-Schriftenreihe für Geschichte der Naturwussenschaften*, "Technik und Medizin," Leipzig, **20**, 1983, pp.25-28.

[7]   H.A. Lorentz, *Versuch einer Theorie elektrischen und optischen Erscheinungen* in *bewegten Korpern*, Brill, Leiden 1895.

[8]   E. Mach, *Die Mechanik in ihrer Entwiklung historisch- kritisch dargestellt*, Brockhaus, Leipzig 1883.

[9]   W. Ostwald, *Lehrbuch der allgemeinen Chemie*, (in zwei Bänden), Leipzig 1893.

[10] R. Avenarius, *Kritik der reiner Erfarung*, 1889/1890.

[11] M. Solovine, *Introduction*, [in:] A. Einstein, *Briefe an Maurice Solovine (1906-1955)*, Veb deutscher Verlag der Wissenschaften, Berlin 1960, pp. VII-XV.

[12] A. Einstein, Die Grundlagen der allgemeine Relativitätstheorie, *Annalen der Physik*, **49** (1916), pp.769-822.

[13] A. Einstein, *Letter to M. Schlick*, December 14th, 1915 (EA 21-610).

[14] A. Einstein, *Letter to P. Ehrenfest*, December 26th, 1915, (EA 9-363).

[15] A. Einstein, *Letter to E. Mach*, December 1913 or January 1914, (EA 16-454).

[16] A. Einstein, Zum Relativitätsproblem, *Scientia*, **15** (1914), pp. 337-348.

[17] A. Einstein, *Letter to H.A. Lorentz*, June3 17, 1916 (EA 16-453).

[18] Ph. Lénard, Uber Relativitätsprinzip, Äther, Gravitätion, *Jahrbuch der Radioaktivität und Elektronik*, **15**, (1918), pp.117-136.

[19] A. Einstein, Dialog uber Einwande gegen die Realtivitätstheorie, *Die Naturwissenschaften* **6**, (1918), pp.697-702.

[20] A. Einstein, *Äther und Relativitätstheorie*, Springer, Berlin 1920.

[21] A. Einstein, *Grundgedanken und Methoden der Relativitätstheorie in ihrer Entwiklung dargestellt*, Unpublished manuscript called *"Morgan Manuscript"* (EA 2-070).

[22] M. Schlick, *Raum und Zeit in der gegenwärtigen Physik*, Springer, Berlin 1920.

[23] A.Einstein and L. Infeld, *The Evolution of Physics*, Simon and Schuster, New York 1938. The quotation is an English translation of the German original entitled *Physik als Abenteuer der Erkentnis*, A.W Sijthoff' Witgevermaatschappij N. V., Leiden 1938.

[24] F. Selleri, Space-time Transformations in Ether Theories, *Z Naturforsch.* **46a**, (1990) pp. 419-425.

[25] J.S. Prokhovnik, *Brit. J.Phi l. Sci.* **14**, (1963), p.195.

On the Question of Physical Geometry

N.A. Tambakis[*]
Interdisciplinary Research Group
Department of Physics
University of Athens

Introduction

For any discussion of the foundations of the Theory of Relativity it seems not only natural but also necessary to look at the art of building foundations itself. I also think that building foundations means constructing an axiomatic physical theory

Certainly such a view doesn't come as a surprise! However, it seems to me that it has been a bit neglected since 1900 when D. Hilbert pointed out, as his 23rd problem, the need "to treat those physical sciences, in which mathematics play an important role, by means of axioms, like geometry."

Of course, I mean the need for a strict axiomatic theory with a rigor somewhere between that of Euclid's "Elements" and a modern formal language. This is certainly a rigor quite above that of the usual presentations of special relativity (SR) in textbooks or of formulations such as "SR from only one axiom," *etc.* [1]

In this axiomatic spirit I'm going to briefly present three matters of the Theory of Physical Geometry, *i.e.* the Theory of Physical Space and Time.

1. The Axiomatic Method in Physics

The axiomatic method in general is usually presented today through the concept of Formal Language [2], *i.e.* of a set L (D, A,Q), where

D: definitions (undefined terms, such a "point," included),

A: axioms,

Q: logical rules of deduction (usually those of classical logic and common to most scientific theories).

From L follow the propositions P of the theory represented by

$$L: L(D,A,Q) \rightarrow P.$$

---

*   Mailing address: N. A. Tambakis, Lysicratous 3, 16343 Athens, Greece Fax +30-1-5546952
Open Questions in Relativistic Physics
Edited by Franco Selleri (Apeiron, Montreal, 1998)

In a purely *mathematical* theory both D and A are, of course, arbitrary, *i.e.* its axioms are neither true nor false. What matters is only some *logical* requirements for them, such as: consistency, independence, uniqueness of meaning (completeness, comprehensibility).

Now what about D and A in a *physical* theory? There are certainly differences of which two are crucial:

(i)   Definitions are again arbitrary but now they are *pragmatic*, *i.e.* they refer to actual things, such as the unit of length held in Paris.

(ii) Axioms are now not at all arbitrary but they have to be checked by experiment, *i.e.* they take the status of so-called "physical laws."

It is now evident that from the above mentioned nature of definitions and axioms of a physical theory, some serious epistemological questions are always open:

1)   A great advantage of an axiomatic physical theory, beyond its rigor, is certainly its economy: instead of checking all its propositions we have to check only its axioms. But unfortunately such an ideal is hardly ever achieved: not only does any experiment designed to check an axiom already contain a certain amount of theory, but even worse, seldom can a physical axiom be verified directly by experiment. (Think for example of how far from direct verification stands Hamilton's principle.)

2)   Even in the case of a direct verification—which, as previously said, already contains a certain amount of theory—serious epistemological questions like the validity of induction, causality, *etc.*, always remain open. (Think of the experiments designed to check local realism *vs.* quantum mechanics.)

2.   A Two-Fold Rejection of any *A Priori* Physical Geometry

After this preliminary, let us now turn to the problem of Physical Geometry (PG). A theory of PG is itself a physical theory and as such should be finally an axiomatic one.

On the other hand the historical development of the question of PG seems to be dominated by *a priori* views, such as Kantian, neo-Kantian or others.

Our intention is now to refute any *a priori* view on two grounds, first for methodological reasons and secondly for logical ones.

2.1 Methodology

Our point is the following:

Since our definitions are arbitrary but pragmatic, the choice of axioms cannot any more be free and, particularly, a priori.

In the case of PG there is a crucial definition, that of *congruence*. We will try to show in an elementary way how this definition restricts the axioms and,

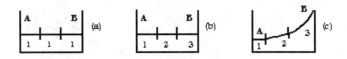

Fig.1 Two ways of recalibrating a segment: (b), (c)

hence, the kind of PG. In other words we will show that the choice of PG is not free but it depends upon the definition of congruence.

By the term "congruence" I mean coincidences either in space or in time. Also, we shall use the concept of *uniformity*, which underlies that of congruence.

"Uniformity" (U) means the possibility of juxtaposing the actual unit of length or repeating the actual unit of time. To the notion of U let us also bring in its negation (-U), which actually means nothing more than the possibility of *recalibration*.

The concept of recalibration is closely connected to that of congruence in the sense that every time we redefine congruence we are actually doing a recalibration.

This fact will become clear with an elementary example:

Let AB be a straight bar measured by a standard unit rod and its length be found equal to 3; the bar is known to be in a box with rigid walls (Fig 1a). Let us then suppose that some unscrupulous assistant measures AB again in three stages, each time leaving the laboratory for a while and committing the following blunder: although each time he uses the standard unit rod, he mistakenly thinks that in the second step he has got a double-unit rod and in the third one a triple-unit rod. In this way he draws his own version of the bar AB (Fig.1b).

Let us finally suppose that this sketch is given to a meticulous guy. This fellow, knowing that the box containing the bar AB has a width of 3 standard units, draws his own version which necessarily shows the bar as curved (Fig.1c).

What is the moral we can draw from this story? It is clearly that our assessment of physical geometry depends crucially on our definition of congruence. The third observer judged the bar as curved because he used a new definition of congruence, that of Fig.1b (Of course, his judgment can equally well be interpreted as a model of a "straight" bar in a non-Euclidean space.)

For this moral to be drawn, there is, however, an implicit assumption we must not miss: we have supposed that the third observer knew nothing of the mess the second observer has made of the units of measurement. If he knew about that, he would have made appropriate corrections and he would have concluded that the bar is a straight Euclidean one, 3 units long. This question

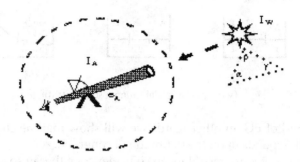

Fig. 2. Experiment designed to check a Euclidean property of the world.

of knowing whether or not some recalibration did indeed take place brings up a new problem, both physical and epistemological.

There are certainly cases where the cause of recalibration is known and its effects can be eliminated, *e.g.*, a surveyor checks the temperature to which his measurement instruments are exposed and he makes necessary corrections. But there also cases where we are actually in the dark as to whether some recalibration has been done or not. This occurs when some force affects all objects equally without it being possible to protect them from it—and, of course, we all know such a force, gravitation! But a discussion on the important issue of differential and universal forces would lead us astray [3].

2.2 Logic

Let X be an experiment designed to check a proposition E which would confirm the Euclidicity of the 3-space (*e.g.*, E:$\alpha + \beta + \gamma = 180°$, for any triangle; Fig. 2).

From a logical point of view, X is a transformation:

X $(I_W, I_A, e_A) \rightarrow$ E, where,

I_W: information from the outside world,

I_A: information from the apparatus (observer included)

e_A: inevitable Euclidean properties of the apparatus.

Let us now suppose that the relation E is found valid; may we then conclude for certain that the 3-space is Euclidean? Certainly not, for the following simple reason: If the 3-space was actually not Euclidean, then I_W would have the form $I_W = (I'_W, h_W)$, where h_W would be some hyperbolic property, like that of $\alpha + \beta + \gamma < 180°$. But then the experiment X, considered as a deductive system, would take the form X (I'_W, h_W, I_A, e_A); but this last system includes the contradictory inputs h_W and e_A and, therefore, anything could be proved by it (including the proposition E!).

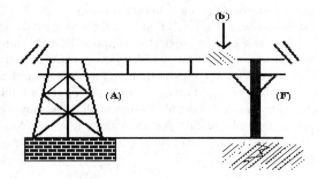

Fig.3. Fracture of a beam (b) does not necessarily mean the failure of the structure (A) but perhaps of another one (F).

3. Can a Physical Theory be "Falsified" because of Wrong Physical Geometry? (Einstein *vs.* Whitehead)

I think that the previous arguments suffice to persuade us that there is enough freedom in the choice of PG (at least on a large scale), once the rules of defining congruence have been fixed.

If so, how valid is a reassurance that a certain PG is the "true" one? Is there indeed any value in the assertion that a physical theory has been falsified because it used the wrong kind of PG?

I think a good example of such an assertion presents the case of A.N. Whitehead's relativity theory *vs.* Einstein's general relativity (GR). Whitehead[4] constructed his own GR on the basis of two geometrical assumptions:

1)   Space is the traditional Euclidean manifold of three dimensions;
2)   Time is not uniform if the space is permeated by a gravitational field.

He further accepted the principle of general covariance which is fulfilled if physical laws are expressed in tensorial form. He actually used two tensorial fields: a "physical metric" g in which material particles and e.m. waves propagate along its geodesics and get deflected by the gravitational fields of the stars, and a flat "background metric" n in which gravitational waves propagate along its geodesics and are thus unaffected by the stars.

Whitehead's relativity theory fully satisfied the four standard tests of GR and, therefore, it became, for reasons related to the sociology of science, an example of a theory for which more effort was made to falsify it than to verify it. This urge for falsification was, however, relieved only in the 70's, and even then in a rather marginal prediction that this theory was making regarding ebbs and floods [5]. It seems to me, however, that such a falsification has to be examined more as an epistemological question than as a physical one. Let me sketch this point with the help of a structural analogy.

To assert that a theory (A) is falsified by some wrong prediction (b) of it is only based on the false epistemological assumption that (A) stands alone by itself as a model of the physical world. However, (A) is actually only a member of a network of theories and, therefore, the empirically falsified prediction (b) may also be based on another theory (F). (I think this is the case for Whitehead's theory but this detailed subject will be examined elsewhere)

This epistemological situation has an analogy to a construction scheme: the fracture of a beam (b) may occur from a fault in the foundation of pillar (F) and not from a failure of (A) (Fig. 3).

4. Conclusions

From the three matters—axiomatics, refutation of *a priori* and falsification—already discussed in short, we may now draw our conclusions concerning the objectivity of PG:

(1)   There is a freedom in the choice of the definition of congruence (in space and time).
(2)   Once this definition is chosen, we are not free any more in the choice of PG: we are necessarily confined within certain axioms, in the sense that only these axioms are empirically verified; another definition of congruence would give rise to another set of axioms, again empirically verified, i.e to another PG [6].
(3)   These two PGs can be transformed into one another, at least locally, by some metrical transformation.

It is the existence of such transformations between possible geometries that points out the objective existence of the physical reality we call "space-time." The different possible PGs are merely possible *representations* of this underlying reality.

It seems to me that in this way we can confirm the well-known epistemological assumption that space and time are not fictions but rather modes of the dynamic existence of matter.

Notes and References

[1]   See, for example, A. Lee, T. Kalotas, *Am. J. Phys.*, *43*, 5 (1975). Also, H. Schwartz, *Am. J. Phys.*, *43*, 4 (1975).
[2]   See, for instance, I. Copi, *Symbolic Logic*, (Macmillan, 1973), 6.5. Also, Blumenthal, *A Modern View of Geometry*, (Dover, 1961), Ch. III.
[3]   A principle of elimination of "universal forces" has been introduced by H. Reichenbach "but has so far not found the attention it deserves" (R. Carnap); see H. Reichenbach, *The Philosophy of Space and Time*, (Dover, 1958), §6.
[4]   A.N. Whitehead, *The Principle of Relativity*, (Cambridge University Press, 1922).
[5]   C.M. Will, *Astrophys. J, 169* (1971), 141-156. For a discussion, see the classic C. Misner, K. Thorne, J. Wheeler, *Gravitation*, (Freeman, 1973), pp. 1067,1124.

[6]   At this point the question arises of which PG is preferable on the grounds of simplicity but, certainly, it makes no sense to talk about the "falsity" or "truth" of some PG. The choice of PG may also be restricted, however, for some physical reasons. (*cf.* H. Reichenbach, *ibid* [3], pp.34-5, 263)

Nonlocality, Relativity, and Two Further Quantum Paradoxes

G. Tarozzi
Istituto di Filosofia, Università di Urbino
Centro Interuniversitario di Filosofia e Fondamenti della Fisica
(Bologna-Urbino)

1. Introduction

In his ongoing struggle against the Copenhagen interpretation of quantum mechanics, in several circumstances Einstein appealed to the principles of his relativity theory to prove either the inconsistency or the incompleteness of the standard point of view.

In the paradox proposed at the Solvay Congress of 1927, he showed that Born's probabilistic interpretation of the quantum mechanical wave function was unacceptable, since in the case of the reduction process, it implied the possibility of instantaneous actions at a distance, in contrast with one of the two postulates of special relativity. In the photon-box experiment of 1930, he stressed how, starting from the principle of equivalence between mass and energy, one was led to a violation of the indeterminacy relations between time and energy. In the famous paper written with Podolsky and Rosen in 1935, the celebrated EPR paradox or argument, he finally showed that the locality condition plus the reality criterion led to the incompleteness of quantum mechanics.

Nowadays after the experimental tests of Bell's inequality, the situation has completely changed and several authors are investigating the consequences of the violation of locality for relativistic theories: for example, the propagation of superluminal signals, the possibility of retroaction in time, the introduction of some new kind of ether, tachyons, Lorentz interpretation *vs.* Einstein interpretation. So whereas for Einstein the presence of a conflict between standard quantum mechanics and a relativistic principle was a proof of the incompleteness of the Copenhagen interpretation, at the present time the relativity theory is under critical examination to clarify what effects the violation of the locality condition has for its axiomatic formulation.

What I intend to discuss now in a sense represents a return to Einstein's original perspective. I propose, in fact, to investigate the consequences of the violation of the locality postulate not for relativity theory—even though I think that this represents an important field of research in the foundations of physics—but for quantum mechanics itself.

In particular, I first discuss my argument that in standard quantum mechanics the violation of the locality condition implies the inconsistency of its present formalism, and show[1], moreover, how the macrorealistic interpretations of the theory of measurement, where the reduction postulate has been abandoned and the measurement process is viewed as an interaction between microsystem and macro apparatus, are plagued by serious paradoxes, among them the possibility of *superluminal long-distance macroscopic effects.*[2]

2.   The photon-box experiment and the conflict between the principle of equivalence and the indeterminacy relations

During the 1930 Solvay Congress [1], Einstein discussed the so-called photon-box experiment, one of his most famous objections against the Copenhagen formulation of quantum mechanics. He proposed to consider a radiation-filled box, suspended on a spring, endowed with perfectly reflecting walls and a shutter, which was opened by a clockwork mechanism enclosed in the box itself. He assumed that the clock was set to open the shutter at time $t = t_1$, for an arbitrarily small interval $t_2 - t_1$, so that a single photon could be released.

Einstein then pointed out that, by measuring the weight of the whole box and observing the length of the spring *before* and *after* the emission of the accurately timed radiative pulse of energy, the difference in the energy content of the box could be determined with an arbitrarily small error ΔE, from the principle of equivalence between mass and energy $E = mc^2$ of the theory of relativity. This energy difference, in accordance with the principle of energy conservation, would then be the energy of the emitted photon.

In this way both the energy of the photon, and its time of arrival at the distant screen can be predicted with arbitrarily small uncertainties ΔE and Δt, in contradiction with Heisenberg's relations, according to which

$$\Delta E \cdot \Delta t \geq \frac{h}{2\pi}. \tag{1}$$

In his famous rebuttal, Bohr emphasized how the reading of the length q of the spring is possible, because of the indeterminacy relations between position and momentum, only with an uncertainty Δq, given by

$$\Delta q \geq \frac{h}{\Delta p} \tag{2}$$

and that the uncertainty Δp in the momentum of the box "must obviously... be smaller than the total impulse which, during the whole interval T of the balancing procedure can be given by the gravitational field to a body of mass Δm." Hence

$$\Delta p \approx \frac{h}{\Delta q} \leq T g \Delta m ,\qquad (3)$$

where g denotes the gravity acceleration. From these considerations Bohr concluded that "the greater the accuracy of the reading q of the pointer, the longer must, consequently, be the balancing interval T, if a given accuracy Δm of the weighting of the box with its content shall be obtained."

Now, from general relativity theory, a displacement of the box by an amount Δq in the direction of a gravitational field causes a change ΔT in the clock reading, in accordance with Einstein's formula for the redshift in a gravitational field

$$T = T g \frac{\Delta q}{c^2} .\qquad (4)$$

Therefore, from the uncertainty Δq of the position reading, on the one hand, and the condition (3) on the other, one obtains

$$\Delta T \geq T g \frac{1}{c^2} \frac{h}{\Delta m} ,\qquad (5)$$

which, again using Einstein formula $E = mc^2$, re-establishes the validity of the indeterminacy relations expressed by (1).

Thus, just as Einstein had succeeded in finding a violation of the indeterminacy relations between energy and time starting from the (relativistic) principle of equivalence between mass and energy, so Bohr seemed to have preserved the validity of such relations by appealing to other relativistic principles, such as the principle of equivalence between inertial and gravitational mass, introduced with relation (4)

Bohr's counter-argument was severely criticized by Popper, who underlined its methodological error: the appeal to another different physical theory, *i.e.* general relativity instead of special relativity, as in Einstein's argument, to guarantee the consistency of quantum mechanics, "changed illegitimately the rules of the game." [2]

We shall now show how Popper's previous objection to Bohr's answer appears more compelling in the light of the fact that Einstein's criticism of the Heisenberg principle is not based on any (even special) relativistic principle, and we will do this by stressing the possibility of deriving the principle of equivalence between mass and energy, or the theorem of the inertia of energy, on the grounds of purely classical assumptions.

3. A non-relativistic derivation of the principle of equivalence between mass and energy

In order to prove the possibility of a strictly classical derivation of the principle of equivalence between energy and mass, let us examine another thought experiment proposed in 1905 by Einstein, in this case with foundational and not refutational purposes. He considered a closed box, in the sides of which there were two exactly similar instruments 1 and 2, *which were constructed so as to send* a light signal in a definite direction or completely absorb an incoming light signal. Now, let the transmitter 1, at a definite instant, send out a signal in the direction of the receiver 2. As a consequence of this process, the instrument 1, and with it the whole box, undergoes a recoil.

According to classical electrodynamics, the momentum p transferred by light of energy E to a given surface corresponds to the radiation pressure

$$p = \frac{E}{c}. \tag{6}$$

In the case of this thought experiment, the momentum transferred by recoil to the box during the emission from 1 is of the same magnitude. If the box considered has the total mass M, it acquires a recoil velocity v to the left, according to the law of conservation of momentum, by

$$Mv = \frac{E}{c}. \tag{7}$$

The box continues its motion during the time employed by the light signal to travel the distance l between 1 and 2. If we neglect terms of higher order this time t is given by

$$t = \frac{l}{c}. \tag{8}$$

During this time interval, the box moves a distance

$$x = vt = \frac{E}{cM}\frac{l}{c} = \frac{El}{Mc^2}. \tag{9}$$

In order to avoid a conflict with the fundamental principle of the centre of gravity of classical mechanics, we must assume that the transfer of energy from 1 to 2 is accompanied by a simultaneous transfer of mass in the same direction. If we denote by m this unknown transferred mass, then we have, on account of the principle of the centre of gravity:

$$Mx - ml = 0 \text{ , i.e. } m = M\frac{x}{l}. \tag{10}$$

Substituting for x the value given by (9), we obtain Einstein's formula for m,

$$m = \frac{E}{c^2}$$

We have thus seen how the theorem of the inertia of energy can easily be derived on the basis of some purely classical principles of mechanics, with the sole exception of the radiation pressure formula of Maxwell's electrodynamics.

One might object that this latter theory is invariant for Lorentz transformations and consequently consistent with the theory of relativity, and thus would not be concerned with a strictly classical derivation. We could reply, first of all, to such an objection that the assumption of the validity of a theory invariant for Lorentz transformations—like classical electrodynamics—, and consequently also the validity of this group of transformations, is a much weaker hypothesis than the acceptance of the validity of the theory of relativity—which corresponds to Einstein's physical interpretation of these transformations themselves—, as required by Bohr's rebuttal.

Against the possibility of preserving Heisenberg's indeterminacy relations between energy and time, once relativistic postulates are violated, there is, moreover, a stronger argument, represented by a variant of the photon box experiment [3]. In this new thought experiment, proposed by Treder, he showed that the uncertainty Δq in the position of the box does not imply any time delay, and consequent uncertainty ΔT of the interval T, if the gravitational field introduced by Bohr is replaced by a classical electrostatic field. Of course, Bohr's appeal to a more general theory than classical electromagnetism to support the validity of the indeterminacy relations, even if methodologically questionable, might still have appeared acceptable, but it becomes absolutely inconsistent once we have established that relativistic principles are contradicted by the quantum mechanical description based on state vectors of the second kind.

That a refutation of the locality postulate in turn entails a violation of the indeterminacy relations between energy and time can also easily be seen by examining relations (4), where, substituting for the velocity of light c an infinite value would amount to determining the time interval ΔT with an absolute precision ($\Delta T = 0$), directly implying the invalidity of formula (1).

4. The undefined state of the unobserved Universe

In the following, I will discuss an argument deriving from the combination of three different paradoxes of quantum mechanics: the first, which is not properly a measurement paradox, was proposed by de Broglie [4] and concerns the problem of the localization of a microobject, whereas the others are connected, the former with negative result measurements discussed by Renninger and Wigner [5], and the latter with the quantum mechanical violation of a very general form of macrorealism, expressed through the realistic hypothesis by Lewis-Carnap "If all minds disappear from the universe, stars still go on on their courses."[6]

I will then discuss this argument in relation to three different interpretations of measurement: the Copenhagen, the macrorealistic and also a microrealistic one, with the aim to show that the macrorealistic appears in some particular situations either even less objectivistic than the Copenhagen one, or not able to forbid the propagation of superluminal signals.

Let us then consider, in a laboratory, a box B, with perfectly reflecting walls, which can be divided into two parts B_1 and B_2 by a double sliding wall. Suppose that B initially contains an electron, whose wave function $\phi(xyzt)$ is defined in the volume V of B. The probability density of observing the electron at point x, y, z at time t is then given by $|\phi(xyzt)|^2$.

Next, B is divided into the two parts B_1 and B_2: B_1 is delivered to the observer O_1, who remains in the laboratory on the Earth, whereas B_2 is connected with an amplification device A_2 in such a way that the presence of the electron in B_2 activates the retarded explosion of a 100,000 megaton nuclear bomb. Then, everything is placed inside a missile which is immediately launched toward the planet Venus. The explosion would cause a disturbance in the orbit of Venus, which would, in turn, produce a (small) displacement of the entire planetary system: if the set of the macroscopic observables $P = F[q_i(t), p_i(t), t]$ corresponds to the ordinary configuration of the planetary system at time t, $P' = F[q_i'(t), p_i'(t)]$ will express the perturbed one.

After the division of the box, the physical situation is described by quantum mechanics with two wave functions, $\phi_1(xyzt)$ defined in the volume V_1 of B_1 and $\phi_2(xyzt)$ defined in the volume V_2 of B_2. The probabilities ω_1, ω_2 of finding the electron in B_1 and B_2, respectively, are given by

$$\omega_1 = \int_{V_1} |\phi_1(xyzt)|^2 \, dV$$
$$\omega_2 = \int_{V_2} |\phi_2(xyzt)|^2 \, dV \tag{11}$$

with $\omega_1 + \omega_2 = 1$.

So we will have the state of the electron described by the initial superposition state:

$$|\psi\rangle = \sqrt{\omega_1}|\phi_1\rangle + \sqrt{\omega_2}|\phi_2\rangle \tag{12}$$

which, if, following standard quantum mechanics, we attribute two wave functions $|P\rangle, |P'\rangle$, respectively, to the normal and perturbed state of the planetary systems, the initial state of the global systems "de Broglie's box + planetary system" will become

$$|\Psi\rangle = \sqrt{\omega_1}|\phi_1\rangle|P\rangle + \sqrt{\omega_2}|\phi_2\rangle|P'\rangle \tag{13}$$

The state (13) corresponding to a very strange superposition between the states of a disturbed and not disturbed universe, is not accepted as a description of "de Broglie's box + planetary system" by all macrorealistic interpretations of the theory of measurement, which maintain the disappearance of the superposition between macroscopic states. More precisely, according to the GRW theory, superposition (13) can describe the physical situation for an interval not larger than 10^{-15} sec. These interpretations assume, however, that the description (12) is the correct one for de Broglie's box, before the occurrence of any physical interaction between the electron and a macrosystem, such as a measuring apparatus, for its detection.

Let us now consider an apparatus A_1 controlled by the observer O_1, who, at any instant preceding the one in which the nuclear explosion might occur on Venus, can connect it with B_1, in this way detecting the electron if it is present in this box. The absence of a detection by A_1 will, instead, inform us that the electron is contained in B_2.

As a consequence of the measurement on B_1, according to the Copenhagen interpretation we can therefore have the reduction of (13) to one of the states:

$$|\Psi_1\rangle = \sqrt{\omega_1}\,|\phi_1\rangle|P\rangle \tag{14}$$

$$|\Psi_2\rangle = \sqrt{\omega_2}\,|\phi_2\rangle|P'\rangle \tag{15}$$

where (14) is a consequence of the detection of the electron in B_1, while (15) is due of the absence of any detection, *i.e.* a typical case of negative result measurement. The only difference between von Neumann's and Bohr's approach is that the reduction of (13) to (14) or (15) occurs for the latter at the level of the measuring apparatus A_1 and for the former at the level of the observer O_1.

We are dealing in both cases with very strange consequences:

(a) in the first case, the detection of the electron by A_1, or at least the observation of this event by O_1, modifies, through an instantaneous action at a distance, the physical situation inside B_2, which is, a spontaneous evolution separated by a few million kilometres from B_1, in this case producing the collapse to the state of the non-perturbed Universe: we are faced therefore with a very *strong form of macroscopic nonlocality;*

(b) in the second case it is the absence of any detection by A_1, which informs O_1 that the electron is contained in B_2, which produces the reduction: in this way it is the non-occurrence of any physical process that generates the transition from (13) to the state of perturbed universe given by (15).

The observation or non observation of the electron on the Earth thus changes the wave-function on Venus, reducing it to zero or to unity.

But the more paradoxical situation, for the orthodox approach, is the one connected with the impossibility of making any measurement or observation, implying the persistence of a state of superposition between the states of the universe. We have, therefore, a direct conflict of this interpretation with the macro-realistic hypothesis of Lewis Carnap: if all minds disappeared from the universe and, as an obvious consequence of such an event, no measurements or observations could be performed, the stars would not continue on their courses, but would remain in the undefined state expressed by (13).

5. Macrorealistic interpretations and superluminal long-distance effects

The most remarkable attempt to give a consistent description of quantum measurement is provided by a large class of theories according to which the process of reduction from the initial state of superposition to a well-defined final state occurs in the transition from the microscopic to the macroscopic level. According to these theories the breaking off of the von Neumann chain is due neither to the privileged status of the measuring apparatus, differentiating it from all other ordinary macroscopic systems, as in Bohr's perspective, nor to the intervention of the observer's consciousness, as in von Neumann's theory, but is simply produced by the macroscopic nature of the measuring apparatus.

Such an hypothesis involves the restriction of the domain of application of quantum formalism to the atomic objects, assuming that macroscopic *apparatus* are complex systems, whose description requires either recourse to classical and semiclassical theories, including, of course, thermodynamics, or the elaboration of a new quantum macrodynamics.

According to several authors, the measuring apparatus is to be considered a thermodynamical system, and the measurement act an irreversible recording process in a macroscopic apparatus triggered by a microscopic event. This hypothesis, first investigated by Jordan, has lead to the theories of measurement of Ludwig, Prigogine and Daneri-Loinger-Prosperi, in which the problem of measurement is identified with the problem of the evolution of a complex macroscopic system toward its state of thermodynamical equilibrium[7].

The fundamental idea on which the previous approaches are based is that in an apparatus the state preceding the measurement must be metastable, in such a way that even a very small perturbation, like the one produced by the interaction with the measured atomic system, causes it to evolve towards a stable state dependent on the one of the measured system.

Another proposal for a realistic solution to the measurement problem, which is also based on the assumption of the macroscopic nature of the measuring apparatus, but which presents several notable elements of novelty with respect to the previous interpretations, has been proposed by Ghirardi-Rimini-Weber (GRW)[8]. In the GRW theory the process of reduction of the initial state of superposition to a well defined final state is replaced by a spontaneous evolution, which can be obtained through a significant modification of the linear nature of the basic quantum laws by adding a non-linear, though very small, term to the Schrödinger equation. In this way not only the dynamics of macroscopic systems, but also the dynamics of the microobjects is modified. For this reason, the GRW approach cannot be viewed as an alternative (macrorealistic) interpretation of quantum measurement, restricting the validity of the quantum laws to the world of microobjects, but as a completely new theory abandoning one of the essential features of quantum formalism.

The previous macrorealistic approaches, which treat the apparatus as a macroscopic system not describable by the quantum formalism and subjected to an irreversible evolution, not only appear the most exhaustive attempts to provide an interpretation of the measurement process able to limit, to the microscopic level, the subjectivist consequences of the standard interpretation, but also have the merit that they have clarified the impossibility of reconciling the idea of a reversible evolution, like the one implied by the Schrödinger equation, with the notion of a disturbing measurement of quantum mechanics. Such an incompatibility already emerged in connection with the paradoxes of thermodynamics, where both the postulate of the existence of Maxwell's demon, i.e. of an ideal non-disturbing measuring apparatus, and the assumption of the general validity of the recurrence theorem by Poincaré, maintaining the intrinsic reversibility of any mechanical process, imply a violation of Boltzmann H-theorem and a consequent conflict with the irreversible nature of macroscopic processes.

If we analyze the previous situation in the light of a macrorealistic theory of measurement, we can easily reconcile quantum mechanics with the Lewis-Carnap realistic hypothesis. As a matter of fact, the de Broglie box, even after the separation of the two parts, is described by superposition (12), but as soon as we include the wave functions $|P\rangle, |P'\rangle$ associated with the states of the planetary system there will be a very rapid spontaneous transition of (13) either to state (14) or to state (15), also in the absence of any measurements on B_1.

Very serious difficulties arise, however, for macrorealistic interpretations in the case of measurements with negative results, i.e. physical situations in which the reduction of the wave function occurs even in the absence of any detection process by the measuring apparatus. So, according to these theories no reduction can occur, since in the absence of any detection by A_1, there is no mechanism of amplification or interaction of the electron with a macroscopic system.

One has, therefore, according to these theories, no reduction of superposition (12), representing the state of the de Broglie box, to the well defined state $|\phi_2\rangle$ as a consequence of the fact that the electron has not been found inside B_1, $\omega_1 = 0$ and $\phi_1(xyzt) = 0$, which in turn implies $\omega_2 = 1$ and $\phi_1(xyzt) = 1$.

Such an absurd conclusion can, however, be partially avoided in the GRW approach, where the reduction is a spontaneous process not necessarily related to a physical interaction of the micro-object with a measuring device, but which is merely due to the increased number of wave functions representing the state of a macroscopic system. However, in the case of GRW, one is not dealing with a different interpretation of the measuring process, but with a new theory involving a radical (though very small) modification in the basic linear structure of the standard equation of motion.

The necessity of maintaining state (12) as the correct description of the previous physical situation, even after the negative result measurement on B_1 has been made, is not only in open conflict with the empirical predictions of standard quantum mechanics, requiring the reduction of the superposition to a well defined state (immediately) after every act of measurement or observation, which in the standard perspective does not necessarily correspond to an interaction between a microsystem and a macrosystem, but in the description of the microscopic world also imposes a higher degree of uncertainty and vagueness than in the Copenhagen interpretation.

As a matter of fact, if the most natural attitude in the interpretation of the paradox is a microrealistic point of view, according to which the electron is already localized in one of the two boxes from the instant of their separation, whereas according to standard quantum mechanics we have a localization as soon as we know where the electron is, this is not a sufficient condition for the usual macrorealist perspective, according to which one must wait for an interaction of the electron with a macroscopic device.

5.   Conclusions

The microrealistic assumption of the localization of the electron before any act of measurement or observation is the only interpretation that would also allow us to get rid of nonlocality, affecting both the standard interpretation and the macro realistic theories, in which the detection of the electron in B_1 destroys "half an electron" on Venus, but requires the introduction of a new additional parameter λ describing such a localization within B_1 and B_2. If $\lambda = +1$, we say that the electron is within B_1, and if $\lambda = -1$, that it is in B_2. This implies, of course, that standard quantum mechanics, which knows nothing about λ, is incomplete.

It can, moreover, be easily shown that it is not merely a question of incompleteness, but that the quantum description becomes ambiguous if one introduces localization through the hidden variable λ.

Let us consider, to this extent, a statistical ensemble of N similarly prepared pairs of boxes B_1 and B_2. Depending on the values of λ, this ensemble can be divided into two sub ensembles, the first composed of about

N/2 systems all with $\lambda = +1$, and the second of about N/2 systems with $\lambda = -1$. For the elements of the first (second) sub ensemble an electron is to be found with certainty in the box on the Earth (on the planet Venus).

But according to this completed version of quantum mechanics, one must necessarily conclude that *even before* any observation, or measuring operation takes place, N/2 elements of the ensemble are described by the state $\phi_1(xyzt)$, and the other N/2 by $\phi_2(xyzt)$, in contrast both with the standard theory, which asserts that all the N elements, before measurement, are described by superposition (12) and macrorealistic theories which imply the persistence of superposition also before certain kinds of measurements, such as the ones giving negative results.

We have seen how macrorealistic theories can reconcile the quantum mechanical description with the Lewis-Carnap realistic hypothesis, *i.e.* with a very weak non-metaphysical formulation of macrorealism, when no measurement is performed, but are in conflict both with the standard theory and a microrealistic interpretation, in the case of a negative result, and with the locality condition, when a positive result is obtained, implying superluminal long distance effects in the latter case. Our argument shows therefore that if one starts from a minimal program of restoring an objectivistic description, at least at the macroscopic level, the need to explain negative result measurements and to avoid nonlocal interactions leads to the introduction of hidden variables at the microscopic level, a possibility that must, nevertheless, be investigated, taking into account the severe limitations imposed in this context by Bell's theorem.

The nonlocal consequences of quantum mechanics, which have been discussed both in relation to the Einstein-Bell correlations and with respect to the macro-objectivistic interpretations to the measurement problem, in our opinion, make research in the field of non-standard approaches to special relativity based on a reformulations of its fundamental postulates, and in particular of the principle of the constancy of light velocity, extremely significant also for the very foundations of quantum theory.

References

[1]   G. Tarozzi, Bell's Theorem and the Conflict between th eTwo Basic Theories of Modern Physics, in: A. Van der Merwe, F. Selleri, G. Tarozzi, *Bell's Theorem and the Foundations of Modern Physics*, World Scientific, Singapore (1992), pp. 448-457.

[2]   G. Tarozzi, *Foundations of Physics* 26, 1996, pp. 907-917.

[3]   *Le Magnetisme*–Rapports et Discussions du Sixième Conseil de Physique de l'Institut International de Physique Solvay.

[4]   K. Popper, *The Logic of Scientific Discovery*, Basic Books New York. 1959; p. 447.

[5]   H.J. Treder, The Einstein-Bohr Experiment, in *Perspectives in Quantum Theory*, eds. W. Yourgau and A. van der Merwe, Dover. New York, 1971.

[6]   L. de Broglie, *J Phys. Radium*, 20, 1959, p. 963.

[7]   M. Renninger, Z. *Phys.* 158, 1960, p. 417; E. P. Wigner, *Am. J. Phys.* 31, 1963, p. 6.

[8] G. Tarozzi, *Epistemologia, III,* 1980, p. 13.
[9] P. Jordan, *Philosophy of Science,* 16, 1949, p. 269; G. Ludwig, *Die Grundlagen der Quantenmechanik,* Springer, Berlin, 1954, p. 122-165; A. Daneri, A. Loinger, G. M. Prosperi, *Nucl. Phys.,* 33, 1962. p.297; I. Prigogine, *From Being to Becoming,* Freeman, San Francisco, 1980.
[10] G.C. Ghirardi, A. Rimini and T. Weber, *Phys. Rev.* D34, 1986, p. 470.

On the History of the Special Relativity Concept

Alexei A.Tyapkin
Joint Institute for Nuclear Research
141980, Dubna, Moscow Region, Russia
atyapkin@vxjinr.jinr.ru

This report contains a short excursus on the origin of relativity concept, not only to supplement a widespread one-sided image of this important stage of the history of natural science, but also to recall those forgotten approaches to relativity theory that cast light on its close relation to the concepts of classical physics. We emphasize certain statements of A. Poincaré and H.A. Lorentz which help us to penetrate deeper into the essence of relativity theory.

1. Introduction

"Relativity burst upon the world, with a tremendous impact... The impact that relativity produced, I think, has never been equaled either before or since by any scientific idea catching the public mind."

Paul A. M. Dirac (1977)

The special relativity concept created in the the first years of our century, initiated a radical transformation of the earlier physical images and became one of the foundations of modern physics. But in spite of its significant place which this theory occupies in the system of modern scientific knowledge, in the historical description of its origin a one-sided approach with substantial gaps became, unfortunately, traditional. In this historiography the period preceding the creation of the relativity theory turned out to be especially underestimated, *i.e.*, it happened when the principle grounds of the new physical theory were put forward to solve contradictions existing in those times.

Such unattentive attitude to the appearance of the principle grounds of the new theory is impossible to explain by to loss of scientific interest to the historical details of the origin of the new scientific concepts. At the same time the principle grounds of a more radical physical theory—quantum mechanics, were developed in physics. But historiographers of quantum mechanics have always regarded this period as a most important element of the deviation from the old ideas of classical physics. The actual era of quantum mechanics is considered to have originated in 1900, the year when M. Planck put forward the hypothesis of discrete energy states of a oscillator and using it derived his formula for the equilibrium black-body radiation spectrum. The subsequent A. Einstein's idea (1905) of photons and L. de Broglie's idea (1923) of a hypothetical wave with a phase velocity related to the velocity of a microparticle were also judged accordingly. In any case, precisely these ideas were always stressed to underlie the wave mechanics created by E. Schrödinger (1926). For a revelatory illustration of the flaws of the historiography of special relativity it is useful to compare the respective presentations of equivalent periods in the development of the two theories, both of which form the foundation of modern physics.

No other physical doctrine excited such widespread interest, as the theory of relativity. The unusual conclusions of the theory on issues seeming most simple always aroused great interest outside the scientific community. Most likely, it was actually because of this widespread popularity of the theory of relativity, organized in the main by men of letters far from science, that its historiographers deviated from an exact and objective description of the history of this most outstanding discovery.

This story, how the historical gaps in the origin of the relativity concept were eliminated in the second half of our century, is the subject of the present report. We also consider those important statements A. Poincaré and H. A. Lorentz, which promoted the development of the deeper understanding of the essence of this theory.

2.   The Origin of the Initial Ideas of Special Relativity

"Experiment has provided numerous facts admitting the following generalization: it is impossible to observe absolute motion of matter, or, to be precise, the relative motion of ponderable matter and ether."

Henri Poincaré (1985)

The descriptions of the history of special relativity, at least those published before 1953, contained no mention whatever of now the initial ideas were formulated during the period preceding its creation. Only the formal utilization was noted, in the works by W. Voigt (1887) and H.A. Lorentz (1892 and 1895), of "local" time in a moving system with the origin of time depending linearly upon the space coordinate.

A truly novel contribution to the historiography of special relativity appeared in 1953 in second volume of the historical work [1] by well-known British mathematician E. Whittaker (the first volume was published in 1910). Whittaker was the first to point out that in 1899 the outstanding French mathematician and theoretical physicist Henri Poincaré expressed firm belief in it being essentially impossible to observe absolute motion in optical experiments owing to the relativity principle being obeyed strictly in optical phenomena, also. The scientist confirmed his idea in a talk at the Paris International Physical Congress held in 1900. E. Whittaker also presented next excerpt about prediction new relativistic mechanics from the talk delivered in St Louis Congress of Arts Science by Poincaré in 1904 stating: "From all these results there must arise an entirely new kind of dynamics, which will be characterised about all by the rule, that no velocity can exceed the velocity of light." [1, p.31].

The chapter of the book by Whittaker on special relativity gave rise to lively discussions and, doubtlessly, aroused the big interest of many scientists in independent historical investigations of the period preceding the creation of this theory. As a result, not only was a more detailed investigation of the works by Poincaré indicated by Whittaker carried out, but several of his publications [4, 5] were also saved from oblivion. It turned out that the principle of relativity for electromagnetic phenomena was proposed by Poincaré even earlier. Thus, the words of Poincaré, used as a epigraph to this chapter, were taken by us from his article of 1895 (4). Further, quoting the Michelson experiment, Poincaré stressed that theory must satisfy the above law without any restrictions related to precision.

In paper [6] I personally drew attention to the fact that in the article "Measurement of time" [5] published in 1898 Poincaré, in discussing the issue of determining the quantitative characteristics of physical time, arrives at important conclusions, on the conventional essence of the concept of simultaneity, not only representing historical interest, but also permitting to clarify the limited nature of the existing interpretation of the space-time aspect of special relativity. Poincaré notes that the postulate of the constant velocity of light "provided us with a new rule for searching for simultaneity," but concerning the assumption made use of here on the independence of the speed of light for the direction of its propagation the author makes the following categorical assertion: "This is postulate without which it would be impossible undertake any measurement of this velocity. The said postulate can never be verified experimentally." [5]. These profound arguments justified Poincaré his article the following no less categorical statement: "The simultaneity of two events, or the sequence in which they follow each other, the equality of two time intervals should be determined so as to render the formulation of natural laws as simple as possible. In other words, all these rules, all these definitions are only the fruit of implicit convention." [5]

These precious ideas of the great thinker were not applied in any explicit form in the creation of the special theory of relativity, unlike his assertion concerning the principle of relativity being rigorously obeyed by electromagnetic phenomena. Later, also; they were not realized; thus, for instance, the conclusion was not comprehended that the concept of simultaneity, for events occupying different sites, was based on measurement of the speed of light in one direction being essentially impossible without the adoption of a convention on the equality of velocities of light for processes propagating in opposite directions. Convincing evidence that the essence of the above issue was not fully realized by specialists is presented, as it is shown in my article [7], by the publication in several central physical journals of proposals, based on false grounds, to measure the speed of light in a sole direction. Such proposals always implicitly contradict the fundamental principle of causality, and their publication in journals is just as inglorious for the publishers of respectable scientific journals, as discussion in the scientific press of proposals aimed at constructing devices experiencing perpetual motion.

A further development of the idea of determining time on the basis of the postulated constancy of the velocity of light was presented by Poincaré in 1900 in an article on the Lorentz theory (8]. In this work the first physical interpretation was given of "local" time introduced by Lorentz as the time corresponding to readings of two clocks synchronized by a light signal under the assumption of a constancy of the velocity of light. This work was ignored by traditional historiography, even though the explanation given by Poincaré of the essence of the proper time was repeated literally in 1905 in a work by A. Einstein.

The works, in which the new transformations of space-time coordinates that subsequently occupied the central place in the theory of relativity, should also be attributed to the period preceding the creation of this theory. In the literature the opinion is widespread that these transformations were obtained in their final form by Lorentz in 1904. The fact is less known that they appeared in the book "Ether and matter" by the British theoretical physicist J. Larmor in 1900 [9].[‡‡‡] And what is totally unknown to historians is that Lorentz first derived the transformations, that subsequently became known, upon the proposal of Poincaré, as Lorentz group, in a work of 1899 [10]. In this article were supplemented by factor $\gamma = \left(1 - v^2/c^2\right)^{-\frac{1}{2}}$ to the transformations of coordinate $x' = x - vt$ and time $t' = t - vx/c^2$ introduced earlier in the work of 1895. Only after this supplement new transformations were brought in strict accordance with the invariance of the Maxwell equations and made to satisfy the requirements of a group.

Thus, by the end of the past century the problem of explaining absence of "ether wind" was quite ready for its ultimate solution by the above works by Poincaré, Lorentz and Larmor.

[‡‡‡] To the presented historical information one must add that the relativistic relation for adding velocities was first obtained by Larmor (see Chapter XI, item 113 of the book [9]) and that the author discussed also the relativistic effect of deceleration of time for electromagnetic processes in a material system travelling through ether (see item 114 in [9]).

3.   The Creation of Special Relativity

"The special theory of relativity is not the creation of a single individual, it is due to the joint efforts of a group of great investigators Lorentz, Poincaré, Einstein, Minkowsky."

Max Born (1959)

The history of the concluding stage in the creation of the special theory of relativity was only complicated by discrepancies in the estimation of the significance of well-known parallel works and, hence, by the insufficient attention subsequently paid to alternative approaches. These discrepancies reflected, first of all, the objective difficulties in comprehending the theoretical constructions, in same cases, and of apprehending the logic of reasoning, in others. But, regretfully, the tendentious attitude in singling out the recognized as the first one hindered objectiveness in estimating the significance of various publications.

In 1921 an extensive article (about 230 pages volume) (11), written by the future eminent theoretical physicist, at the time a twenty-years-old student of the Munich university, Wolfgang Pauli, was published in the German edition of the Encyclopedia of Mathematical Sciences. This article, later published as book in various languages, still remains one the best expositions of the fundamentals of the special and the general relativity. The article began with a short historical study, before the publication in 1953 of the book by Whittaker was the most complete and objective review of history of special relativity.

In concluding a incomplete list of works were published during the period preceding the creation of the theory Pauli singled out for further discussion "three contributions, by Lorentz [12], Poincaré [13] and Einstein [14], which contain the reasoning and the developments that form the basis of the special theory of relativity." Indeed, the grounds do exist for considering the three authors of these fundamental works the creators of the special theory of relativity, even though the contribution of each scientist differs from that of the others. But, in spite of the great success of the article and book by Pauli, many scientists subsequently ignored his historic estimation and adhered in their scientific publications to the widespread version, presented in popular literature, that the sole creator of the theory was Einstein.

The publication in 1935, in the Russian language, of a collection of the classics of relativity, edited by V.K. Frederiks and D.D. Ivanenko [15] turned out to be a digression from the obvious hushing up of the work of H. Poincaré [13]. Unlike the collection of the first works on relativity theory, published in Germany in 1913, the Russian edition contained the principal work written by H. Poincaré in 1905 [13,b]. The editors pointed out, in the comments to the articles included in the collections, that the main article by Poincaré "not only contains Einstein's parallel work, but in certain parts also the more recent—by nearly three years—article by Minkowskii, and partly even exceeds the latter"

[15, p. 367], while the fact that this fundamental work had been forgotten was classified as not having analogs in modern phisics. But this high estimate of the work by Poincaré only had some influence among theoretical physicists, and did not become known to the historians of science even in Russia. It is no chance that the high estimate of the work by Poincaré, given by the editors of the collection in the concluding remarks, was supported and acquired further development in Russia in the work of the next generation of physicists. Thus, in 1973 I compiled and submitted for publication by *Atomizdat* the most complete collection of pioneer works in special relativity theory, which included translations into Russian of articles written by Poincaré in 1895-1906 [16]. Subsequently, in 1984, A.A. Logunov published a book under the title "On the works of Henri Poincaré *On the Dynamics of the Electron*" [17].

My proposal to publish a more complete collection of works of the classics of relativity is based on the example of the 1935 collection "The Principle of Relativity," which reveals that the publication of translations of the original texts of forgotten early works by A. Poincaré and G. Larmor would serve as the most objective and effective way to convince the readers of the decisive role of these scientists in creating the concept of relativity and in preparing a scientific atmosphere for final solution of the problem.

In his book, dedicated to two 1905(06) publications by H. Poincaré [13], A.A. Logunov chooses a non-traditional form of exposition for analyzing these works. Instead of usual quotations of fragments from the originals under discussion, the book includes the complete texts of these two articles, published by H. Poincaré under the common title of "On the dynamics of the electron," which are time to time interrupted by detailed comments written by A.A. Logunov. These comments, in the main, serve a sole purpose: to show the profound physical meaning and the essential novenlty of particular points and relations established by H. Poincaré. Here, A.A. Logunov often inserts into the text of his explanations quotations from earlier articles by Poincaré. From these additions it becomes quite clear that the main points of the new theory were put forward by the French scientist long before 1905, while certain new concepts such as "local" time were given a clear explanation of their physical meaning in his earlier articles. At the same time, it becomes clear, how much better, from the point of view of physicists, could the main article of Poincaré, intended for mathematical journal *Rendiconti del Circolo Matematico di Palermo* have become, had the earlier explanations or, at least, references to his articles on such explanations of the physical meaning, been utilized.

It is important to note that all the formulae in the articles by Poincaré, that are presented in A.A. Logunov's book, are given in accordance with modern notation, which essentially simplifies understanding the theoretical relations.

To conclude this section we note that the history of the creation and development of novel scientific concept is best studied making use of the originals of scientific articles, access to which is significantly simplified owing to the publication of topical collection of old original articles. I have no doubt that, upon acquaintance with the original works of the classics of relativism, any benevolent reader will arrive at the conclusion that special relativity was created by a whole of eminent scientists, Poincaré, Lorentz, Einstein and Minkowski. I discussed detail principal significanceof a contribution of every founder of this theory in concluding article in the Collection [16].

We now terminate the above fragmentary historical sketch the aim of which was to draw attention to the ideas of Einstein's predecessors, the falling of which into oblivion doubtlessly impoverished the understanding of special relativity for many years. The same idea concerning the limitation of the understanding of this theory was expressed by A.A. Logunov in the preface to his book [18]. by following words: "However, dogmatism and faith, alien to science, but always accompanying it, have done their business. Nearly up to our time have they limited the level of understanding and, consequently, reduced the range of applications of the theory of relativity."

Now we consider question about of the more profound conception of the special relativity, following my book [19] which was published in Italy into the encyclopedic series.

4. The Essence of Special Relativity

"The true relation between real objects are the only reality we are capable of apprehending."

Henri Poincaré (1902)

Further we must realize that the relations are preserved plays a decisive role here, totally in accordance with the simple, but extremely profound assertion made by Poincaré [20], adopted as an epigraph for this section of the present article. The term "relativity" occurring in the title of the theory has a second unexpected justification. Besides the conventional meaning used for establishing in the theory new quantities depending on the relative velocity of motion of reference frames, the term "relativity" may be justified, also, in that the new absolute quantities and invariant relations established by this theory signify conservation of the relations between quantities depending on the respective velocities.

Indeed, the main content of special relativity resides in the general properties of physical phenomena corresponding to the pseudo-Euclidean geometry of the four-dimensional world1 in which space and time join in a certain entity, independent of the relative motion of inertial reference frames. However, this extremely concise formulation, naturally, requires some decoding, separation of the physical essence from the adopted form of its mathematical expression. It is even useful to digress some time from a form adequate to the content and deal with another plausible expression, so as to reveal in a clear manner the physical essence of the new theory.

The idea of the main content of special relativity was expressed by Minkowski in his famous talk "Space and time" by the following statement termed by the author the postulate of the absolute world: ."... the postulate comes to mean that only the four-dimensional world in space and time is given by phenomena, but that the projection in space and in time may still be undertaken with a certain degree of freedom..." [21]. Minkowski's talk began with even sharper words concerning the arbitrariness that arose in the new theory, when space and time quantities were considered separately: "Henceforth space by itself and time by itself are doomed to fade away into mere shadows, and only a kind of union of the two will preserve an independent reality." I do not think Minkowski termed these quantities shadows, because in the new theory they became relative, dependent upon the velocity of relative motion. Most likely, Minkowski implied arbitrariness to signify the apparent contradiction of the obtained results: lengths in each considered reference frame exhibit contraction with respect to any other frame, clocks in each frame slow down relative to other frames. But, anyhow, enrolling quantities in the category of fictions does not free one from the necessity of clarifying the essence of the corresponding effects.

These "miraculous" reversals of quantities, resulting from comparing lengths and time intervals, are due to transition from the simultaneity of one frame to the simultaneity of anower frame. We see that the point is that proper simultaneities adopted in different inertial frames differ from each other. It remains for us to clarify the meaning of the central provision of all the theory, the relativity of simultanety, in any words, to understand which common properties of physical processes are reflected in the artificially chosen shift of origins of time at differing points of a moving inertial reference frames. For ultimate clarification of this issue without renouncing arguments based on common sense it is best to turn to the description of velocities of physical processes in a moving frame within the Galilean approach utilizing a unique simultaneity for the two frames being considered. But before presenting the results of such an analysis we shall recall the main advantage achieved by introducing a shift in the simultaneity along the direction of relative motion of the frames. The shift in simultaneity was introduced under the condition of constancy of the speed of light, and as a result the independence upon direction is obtained of the velocities of all physical processes in each inertial frame from sources at rest in these frames. The calculus of space-time coordinates in each inertial frame was also chosen under the condition that the principle of relativity be satisfied,, and therefore the laws of physics turn out to be invariant with respect to relativistic transformations of coordinates. Precisely this represents the content of the correspondence, noted above, of the chosen relativistic metric to the properties common to physical processes.

Now let us ponder over the main question: what significance has Nature being consistent precisely with the special principle of relativity, and not with the Galileo-Newton-Hertz principle of relativity? Clearly, it means conservation of the form of mathematical equations expressing physical laws only under the condition that a relative shift in simultaneity be introduced, when time coordinates of events are calculated in two inertial reference frames moving relative to each other. This means that relative to the simultaneity in the initial frame $K(x, t)$ the reading of a clock in the frame $K'(x', t')$ is ahead by a quantity, that increases linearly along the x'-axis. This shift oa simultaneities does not violate the equivalence of the reference frames, since the reading of the clock in frame $K(x, t)$ will be ahead relative to the simultaneity in frame $K'(x', t')$ by a quantity increasing linearly along the direction opposite to the x-axis.

Hence it should be clear how unjustified it would be to interpret the spetial principle of relativity as the assertion of identity of how physical processes proceed in different inertial reference frames moving relative to each other, if the identity of mathematical expressions for the respective physical processes is achieved by taking advantage in these reference frames of noncoinciding times, t and t'. The point is that their main difference consisting in the relative shift of simultaneities means taking into account the general delay of processes along the direction of relative motion of the frames. The principle of relativity being satisfied signifies conservation of kinematical similarity while all processes experience a common delay along the x'-axis. This can be ultimately verified by considering the velocities of processes in amoving inertial reference frame $\hat{K}'(\hat{x}',t)$, the coordinates of which are related to coordinates in the initial frame $K(x, t)$ by the Galileo transformations (1).

Indeed, for the absolute velocity of an arbitrary physical process reproduced at an angle 0 in a moving reference frame, utilizing the coordinates $\hat{x}' = x - vt$, $\hat{y}' = y$, $\hat{z}' = z$, $\hat{t}' = t$ we obtain, in accordance with refs. [6,19], the following relation:

$$\hat{u}'\left(\hat{\theta}'\right) = \frac{u_o\left(1 - v^2/c^2\right)}{\left[1 - \left(v^2/c^2\right)\sin^2\hat{\theta}'\right]^{\frac{1}{2}} + \left(u_o c/c^2\right)\cos\hat{\theta}'}, \tag{1}$$

where $u_o = \text{const}(\theta')$ stands for the absolute velocity of the same process, if the coordinates x', y', z', and t' are used.

For the direction along the x'-axis ($\hat{\theta}' = 0$) and the opposite direction $\hat{\theta}' = \pi$ we obtain from (1) the respective velocities

$$\hat{u}'(0) = u_o \frac{1 - v^2/c^2}{1 + u_o v/c^2}, \quad \hat{u}'(\pi) = u_o \frac{1 - v^2/c^2}{1 - u_o v/c^2}$$

Hence for light ($u_o = c$) we obtain the velocities

$$\hat{u}'(0) = c - v \text{ and } \hat{u}'(\pi) = c + v$$

which correspond to the expressions of classical physics and to the problem of "ether wind" that arose in this connection.

Consequently, relativistic theory introduced no changes directly into the motion of a light front in a moving reference frame, while substitution of the constant "c" for the velocities (1) for all direction is due to transition in the moving reference frame from space-time coordinates, $\hat{K}'(\hat{x}',t)$, to the new calculus of coordinates in the same inertial frame, $K'(x',t')$. This corresponded to the primary provision of the Lorentz theoretical construction concerning the conservation, in an intact form, of classical electrodynamics and optics. The same result transition from the velocities of light $c - v$ and $c + v$ for opposite direction to the constant speed of light "c" was interpreted by Einstein as the result of clarification of the true course of time in a moving frame. We introduce into this assertion only a small, but exstremely significant correction: chanding the notion of the true course of time in some frame signifies a corresponding change of the general course of physical processes in this frame, which can be clearly ilustrated within the preceding approach involving a unique time $\hat{t}' = t$, or $\hat{t} = t''$, for two inertial reference frames.

In the first case ($\hat{t}' = t$) we have isotopic velocities of physical processes in the frame $K(x, t)$ and we fix anisotropic velocities of similar physical processes reproduced in indentical conditions in another inertial frame $\hat{K}'(\hat{x}',t)$. This dependence of the velocity upon the angle, represented by relation (1), exhibits a remarkable peculiarity: in no real experiment can it be distinguished from the case $u = \text{const}(\theta)$, if in Nature there exist no processes with velocities exceeding, within this version of the description, the speed of light in vacuum, i.e. $u_o < c$. Relation (1) is, naturally, implied to apply to all processes, without exception. The noted remarkable feature of relation (1) follows formally from the fact that the simple transformation of the coordinates of events from~the fact that the simple transformation of the coordinates of events from $\hat{K}'(\hat{x}',t)$ to $K'(x',t')$ [§§§] realizes transition to the isotropic velocities $u' = \text{const}(\theta)$. Doubtless, it is of interest, however, to consider in detail the physical reasons underlying the indistinguishability of the obtained angular dedependence (1) and the isotropy of velocities.

It lies in the general property of conservation of kinematical sirnilarity for all physical processes. The velocity angular dependence (1) exhibits the same peculiarity consisting in that the relation between different processes are essentially indistinguishable from the relations between the processes, when the velocities of the processes are independent of the angle. Thus, included in the general, and therefore nonobservable, effects is the difference

---

[§§§] These event coordinates are related by [18, p. 28] as follows:

$$x' = \gamma \hat{x}'; \ y' = \hat{y}; \ z' = \hat{z}'; \ t' = \gamma\left[\hat{t}'\gamma^{-2} - \hat{x}'\frac{v}{c^2}\right].$$

Here $\gamma = \left(1 - v^2/c^2\right)^{-\frac{1}{2}}$.

between the velocities of processes in opposite directions along the $\hat{x}'$-axis. The time required for a certain length to be translated in one direction differs from the time required for going in the opposite direction by the same quantity for all physical processes. In other words, the difference between the velocities of light in the positive and in the opposite directions for a moving frame, (1), encountered by classical physics is essentially nonobservable in experiments performed in this frame, only because any other physical process exhibits the same propagation delay in the positive direction with respect to propagation in the opposite direction. Now, does such a nonobservable delay exist for all processes? It exists objectively with respect to processes reproduced in similar conditions in another frame, conventionally regarded as the primary frame. The physical meaning of this delay is totally equivalent to introduction in the moving frame of a proper simultaneity differing from the simultaneity of the primary frame. The limitation of the orthodox interpretation of special relativity consists precisely in that it actually does not reveal the true meaning of the relativity of simultaneity.

The orthodox interpretation of special relativity concentrated on substantiation of a proper basis for calculating space-time coordinates in each individual inertial reference frame. Set aside was the approach initiated by Lorentz, that was based on parallel consideration of two bases for the calculus of coordinates in each of the two inertial reference frames being considered: $K(x,t)$ and $\hat{K}'(\hat{x}',t')$ for one, and $K'(x',t')$ and $\hat{K}'(\hat{x}',t)$ for the other. As a result of this economic approach the problem of substantiation acquired a formal solution involving an essential rupture of the common-sense logic. Preliminary consideration of the velocities of physical processes expressed in unified Galilean scales in two inertial frames permits to verify in the simplest manner the relative difference between the courses of processes in the direction of relative of the reference frames. All processes proceed slower in frame K', that in frame K, along the x-axis, but this does not violate equivalence of the frames, since the opposite direction: all the processes in frame K are delayed with respect to the processes in frame K'. The assertion concerning the relative delay of velocities of processes only reveals the physical meaning of the relativity of simultaneity. The role played by the utilized Galilean scales on rulers and faces of clocks is the same as that of reduction to common measurement units of the quantities being compared.

The constancy of the velocity of light exhibits two different aspects. Thus, the initial provision on the independence of the velocity of light of the motion of the source is something that can be checked experimentally. The assertion of independence of the velocity of the motion of the reference frame has another foundation. Here, instead of the velocities of other physical processes in the given inertial frame is preserved. Precisely because of the relation between the velocities of processes remaining unchanged the proper time introduced in the given frame acquires the status of real time singled out among all possible calculated times by a sole indisputable advantage:

iprovides for the absolute values of velocities of physical processes originating from sources at rest in the given reference frame being independent of the direction of propagation.**** But this advantage of choosing for each inertial reference frame its proper basis for calculating space-time coordinates must not, however, overshadow the objective relative difference between the velocities along the direction of relative motion of two reference frames. It is merely this fact that is expressed by the difference between the proper simultaneities in these frames leading to the delay of all processes by one and the same quantity depending only on the distance along the x-axis.

In the spirit of the ideas of Lorentz with respect to ether the "ether wind" being nonobservable in the case of light could be explained by the corresponding motion through ether influencing all physical processes. However, imposition by such an explanation of the motion of ether secretly at rest in the initial frame $K(x, t)$ has no sufficient foundation, since, in considering the propagation velocities of the corresponding processes, we obtain, utilizing a unique time $\hat{t} = t'$, asymetric velocities for the nonobservable "ether wind" in the opposite direction in the initial frame $\hat{K}(\hat{x}, t')$. Therefore we are justified in relating the discussed kinematical effects only to the fact itself of relative motion, while their appearance should be explained by the universal dependence of the dynamics of any whatever interaction upon the velocity of relative motion.

5. Conclusion

"A problem arises only when we assume or postulate that the same physical situation admits of several ways of description."

Albert Einstein (1949)

Revolutionary transformations of basic physical conceptions never proceed smoothly. Giving up conventional views is always painful. Smoothing out the uneven development of knowledge proceeds gradually as the essence of novel concepts is penetrated. Bridges across abysses and crevices separating levels of knowledge are most often built by new generations of scientists, much later than when the new physical theory originates. The process of extending the understanding of a fundamental theory lasts many decades and develops along several main directions. One of these involves revelation of the relation to preceding physical opinions and clarification of the actual degree of novelty inherent in the primary provisions

---

**** Precisely for this reason, to determine the proper time in some inertial reference frame any physical process from a source at rest in the given frame can be utilized, under the assumption that the velocity of the process be independent of the direction in which it propagates. Besides this, clocks previously synchronized at the same point of the frame and then slowly taken apart to different points exhibit readings corresponding to the proper simultaneity of the given inertial frame.

of the discussed theory. Another approach is to clarify the limits justifying application of the theory, based on further development of the understanding of the physical theory.

The latter type of development of the interpretation of a fundamental theory lasts the longest, since it is completed only by the creation of a more general theory ultimately establishing the limits of the given physical theory. Thus, comprehension of classical mechanics, in this respect, was completed only upon creation of the special and general theory of relativity and of quantum mechanics, that imposed limits on its application and explained the reasons of this limitation. The example of classical mechanics also clarified the significance of the criticism, initiated by E. Mach, of the formulation of its laws, originated with Newton, for the subsequent devitation from the conceptions of classical mechanics.

It is no chance that these general issues, related to the knowledge of the essence of physical laws, have been touched upon in the concluding part of my paper on special relativity. I hope to convince the readers that further development of the interpretation of the existing theoretical foundation of the physical science represents a most interesting sphere of scientific activity. The scope of such activities enhancing the profundity of scientific truths, actually already established in physics, can be termed "Foundation of physics," after the title of the international journal that organizes successful discussions of the investigations in this fascinating, and important for the further development of physics, field of scientific activity.

The author sincerely hopes the analysis performed in this article and the critical discussion of the simplest of modern physical theories will convince the readers of the existence of more significant possibilities of fruitful activity aimed at the developing the interpretation of other modern theories. Thus, for example, in physics great efforts are still required for clarifying such most important issues, as the reasons underling the appearance of energy nonconservation in the formalism of the geometrized relativistic theory of gravity, and to explain the astonishing interference phenomenon in experiments involving individual quantum objects for which the theory till now provides a formal description.

Truly, for fruitful activity in the indicated field it is important to free oneself from the prejudice that a physical theory is completed, when a set of mathematical relation is established that describes experimental facts in the respective range of physical phenomena. It must become quite clear that penetration of the essence of profound truths of truly scientific knowledge of Nature merely originates with the establishment of rigorous quantitative laws.

References

[1]   Whittaker E. A *History of the Theories of Aether and Electricity*, v. 2, N.Y. 1953, p. 27.

[2]   Poincaré H. *Rapports du Congres de Physique*, Paris, 1900, t.1, p.22.

174   Open Questions

[3]   Poincaré H. *Bull. des. Sci. Math.*, ser. 2 v. 28 p. 302 (1904); *The Monist of January*, v. 15, p. 1(1905).
[4]   Poincaré H. *L'Eclairage Electrique*, t. 5, p. 5 (1895).
[5]   Poincaré H. *Revue de Metaphysique et de Morale*, t.- 6, p. 1, (1898).
[6]   Tyapkin A.A. *Usp. Fiz. Nauk* (USSR), vol. 106, no 4, p. 617-65, (April 1972), (in Russian); Translation into English: *Sov. Phys. Usp.* (USA), vol. 15, no 2, p. 205-29 (Sept.-Oct. 1972).
[7]   Tyapkin A.A. *Lett. Nuovo Cimento*, v. 7, p. 760 (1973).
[8]   Poincaré H. *Archives Neerland*, v. 5, p. 252 (1900).
[9]   Larmor J.J. *Aether and Matter*. Cambridge, 1900, (see p. 167-177).
[10]   Lorentz H.A. *Zittingsverlag. Acad.Wet.*, v. 7, s. 507 (1899); *Amsterdam Proc.*, 1898-1899, p. 427.
[11]   Pauli W. Relativitatstheorie. – In: *Encyclopadie der math. Wissenschaften*, b. 5, t. 2, Leipzig, 1921, s. 539-775;
[12]   Lorentz H.A. *Versiag. Konincl. akad. wet. Amsterdam*, v. 12, s. 986 (1904); *Proc. Acad. Sci. Amster.* v. 6, p. 809 (1904).
[13]   Poincaré H. a) *Comptes Rendus*, t. 140, p. 1504 (1905); b) *Rendiconti del Cir. Matem. di Palermo*, v. 21, p. 129 (1906).
[14]   Einstein A. *Annalen der Physik*, b. 17, s. 891(1905).
[15]   *The Principle of Relativity. A Collection of Papers by the Classics of Relativity* (H.A. Lorentz, H. Poincaré, A. Einstein, H. Minkowski) (Russian translation edited, and with comments, by V.K. Frederiks and D.D. Ivanenko), Leningrad: ONTI, 1935.
[16]   *The Principle of Relativity. Collection of Papers on the Special Theory of Relativity*, (compose by A.A.Tyapkin) Moscow: Atomizdat, 1973 (in Russian).
[17]   Logunov A.A. *On the articles by Henri Poincaré on the dynamics of the electron*, Dubna: Publishing Depart. of the JINR, 1995 (English translation by G. Pontecorvo from the last 1988 Russian edition).
[18]   Logunov A.A. *Lectures on the Theory of Relativity*. Modern Analysis, Moscow: Moscow University Press, 1984.
[19]   Tyapkin A.A. *Relativita' Speciale* – Milano: Jaca Book, Un Enciclopedia EDO D Orientamento, 1992; FISIC A, Enciclopedia Tematica Aperta, Prolusioni di J.V. Narlikar, H.G. Owen, F. Selleri, A.A. Tyapkin (p. 101-128), Milano: Jaca Book, 1993.
[20]   Poincaré H. *La Science et l'Hypothese*, Paris: Flammarion, 1902 p. 103 see in book (Russian translation) *On Science*, Moscow: Nauka, 1983.
[21]   Minkowski H. *Phys. Zs.* v. 10 (1909) S. 104.

Structures in
Space and Time

The Problem of Surface Charges and Fields in Coaxial Cables and its Importance for Relativistic Physics

A. K. T. Assis[††††‡‡‡‡] and J. I. Cisneros
Instituto de Física 'Gleb Wataghin'
Universidade Estadual de Campinas – Unicamp
13083–970 Campinas, São Paulo, Brasil

We calculate the surface charges, potentials and fields in a long cylindrical coaxial cable with inner and outer conductors of finite conductivities and finite areas. It is shown that there is an electric field outside the return conductor.
Key Words: Surface charges, coaxial cable, classical electrodynamics.
PACS: 41. 20. –q, 41. 20. Gz

1.   Coaxial Cable

The possible existence of a second order motional electric field arising from steady conduction currents and its implications to the theory of relativity has been discussed recently by a number of authors: [1-7]. This field is of the second order in v_d/c, v_d being the drifting velocity of the conduction electrons, and is supposed to exist outside the wire. However, most of these authors do not consider the first order coulombian electric field (proportional to the current or to the drifting velocity) which should arise outside resistive wires carrying a steady current. As this first order electric field is relevant to the interpretation of some experiments, we decided to consider it here in a particular geometry. We discussed the second order electric field in [8], [Section 6. 6] [9] and [Section 5. 4] [10].

Before discussing the first order electric field we want to call attention to Ivezic's work: [11], [12], [13] and [14]. Although discussing the second order electric field, he was aware of the coulombian electric field. According to him the second order field might be due to a relativistic contraction of the average

---

[††††] E-mail: assis@ifi.unicamp.br; home page: http://www.ifi.unicamp.br/~assis
[‡‡‡‡] Also Collaborating Professor at the Department of Applied Mathematics, IMECC, State University of Campinas, 13081-970 Campinas, SP, Brazil.

distance between moving electrons as
compared with their average distance
when there is no current. He could then
explain several experiments based on this
approach. In this connection it would be
relevant to analyse Selleri's recent
proposal of a new set of spacetime
transformations between inertial systems
in order to see if this new approach might
give theoretical support to Ivezic's ideas
when there is curvature in the wires (and
then centripetal acceleration of the
electrons) and explain the same set of
experiments, [15] and [16]. Selleri's
transformations don't present the
discontinuity between accelerated and
inertial reference frames which exists in
Einstein's theory of relativity.

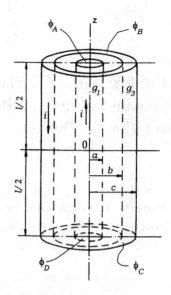

Figure 1. Geometry of the problem.

In the study of dc and low frequency
ac circuits, the following subjects are seldom analysed in electromagnetic
textbooks: electric fields outside the conductors, surface charges on the wires
and energy flow from the sources to the conductors where energy is
dissipated. There are two main reasons for this: (I) The scalar electric potential
is the solution of Laplace's equation with frequently complicated boundary
conditions; and (II) the solution of elementary circuits, based on Ohm's law, is
obtained by the application of Kirchhoff's rules. As these rules utilize only the
values of current and potential inside the conductors, the discussion of the
subjects listed above is unnecessary. However, some authors have treated
these topics in the past few years: [17,18] and references therein. The case of a
long coaxial cable has been treated by Sherwood, [18], Marcus, [19],
Sommerfeld, [20, pp. 125–130] (German original from 1948), Griffiths, [21, pp.
336–337] and a few others. All of these works considered a grounded return
conductor either with an infinite area or with an infinite conductivity. Our
main contribution in this work is to generalize these assumptions considering
a return conductor with finite area, finite conductivity and with a variable
electric potential along its length. We calculate at all points in space the scalar
and vector potentials, the electric and magnetic fields and analyse the energy
flow by means of Poynting vector. We also calculate the surface electric
charges.

The geometry of the problem is that of Fig. 1. A constant current I flows
uniformly in the z direction along the inner conductor (radius a and
conductivity g_1), returning uniformly along the outer conductor (internal and
external radii b and c, respectively, and conductivity g_3). The conductors have
uniform circular cross sections and a length $1 >> c > b > a$ centered on $z = 0$.

The medium outside the conductors is considered to be air or vacuum with $\varepsilon = \varepsilon_o = 8.85 \times 10^{-12} \, C^2 \, N^{-1} m^{-2}$. The potentials at the right extremities ($z = 1/2$) of the inner and outer conductors are maintained by a battery at the constant values ϕ_A and ϕ_B, respectively. The potentials at the left extremities ($z = -1/2$) of the outer and inner conductors are maintained by another battery at the constant values ϕ_C and ϕ_d, respectively. Instead of two batteries, the solution presented here can also be applied to the situation of one battery at one extremity and a resistor at the other extremity.

In the previous works quoted above the authors considered only a particular case: a grounded outer conductor ($\phi_C = \phi_d = 0$) with an infinite area (Sommerfeld, $c \to \infty$) or with an infinite conductivity (Griffiths, $g_3 \to \infty$).

We are interested in calculating the potentials and fields in a point $\vec{r} = (\rho, \varphi, z)$ such that $1 >> \rho$ and $1 >> |z|$, so that we can neglect border effects (ρ, φ and z are the cylindrical coordinates). All solutions presented here were obtained in this approximation. With this approximation and geometry we then have the potential as a linear function of z, [22]. In order to have uniform currents flowing in the z direction along the inner and outer conductors, with a potential satisfying the given values at the extremities, we have:

$$\phi(\rho \le a, \varphi, z) = \frac{\phi_A - \phi_D}{\ell} z + \frac{\phi_A + \phi_D}{2}, \qquad (1)$$

$$\phi(b \le \rho \le c, \varphi, z) = \frac{\phi_B - \phi_C}{\ell} z + \frac{\phi_C + \phi_B}{2} \qquad (2)$$

where, by Ohm's law (R_1 and R_3 being the resistances of the inner and outer conductors, respectively):

$$\phi_d - \phi_A = R_1 I = \frac{\ell I}{\pi g_1 a^2}, \qquad (3)$$

$$\phi_B - \phi_C = R_3 I = \frac{\ell I}{\pi g_3 (c^2 - b^2)}. \qquad (4)$$

In the four regions ($\rho < a$, $a < \rho < b$, $b < \rho < c$ and $c < \rho$) the potential ϕ satisfies Laplace's equation $\nabla^2 \phi = 0$. The solutions of this equation for $\rho < a$ and for $c < \rho$ in cylindrical coordinates satisfying the boundary conditions above, Eqs. (1) and (2), and imposing the value $\phi(\rho = 1, \varphi, z) = 0$ to complete the boundary conditions, yield:

$$\phi(a \le \rho \le b, \varphi, z) = \left[\frac{\phi_A - \phi_D + \phi_C - \phi_B}{\ell} z + \frac{\phi_A + \phi_D - \phi_C - \phi_B}{2} \right] \frac{\ln(b/\rho)}{\ln(b/a)}$$
$$+ \left[\frac{\phi_B - \phi_C}{\ell} z + \frac{\phi_C + \phi_B}{2} \right] \qquad (5)$$

$$\phi(\leq \rho, \varphi, z) = \left[\frac{\phi_B - \phi_C}{\ell} z + \frac{\phi_C + \phi_B}{2} \right] \frac{\ln(\ell/\rho)}{\ln(\ell/c)}. \tag{6}$$

The electric field $\vec{E} = -\nabla\phi$ is given by

$$\vec{E}(\rho < a, \varphi, z) = \frac{\phi_D - \phi_A}{\ell} \hat{z}, \tag{7}$$

$$\vec{E}(a < \rho < b, \varphi, z) = \left[\frac{\phi_A - \phi_D + \phi_C - \phi_B}{\ell} z + \frac{\phi_A + \phi_D - \phi_C - \phi_B}{2} \right] \frac{1}{\ln(b/a)} \frac{\hat{\rho}}{\rho}$$
$$+ \left[\frac{\phi_C - \phi_B}{\ell} + \frac{\phi_D - \phi_A + \phi_B - \phi_C}{\ell} \frac{\ln(b/\rho)}{\ln(b/a)} \right] \hat{z} \tag{8}$$

$$\vec{E}(b < \rho < c, \varphi, z) = \frac{\phi_C - \phi_B}{\ell} \hat{z}, \tag{9}$$

$$\vec{E}(c < \rho, \varphi, z) = \left[\frac{\phi_B - \phi_C}{\ell} z + \frac{\phi_C + \phi_B}{2} \right] \frac{1}{\ln(\ell/c)} \frac{\hat{\rho}}{\rho}$$
$$+ \frac{\phi_C - \phi_B}{\ell} \frac{\ln(\ell/\rho)}{\ln(\ell/c)} \hat{z}. \tag{10}$$

Eqs. (5) and (8) had been obtained by Jefimenko, [pages 509–511] [23], who also discussed the flow of energy in this system.

The surface charges densities σ along the inner conductor ($\rho = a$, $\sigma_a(z)$) and along the inner and outer surfaces of the return conductor ($\rho = b$, $\sigma_b(z)$ and $\rho = c$, $\sigma_c(z)$) can be obtained easily utilizing Gauss's law:

$$\oiint_S \vec{E} \cdot d\vec{a} = \frac{Q}{\varepsilon_o}, \tag{11}$$

where $d\vec{a}$ is the surface element pointing normally outwards the closed surface S, Q is the net charge inside S. This yields $\sigma_a(z) = \varepsilon_o E_{2\rho}(\rho \to a, z)$, $\sigma_b(z) = -\varepsilon_o E_{2\rho}(\rho \to b, z)$ and $\sigma_c(z) = \varepsilon_o E_{4\rho}(\rho \to c, z)$, where the subscripts 2ρ and 4ρ mean the radial component of $\vec{E}$ in the regions $a < \rho < b$ and $c < \rho$, respectively. This means that:

$$\sigma_a = \frac{\varepsilon_o}{a} \frac{1}{\ln(b/a)} \left[\frac{\phi_A - \phi_D + \phi_C - \phi_B}{\ell} z + \frac{\phi_A + \phi_D - \phi_C - \phi_B}{2} \right], \tag{12}$$

$$\sigma_b(z) = -\frac{a}{b} \sigma_a(z), \tag{13}$$

$$\sigma_c(z) = \frac{\varepsilon_o}{c} \frac{1}{\ln(\ell/c)} \left[\frac{\phi_B - \phi_C}{\ell} z + \frac{\phi_C + \phi_B}{2} \right]. \tag{14}$$

Jefimenko obtained only Eqs. (12) and (13), but not (14). This last one was calculated here for the first time.

An alternative way of obtaining ϕ and $\vec{E}$ is to begin with the surface charges as given by Eqs. (12) to (14). We then calculate the electric potential ϕ (and $\vec{E} = -\nabla\phi$) through

$$\phi(\vec{r}) = \frac{1}{4\pi\varepsilon_o} \sum_{j=1}^{3} \iint_{S_j} \frac{\sigma(\vec{r}_j)\mathrm{d}a_j}{|\vec{r} - \vec{r}_j|} . \tag{15}$$

Here the sum goes over the three surfaces. We checked our results with this procedure.

We can calculate the vector potential utilizing

$$\vec{A}(\vec{r}) = \frac{\mu_o}{4\pi} \iiint \frac{\vec{J}(\vec{r}')\mathrm{d}V'}{|\vec{r} - \vec{r}'|}, \tag{16}$$

where $\mu_0 = 4\pi\times 10^{-7}$ kg m C^{-2} and $\mathrm{d}V'$ is a volume element. With the approximation above we obtain:

$$\vec{A}(\rho \le a, \varphi, z) = \frac{\mu_o I}{2\pi}\left[\frac{\rho^2}{2a^2} - \frac{c^2 \ln(c/a) - b^2 \ln(b/a)}{c^2 - b^2}\right]\hat{z}, \tag{17}$$

$$\vec{A}(a \le \rho \le b, \varphi, z) = \frac{\mu_o I}{2\pi}\left[\frac{c^2 \ln(c/\rho) - b^2 \ln(b/\rho)}{c^2 - b^2} - \frac{1}{2}\right]\hat{z}, \tag{18}$$

$$\vec{A}(b \le \rho \le c, \varphi, z) = \frac{\mu_o I}{2\pi}\left[\frac{c^2 \ln(c/\rho)}{c^2 - b^2} - \frac{c^2 - \rho^2}{2(c^2 - b^2)}\right]\hat{z}, \tag{19}$$

$$\vec{A}(c \le \rho, \varphi, z) = 0. \tag{20}$$

The magnetic field can be obtained either through the magnetic circuital law $\oint \vec{B}\cdot\mathrm{d}\ell = \mu_o I$, or through $\vec{B} = \nabla\times A$. Both approaches yield the same result, namely:

$$\vec{B}(\rho \le a, \varphi, z) = \frac{\mu_o I \rho}{2\pi a^2}\hat{\varphi}, \tag{21}$$

$$\vec{B}(a \le \rho \le b, \varphi, z) = \frac{\mu_o I \hat{\varphi}}{2\pi \rho}, \tag{22}$$

$$\vec{B}(b \le \rho \le c, \varphi, z) = \frac{\mu_o I}{2\pi}\frac{c^2 - \rho^2}{c^2 - b^2}\frac{\hat{\varphi}}{\rho}, \tag{23}$$

$$\vec{B}(c \le \rho, \varphi, z) = 0. \tag{24}$$

This completes the solution of this problem.

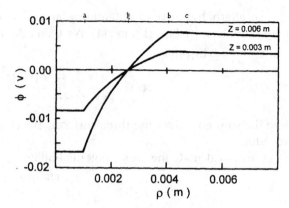

Figure 2. Electric
potential as a
function of ρ.

2.   The Symmetrical Case

We now consider two equal batteries symmetrically located on both
ends, such that $\phi_d = -\phi_A = \phi_1$ and $\phi_B = -\phi_C = \phi_3$. In this case the potential is
simply proportional to z without any additive constant. We can then write it
as $\phi(\rho, \varphi, z) = R(\rho)z$, where $R(\rho)$ in terms of the currents and conductivities is
given by

$$R(\rho \leq a) = -\frac{1}{\pi g_1 a^2}, \tag{25}$$

$$R(a \leq \rho \leq b) = -\frac{I}{\pi}\frac{1}{\ln(b/a)}\left[\frac{\ln(b/\rho)}{g_1 a^2} - \frac{\ln(\rho/a)}{g_3(c^2 - b^2)}\right], \tag{26}$$

$$R(b \leq \rho \leq c) = \frac{I}{\pi g_3(c^2 - b^2)}, \tag{27}$$

$$R(c \leq \rho) = \frac{I}{\pi}\frac{\ln(\ell/\rho)}{\ln(\ell/c)}\frac{1}{g_3(c^2 - b^2)}. \tag{28}$$

A plot of $\phi(\rho) = R(\rho)z$ versus ρ is given in Figure 2. In order to obtain this
plot we utilized the following data: $a = 0.0010$ m, $b = 0.0040$ m, $c = 0.0047$ m,
$I = 50$ A, $g_1 = 5.7 \times 10^6$ m^{-1} Ω^{-1}, $g_3 = 2 \times 10^6$ m^{-1} Ω^{-1} and $1 = 1$ m. There are
two curves, one for $z = 0.003$ m and another for $z = 0.006$ m. We see that the
potential is constant for $0 \leq \rho \leq a$, increases between a and b, is constant
between b and c, decreasing for $\rho > c$. As $\vec{E} = -\nabla\phi$, the z component of $\vec{E}$ is
given by $E_z = -R(\rho)$, so that its behaviour is the same as that of $\phi(\rho)/z$ with an
overall change of sign. The point where $\phi(\rho) = R(\rho) = 0$ is $\rho = \xi$, where

$$\xi = \exp\frac{g_1 a^2 \ln(a) + g_3(c^2 - b^2)\ln(b)}{g_1 a^2 + g_3(c^2 - b^2)}. \tag{29}$$

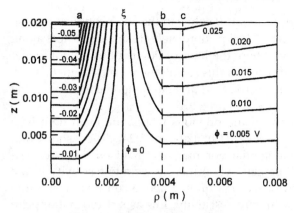

Figure 3. Equipotentials.

Sommerfeld or Griffiths's solutions are recovered taking $g_3(c^2 - b^2) \to \infty$, such that $\xi \to b$, $\sigma_c(z) \to 0$, $\vec{E}\,(\rho > b) \to 0$ and $\phi(\rho \geq b) \to 0$ for any z. The opposite solution when the current flows in an inner conductor of infinite conductivity, returning in an outer conductor of finite area and finite conductivity is also easily obtained from above, yielding $\xi \to a$, $\vec{E}\,(\rho < a) \to 0$ and $\phi(\rho \leq a) \to 0$ for any z.

In Figure 3 we plotted the equipotentials with the same data as above, in SI units. The values of the surface charges at z = 0.001 m obtained from Eqs. (12) to (14) are: $\sigma_a = -6.54174 \times 10^{-12}$ C m^{-2}, $\sigma_b = 2.61670 \times 10^{-11}$ C m^{-2} and $\sigma_c = 4.49027 \times 10^{-13}$ C m^{-2}. As the surface charges vary linearly with z, it is easy to find their values at any other distances from the center of the cable.

3.   Discussion

The distribution of charges given by Eqs. (12) to (14) shows that the facing surfaces $\rho = a$ and $\rho = b$ work as a set of capacitors. That is, the charge at the position $\rho = a$, z, in a length dz, $dq_a(z) = 2\pi\,a\,dz\,\sigma_a(z)$, is equal and opposite to the charge at the position $\rho = b$, z, in the same length dz: $dq_b(z) = 2\pi b\,dz\sigma_b(z) = -dq_a(z)$. The field outside the coaxial cable depends then only on the surface charges at the external wall of the return conductor, $\sigma_c\,(z)$:

$$\phi(c \leq \rho, \varphi, z) = \frac{c}{\varepsilon_o}\sigma_c(z)\ln\frac{\ell}{\rho}. \tag{30}$$

The flux of energy from Poynting vector $\vec{S} = \vec{E} \times \vec{B}/\mu_o$ is also represented in Figure 3. That is, the lines of Poynting flux lie in the equipotential surfaces, as had been pointed out by [17] and [18]. The classical view is that the energy comes from the batteries (not represented in Figure 1). In Figure 3 it would come from the top of the graph moving downwards towards decreasing values of z, along the equipotential lines. It would then enter the conductors and move radially in them. In the inner conductor it would dissipate as heat while moving radially from $\rho = a$ to $\rho = 0$, while in the outer conductor it also moves radially from $\rho = b$ to $\rho = c$, being completely dissipated as heat along this trajectory. The only region where the lines of Poyting flux do not follow the equipotential surfaces is for $\rho > c$. In this region there is no magnetic field. Although we have obtained an electric field and equipotential lines here, the Poynting vector goes to zero.

Beyond the generalizations of the previous works, the main nontrivial conclusion of this analysis are Eqs. (10), (20) and (24). They show that although there is no vector potential nor magnetic field outside a coaxial cable, the electric field won't be zero when there is a finite resistivity in the outer conductor. As the previous works quoted above considered only the case of a return conductor with zero resistivity, this aspect did not appear. The existence of the tangential component E_z of $\vec{E}$ outside the coaxial cable might be guessed from Maxwell's equations. That is, as there is a resisitivity in the outer conductor of finite area and finite g, there must be an electric field at $b \leq \rho \leq c$ balancing Ohm's resistance in a dc current. As the tangential component of the electric field is continuous in any boundary, this means that E_z must also exist outside the external conductor. Although this may seem trivial, it is almost never mentioned for the case of a coaxial cable. More important than this is that we have shown that there will also exist a radial component of $\vec{E}$, $E\rho$ given by Eq. (10). Although it is inversely proportional to ρ, it will have a reasonable value close to the cable and in principle might be measured in the laboratory. To our knowledge the first to mention this external electrical field outside a resistive coaxial cable was Russell in his important paper of 1983, [24]. Our work presents a clear analytical calculation of this field, which Russell could only estimate. Our paper might be considered the quantitative implementation of his insights.

Acknowledgements

One of the authors (AKTA) wishes to thank the European Community and FAEP–UNICAMP for financial support. He thanks also Drs. Franco Selleri, Mark A. Heald, Álvaro Vannucci and Daniel Gardelli for discussions and suggestions.

References

[1]   W. F. Edwards, C. S. Kenyon, and D. K. Lemon. Continuing investigation into possible electric fields arising from steady conduction currents. *Physical Review D*, 14:922–938, 1976.

[2]   D. F. Bartlett and B. F. L. Ward. Is an electron's charge independent of its velocity? *Physical Review D*, 16:3453–3458, 1977.

[3]   G. Bonnet. Electric field arising from a steady current passing through a superconductor. *Physics Letters A*, 82:465–467, 1981.

[4]   J. C. Curé. A modified version of the Millikan oil drop experiment to test the probable existence of a new electrodynamic field. *Physics Letters B*, 116:158–160, 1982.

[5]   D. F. Bartlett and S. Maglic. Test of an anomalous electromagnetic effect. *Review of Scientific Instruments*, 61:2637–2639, 1990.

[6]   D. F. Bartlett and W. F. Edwards. Invariance of charge to Lorentz transformation. *Physics Letters A*, 151:259–262, 1990.

[7]   C. S. Kenyon and W. F. Edwards. Test of current–dependent electric fields. *Physics Letters A*, 156:391–394, 1991.

[8]   A. K. T. Assis. Can a steady current generate an electric field?*Physics Essays*, 4:109–114, 1991.

[9]   A. K. T. Assis. *Weber's Electrodynamics* . Kluwer Academic Publishers, Dordrecht, 1994. ISBN: 0–7923–3137–0.

[10] A. K. T. Assis. *Eletrodinamica de Weber – Teoria, Aplicacoes e Exercicios*. Editora da Universidade Estadual de Campinas – UNICAMP, Campinas, 1995. ISBN: 85–268–0358–1.

[11] T. Ivesic. The relativistic electric fields arising from steady conduction currents. *Physics Letters A*, 144:427–431, 1990.

[12] T. Ivesic. Electric fields from steady currents and unexplained electromagnetic experiments. *Physical Review A*, 44:2682–2685, 1991.

[13] T. Ivesic. The definitions of charge and the invariance of charge. *Physics Letters A*, 162:96–102, 1992.

[14] T. Ivesic. Reply to "Comments on 'Electric fields from steady currents and unexplained electromagnetic experiments'". *Physical Review E*, 44:4140–4142, 1993.

[15] F. Selleri. Noninvariant one–way velocity of light and particle collisions. *Foundations of Physics Letters*, 9:43–60, 1996.

[16] F. Selleri. Noninvariant one–way velocity of light. *Foundations of Physics*, 26:641–664, 1996.

[17] M. A. Heald. Electric fields and charges in elementary circuits. *American Journal of Physics*, 52:522–526, 1984.

[18] J. D. Jackson. Surface charges on circuit wires and resistors play three roles. *American Journal of Physics*, 64:855–870, 1996.

[19] A. Marcus. The electric field associated with a steady current in long cylindrical conductor. *American Journal of Physics*, 9:225–226, 1941.

[20] A. Sommerfeld. *Electrodynamics*. Academic Press, New York, 1964.

[21] D. J. Griffiths. *Introduction to Electrodynamics*. Prentice Hall, Englewood Cliffs, second edition, 1989.

[22] B. R. Russell. Surface charges on conductors carrying steady currents. *American Journal of Physics*, 36:527–529, 1968.

[23] O. D. Jefimenko. *Electricity and Magnetism*. Electret Scientific Company, Star City, 2nd edition, 1989.

[24] B. R. Russell. Surface charges and leaky shields. *American Journal of Physics*, 51:269–270, 1983.

A New Appraisal of the Relativistic Quantum Theories of the Electron

A.M. Awobode
Department of Physics
University of Ibadan
Ibadan, Nigeria

An additional term introduced in a recent modification of the Dirac theory has been shown to produce interesting consequences. Physical quantities, such as the gyromagnetic ratio and the Lamb-shift, which the Dirac equation could not adequately account for, have been calculated (approximately). Though these quantities are obtainable from relativistic QED with higher accuracies, the greater advantage of the modified Dirac equation is exhibited in the consistent calculation of the precessional frequencies of the helicity and angular momenta.

1. The Successes and Limitations of the Dirac Theory

One of the most significant achievements of relativistic quantum mechanics is the account of electron spin given by the Dirac theory which, in addition, predicted the existence and properties of positrons. Several other consequences of the Dirac equation have also been shown to have experimental validity, particularly with regard to the behaviour of electrons and muons [1,2]. Therefore, despite certain inherent difficulties, in view of its many successful applications the Dirac theory constitutes a very important part of atomic physics and quantum field theory.

Foremost among the difficulties of the Dirac equation is its interpretation as a single-particle theory; the existence of negative-energy solutions to the equations according to the hole theory is interpreted as implying that the vacuum is populated with negative-energy electrons, and thus the vacuum carries infinite charge [3]. This appears to be a conceptual difficulty by itself, quite apart from the practical difficulty of constructing a wave function for a single particle, which obviously requires that all the filled negative-energy states be taken into consideration. However, despite its shortcomings, the hole theory allowed the prediction of pair creation, and hence the possibility of vacuum polarization, which appears to have been experimentally established, particularly by the structure of the positronium ground state [4], the Uehling effect and the level shift of mesic atoms.

Another interpretational difficulty usually associated with the Dirac theory is concerned with the fluctuating behaviour of some physical quantities. For example, contrary to expectations, the velocity of a particle which is free from all external fields is not a constant of the motion; The solution of the Heisenberg equation of motion for the velocity operator demonstrates that a time-dependent oscillating function is superimposed on a constant term.

Consequently, the coordinates of the particles exhibit similar oscillatory behaviour known as *Zitterbewegung* [3,4,5]. The spin [6] and the rest mass [7] have also been shown to exhibit such characteristics. In addition, the eigenvalues of the velocity operator $c\alpha$ are $\pm c$ since the eigenvalues of α are ± 1 [8]. Hence, it appears that a particle of non-zero mass can travel at the speed of light contrary to the conclusions of special relativity.

Moreover, the components of the velocity operator do not commute, and hence the measurement of the x-component of the velocity is incompatible with that of the y-component [8]. In order to resolve these difficulties, the Dirac theory was expressed in a different form in which even and odd operators are separated. In the Foldy-Wouthuysen representation, new Hamiltonian, spin and position operators were defined such that the velocity is a constant of the motion and the position coordinate is free from all oscillatory behaviours [2,3,9]. However, among the limitations of the Foldy-Wouthuysen representation was the observation that the position operators are non-local and the position coordinates do not possess the correct Lorentz transformation.

In addition to these problems of interpretation, the Dirac equation was also confronted with a real experimental difficulty in the calculation of the Lamb shift, *i.e.*, the displacement of the $2S_{1/2} - 2P_{1/2}$ energy levels of the hydrogen atom [10]. The Dirac equation predicts a degeneracy between the two levels contrary to experimental observations. Also in conflict with observation is the magnitude of the gyromagnetic ratio g deducible from the Dirac theory; measurement show that $g > 2$ in contradiction with that obtained from the Dirac theory [12].

Nevertheless, both the Lamb shift and the gyromagnetic ratio have been accurately accounted for by quantum electrodynamics (QED) employing a renormalization procedure which however is not entirely free from conceptual difficulties. Another useful modification which deserves mention at this point is the addition of the Pauli moment term [11]. Before the measurement of the anomalous part of the magnetic moment, Pauli introduced terms into the Lagrangian and Hamiltonian of particles interacting with the electromagnetic field on the assumption of a more general value for the magnetic moment [11]. In non-covariant form, the additional term is given as $\mu_B \mu'(\vec{\sigma} \cdot \mathbf{B} + i\vec{\alpha} \cdot \mathbf{E})$ where $(\vec{\alpha}, \vec{\sigma})$ are the Dirac matrices, μ_B is the Bohr magneton, μ' is the anomalous magnetic moment, and $(\mathbf{E}, \mathbf{B})$ are the electric and magnetic field vectors respectively. The weakness of this approach in resolving the difficulties of the Dirac equation lies in the fact that μ' enters as an adjustable parameter. Thus instead of obtaining a theoretical value of μ' from the modification and then comparing with measured values, μ' is input into the theory from experiments. The Pauli phenomenological correction therefore appears to treat the anomalous moment like an intrinsic property of the electron like the mass, spin and charge. Although the inclusion of the term has been found useful for describing the precession of the spin in a magnetic field [13], the approach is severely limited because the relative displacement of the $2S_{\frac{1}{2}} - 2P_{\frac{1}{2}}$ energy levels is neither accounted for, nor the *Zitterbewegung* given a possible interpretation by means of this procedure.

It becomes clear in view of these problems that the Dirac theory, despite these modifications, requires some new improvement. The modifications that have been proposed to resolve these difficulties mentioned above, namely the Hole theory, the Foldy-Wouthuysen representation and field-theoretic quantum electrodynamics, *etc.*, have all been concerned with individual problems or a specific group of problems and are not general enough to resolve them all in a unified and systematic way. Moreover, there exists some problems which these schemes are incapable of solving quite apart from the fact that they themselves are not without some difficulties.

II. A New Modification of the Dirac Hamiltonian

Following the discovery that the rest mass in relativistic quantum mechanics is time-dependent [7], an extension of the Dirac theory which is presently proving successful in the resolution of some of the problems mentioned above was proposed [14]. The Dirac Hamiltonian was extended by adding a term describing the time variation of the rest mass [7]. The resultant time-dependent equation was then used in the interpretation of the fluctuations of the velocity and position coordinates as consequences of the non-stationary states of the electron [14]. The newly proposed equation, consisting of the Dirac Hamiltonian and an extra term has been shown to be Lorentz covariant [15].

More importantly however, is that recently, empirically testable conclusions have been reached on the basis of the newly proposed equation. For example, we have shown that

(1)   In the limit $t \to \infty$, replacing the time-dependent term in the proposed Hamiltonian by an approximate time-independent one, and considering an electron in a static magnetic field, a gyromagnetic ratio $g > 2$ can be successfully inferred in the non-relativistic regime [14]. The additional contribution, *i.e.*, the anomalous part of g is calculated to be about 0.00016

which to four places of decimal is approximately equal to that obtained by QED.

(2) By considering a particle in a centrally symmetric Coulomb potential, we have been able to calculate a relative displacement between the $2S_{\frac{1}{2}}$ and $2P_{\frac{1}{2}}$ levels *i.e.*, the Lamb shift. Although the value of the separation calculated is about 1.5 times higher than the measured value, there is considerable confidence that further refinements will bring the value closer to 1060 MHz. What we need note here however is that it has been satisfactorily shown that, in general, there is a splitting between states with same n and j but different l [15]. These states, according to the original Dirac equation, are degenerate.

(3) Also by evaluating the commutator $[H, \vec{\sigma} \cdot \mathbf{p}]$ of the relativistic Hamiltonian H with the helicity operator $\vec{\sigma} \cdot \mathbf{p}$, and then using the Heisenberg equation of motion, we have demonstrated the observed variation of the electron helicity in a static magnetic field [18]. This approach is the proper quantum mechanical procedure which gives further proof of the general validity of the Heisenberg equation of motion beyond non-relativistic quantum mechanics. The precessional frequency $[(eB/mc)\mu'$ where $\mu' = 0.0012$], which we have calculated is in agreement with the experimentally measured value [2].

(4) We have also calculated relativistic corrections to the precessional motions of the spin, total and orbital angular momenta [19]. It is remarkable that it was only possible to demonstrate the precession of the total angular momentum vector $\mathbf{J}$ about the direction of the magnetic field for a relativistic particle because of the extra term. In the Dirac theory the total angular momentum operator commutes with the Hamiltonian for a relativistic particle in a static magnetic field thus implying that J does not precess.

While QED has been exceedingly successful in resolving the first two problems (1 & 2) mentioned above with unsurpassed accuracy, the conceptual modifications suggested by QED when considered as ammendments to the Dirac theory however, are not capable of solving the other two (3 & 4). Nevertheless, the equation we have proposed as mentioned above, has successfully produced some empirical results which are worth comparing with those obtained from other efforts aimed at removing the defects of the Dirac equation.

Precession of the Spin: First we note that the Heisenberg equation of motion derived for the spin operator ä on the basis of our new Hamiltonian has a form similar to that of the classical Thomas equation with $\mathbf{E} = 0$ and $\mathbf{v} \cdot \mathbf{B} = 0$ [15]. Moreover, the precessional angular velocity which we have calculated from our proposed equation is approximately equal to that obtained from the Thomas equation. The formal similarity between the classical Thomas equation and the Heisenberg equation of motion for is in conformity with Ehrenfest's theorem. According to Ehrenfest's theorem the Heisenberg equation of motion for the average quantities have the same form as the classical equation of motion for the respective quantities which the operators represent. This, however, is only approximately true, particularly for relativistic motion as may be observed with velocity and position coordinates in the Dirac theory. Thus considering the situation more accurately, we find that when the precessional angular velocity is calculated to second-order using our new Hamiltonian it contains a term proportional to $\mu_B \cdot B/mc^2$, while the angular velocity ω_p obtained from the classical equations differ from the Larmor frequency ω_L by ω_c, i.e., $\omega_p - \omega_L = \omega_c$. The classical precessional angular velocity ω_p is also obtainable from the Bargmann-Michel-Telegdi equation [13] which is a relativistic generalization containing the Thomas equation as a special case.

Precession of the Helicity: Of greater importance, however, is the Heisenberg equation of motion which we have derived for the helicity operator $\vec{\sigma} \cdot \mathbf{p}$ using the new Hamiltonian. This operator commutes with the original Dirac Hamiltonian for a particle in a magnetic field thereby implying that the helicity is a constant of the motion, whereas observations show that the helicity varies with time. The additional term in our new Hamiltonian however does not commute with $\vec{\sigma} \cdot \mathbf{p}$ thereby establishing the time-dependence of the helicity. It is remarkable and note-worthy that again, to first approximations, the frequency of precession calculated is the same as that obtained from classical arguments based on the Thomas equation and cyclotron motion. The frequency calculated from classical arguments in this case, however, is completely devoid of all relativistic effects (Grandy, Jr., ref. [2]). It is worth noting that the precessional angular frequency we have calculated quantum-relativistically on the basis of our proposed equation, when taken to second order of approximations, depends on $\mu \cdot B/mc^2$.

Precession of **L** & **J**: In addition to the equations of motion for the spin and helicity discussed above, the Heisenberg equation of motion for the total angular momentum **J**, and the orbital angular momentum **L** were also obtained from the modified Hamiltonian. The original Dirac theory indicated that in a static magnetic field, **J** is a constant of the motion, while again because of the additional term, **J** is shown to precess about the direction of the field with the Larmor frequency. Similarly **L** precesses with the Larmor frequency.

Considering the above, we have thus shown that, despite the approximations, the new equation has achieved notable successes in calculating the anomalous magnetic moment, the Lamb shift and the precession of the helicity which had all been previously observed by experiments. That is to say, the new equation which we are presently introducing has produced equivalent results in the domains where the previous modifications had been separately successful in varying degrees.

Also, it has been possible, using the new Hamiltonian to describe the relativistic precession of **L**, **S** and **J** and to calculate their precessional frequencies. It therefore marks a significant development in quantum relativistic electron theory. The calculated frequencies, when taken to second-order contain a new term $\mu \cdot B/mc^2$ which is appearing for the first time and may therefore be tested for confirmation by suitable experiments. The new equation is Lorentz covariant and thus meets the requirements for a relativistic quantum-mechanical operator. It therefore has the decisive advantage of being useful in describing the dynamics of operators which may have no classical or non-relativistic counterparts.

III. Summary

We conclude therefore, that the newly modified Dirac equation has produced results which are in reasonable agreement with measured quantities, while providing others which can be experimentally verified. The equation has accounted for a range of physical observations which have traditionally taken two or more disparate theories to achieve. More interestingly, it has produced new features of known phenomena, *viz.* the $\mu \cdot B/mc^2$ dependence of the precessional frequencies of and $\vec{\sigma} \cdot \mathbf{p}$, and described entirely new phenomena, *e.g.*, the relativistic precession of the **L** and **J**.

The physical interpretation of the new equation which is implicit in its application shows that it is not a single particle theory. The presence and properties of other particles have to be taken into consideration even in the absence of external fields. The modified equation therefore corrects the defects of the Dirac theory while also providing a basis for the construction of a more consistent field theory.

Thus interestingly an equation arrived at by modifying the mass term in the Dirac equation has successfully corrected notable defects in the Dirac theory in a unified and systematic way. The defects had hitherto been addressed by methods, techniques and procedures which had shown obvious limitations. The newly proposed equation therefore forms the basis of a promising development, which merits further consideration and acceptance as a physical theory of the electron and other similar spin-½ particles.

References

[1]   B. Thaller, *The Dirac Equation*, (Springer-Verlag 1992).

[2]   W.T. Grandy, Jr., *Relativistic Quantum Mechanics of Leptons and Fields* Kluwer, Dordrecht (1990).

[3]   G.L. Trigg, *Quantum Mechanics*, (Van Nostrand, N.Y. 1964); J. D. Bjorken and S. D. Drell, *Relativistic Quantum Mechanics* (McGraw-Hill Book Co., N.Y. 1964).

[4]   S.S. Schweber, *An Introduction to Relativistic Quantum Field Theory* (Row, Peterson & Co., Evanston, Illinois 1961).

[5]   E. Schrödinger, *Sitzber. Preuss. Akad. Wiss. Phys. Math. Kl.* 24 (1924); H. Thirring, *Principles of Quantum Electrodynamics* (Acad. Press. N.Y. 1958).

[6]   A.O. Barut & A. J. Bracken, *Phys. Rev. D*, 23 254 (1983).

[7]   A. M. Awobode, Dynamical Fluctuation of the Rest Mass, *Found. Phys. Lett.* 2, 167 (1990).

[8]   P.A.M. Dirac, *The Principles of Quantum Mechanics* O.U.P (1958).

[9]   L.L. Foldy & S.A. Wouthuysen, *Phys. Rev.* 78, 29 (1950).

[10] W.E. Lamb & R.C. Retherford, *Phys. Rev.* 72, 241 (1947).

[11] W. Pauli, *Rev. Mod. Phys.* 13 (1941).

[12] H. M. Foley & P. Kusch, *Phys. Rev.* 73, (1948). p.412.

[13] V. Bargmann, L. Michel & V. L. Telegdi, *Phys. Rev. Lett.* 2 (1959) p.435.

[14] A.M. Awobode, A Generalized Time-dependent Relativistic Wave Equation, *J. Maths. Phys.* 35 (2) (1994) p.568.

[15] A. M. Awobode, "The Lorentz covariance of a Hermitian Time-dependent Hamiltonian" Preprint 1995.

[16] A. M. Awobode, "The Anomalous magnetic Moment of the Electron" Preprint 1994.

[17] A. M. Awobode, "Atomic Energy Levels in Relativistic Quantum Theory: The Lamb-shift" Preprint 1995.

[18] A. M. Awobode, "The Helicity of a Relativistic Electron in a Static Magnetic Field" Preprint 1994.

[19] A. M. Awobode, "Relativistic Precessional Motion of Angular Momenta" Preprint 1997

[20] A. M. Awobode, "Quantum-relativistic correction to the Motion of the Electron Helicity in a Static Magnetic Field" Preprint 1997.

Nature of Relativistic Effects and Delayed Clock Synchronization

V. S. Barashenkov
Joint Institute of Nuclear Research
Dubna, Russia

E. Kapuscik
Cracow Pedagogical University and Institute of Nuclear Physics
Kraków, Poland

M. V. Liablin
Joint Institute of Nuclear Research
Dubna, Russia

An interference experiment with incoming and refracted light rays is described which distinguishes between the Einstein and Poincaré–Lorentz views on the nature of relativistic effects. The Lorentz transformation rule with a delayed clock synchronization is checked experimentally.

1. Introduction

As is known, at present there are two in principal different views on the nature of relativistic effects. One group of physicists, following Poincaré and Lorentz, views them as a consequence of deformations experienced by bodies moving relative to the preferred (ether) reference frame. Other physicists, like Einstein, regard these effects as a pure kinematic consequence of simultaneity in moving and motionless frames.

Our aim is to try to distinguish these two viewpoints by observing a possible change of the interference of incoming and refracted laser rays depending on the reference frame velocity.

Einsteinian theory states that such a dependence is not present, since otherwise one would be able to discover inertial motion from inside an isolated system. However, from the viewpoint of those who share Lorentz's idea concerning the existence of some preferred reference frame, in the experiments considered one can measure an interference fringe displacement owing to a time delay in the process of the ray refraction.

Open Questions in Relativistic Physics
Edited by Franco Selleri (Apeiron, Montreal, 1998)

In addition, one more question remains undecided. Special relativity is based on the purely classical principle of time synchronization which, in turn, assumes an instantaneous reflection of the sent signal, without any delay. At first sight we encounter here a contradiction with the fact of a finite duration of excitation and deexcitation processes of atoms inside mirrors due to which the reflection is realized. Nevertheless, we shall show that the known Lorentz transformations preserve their form even if the time delay is explicitly taken into account.

In the next Section, the interference experiment and the results are described. Sec. 3 is devoted to a theoretical consideration of the Lorentz transformations with a time delay. In Sec. 4 we discuss the results.

2.   Experiment with delayed refraction

Our goal is to check up the velocity dependence of an interference of two laser rays. The principal idea of such a check–up can be explained by the following *gedanken experiment*. Let us imagine two parallel coherent light beams in a reference frame moving with a velocity v . One of these beams gets immediately into an interferometer but the other beam encountering a mirror is absorbed, *i.e.* is detained for some time t_0 and then is emitted into the interferometer. If Lorentz's idea on the preferred reference frame is true, then turning the plate where a light beam source and the interferometer are located perpendicularly to the frame velocity vector $\vec{v}$ will result in a relative phase shift of the light beams φ and, therefore, the fringes in the interferometer will be displaced.

According to the Lorentz's idea the light velocity in the moving frame is $c + v$, so the times during which two mentioned above rays run from their common source up to the interferometer are

$$t_1 = \frac{x}{c+v}, \qquad t_2 = t_o + \frac{x - vt_o}{c+v} \tag{1}$$

and the corresponding time delay is

$$\tau = t_2 - t_1. \tag{2}$$

In the case of the perpendicular disposition of the installation we have

$$t_1' = \frac{x}{c}, \qquad t_2' = t_o + \frac{x}{c} \tag{3}$$

with the time delay

$$\tau' = t_2' - t_1'. \tag{4}$$

The time difference, determining the interference picture change in the process of the transition from an initial disposition to the perpendicular one, is

$$\Delta \tau = \tau' - \tau = \frac{vt_o}{c+v} \cong t_o \beta \tag{5}$$

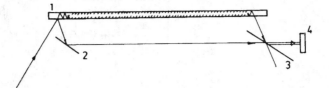

Fig. 1 : Scheme of the interference experiment:
1 – plate – parallel slab,
2 and 3 – mirrors,
4 – photodetector.

with the exactness to within quadratic terms $\beta^2 = (v/c)^2$ stipulated by the Lorentz contraction of lengths. Unlike the well-known Michelson-Morley and other measurements, where the propagation of light in two opposite directions compensates for any linear terms, in the experiment considered here with a single-directed light path these terms are conserved and can be investigated.

The corresponding phase shift of sinusoidal light waves

$$\varphi = \frac{2\pi c \Delta \tau}{\lambda} \qquad (6)$$

creates the fringe displacement in the photodetector equal to

$$\Delta l = \frac{\varphi \lambda}{2\pi} . \qquad (7)$$

Measuring this displacement we may get

$$\Delta \tau = \frac{\Delta l}{v} \qquad (8)$$

and the time delay

$$t_o \cong \frac{\Delta \tau}{\beta} = \frac{2\pi \Delta l}{v} . \qquad (9)$$

The real experiment differs from the one considered here only in the substitution of a mirror for the Mach-Zander interferometer with 125-multiple reflections of the light beam between two plane-parallel silvered plates (see Fig. 1) which makes it possible to extend the time delay considerably. The fringe displacement is recorded by a photodetector and the results are accumulated in a computer. To reduce the random noise level, all equipment placed on the turning plate, after a process of correction, is glued to this plate which, in turn, is attached to a massive base plate by means of a small area contact. As a transfer velocity v we use the Earth's velocity ~380 km/s.

The arrangement is calibrated by introducing a plate-parallel slab with precisely known thickness and refraction coefficient, *i.e.* producing known time delay, into the free arm of the interferometer (the lower arm in Fig. 1). In Fig. 2, where the results of our measurement are plotted, one can see such a

198   Open Questions

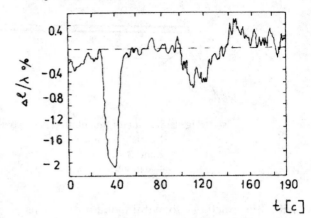

Fig. 2 : Displacement of the interference fringes $\Delta l/\lambda$ as a function of time.

calibration peak corresponding to the interference fringe displacement $\Delta l/\lambda = 0.02$. (In this case $\lambda = 63$ A).

In Fig. 2 the $360°$-turning of the arrangement corresponds to the time interval $90 \leq \Delta t \leq 190$ seconds inside which $\Delta l/\lambda = 0 \pm 0.006$, *i.e.* $t_0 = 0 \pm 8 \times 10^{-17}$ second. (We take into account that t_0 in (9) must be divided by the number of refractions inside the interferometer $n = 125$.) The observed oscillations of $\Delta l/\lambda$ are errors due to small deformations of the turning plate. The corresponding error Δt_0 is much smaller than the duration of the light refraction from metal mirror $t_0 \sim 10^{-14}$ seconds measured in direct experiments [1], and this fact justifies the claim that the Lorentz's hypothesis regarding the existence of a preferred reference frame and the dependence of light velocity on the frame speed v is wrong.

3.   Delayed synchronization of clocks

The Einsteinian synchronization method assumes that if at the instant of time t_1 any observer sends a light signal towards some event and the signal after a reflection at the position of the event comes back to the same observer at the instant of time t_2 then the clock located at the event at the moment of reflection should show up the instant of time t which satisfies the following *classical synchronization condition*

$$c\,(t - t_1) = c\,(t_2 - t) \tag{10}$$

where c is the velocity of light. Such an assumption is quite clear in the domain of macroscopic physics but need an additional discussion on the microscopic level.

To simulate the excitation and the subsequent de-excitation of atomic processes close to the mirror surface, we modify the Einstein synchronization condition (10) into the form

$$c(t - t_1) = c(t_2 - t - \tau) \tag{11}$$

where τ is the *delay time* which is the macroscopic parameter due to quantum microscopic processes in the mirror[§§§§]. From this condition we get the synchronized time

$$t = \frac{t_1 + t_2 - \tau}{2} \tag{12}$$

and the distance to the event

$$x = c\frac{t_2 - t_1 - \tau}{2}. \tag{13}$$

Another observer, operating with times t_1' and t_2' to synchronize clocks and to compute distance, ascribes to the same event coordinates

$$t' = \frac{t_1' + t_2' - \tau'}{2} \tag{14}$$

and

$$x' = c\frac{t_2' - t_1' - \tau'}{2}. \tag{15}$$

where the same invariant light velocity is present and another delay time τ' is used since this quantity need not to be same for all observers.

The relations between times used by different observers are given by the Lorentz transformations

$$t' = kt_1, \quad t_2' = k^{-1}t_2 \tag{16}$$

where k is a dimensionless parameter determined by the relation between observers in question. The transformation rules (16) are the same as in the usual Einsteinian synchronization because they are characteristics of the light frequencies and are therefore independent from the interaction of light with the matter of the mirrors.

The relations (16) can be rewritten in terms of the spacetime coordinates as

$$x' = \frac{(k^2+1)x - c(k^2-1)t + k\tau' - \tau}{2k} \tag{17}$$

and

$$t' = \frac{(k^2+1)t - (k^2-1)\frac{x}{c} + k\tau' - \tau}{2k}. \tag{18}$$

[§§§§] It should be noted that the customary relativistic energy-momentum relation considered usually as the most evident test of applicability of the Einsteinian theory of relativity to microscopic processes is not changed for the considered more general synchronization procedure. The modification introduced by the condition (11) is some kind of an additional translation which do not influence the definition of four–vectors because in this definition only the homogeneous part of space-time transformations is involved. The energy-momentum four-vector has therefore exactly the same properties as in theories with the customary Einsteinian sychronization.

These formulas coincide with the standard Lorentz transformations if the delay time transforms also according to the Lorentz rule for the times t_i :

$$\tau' = k^{-1}\tau = \sqrt{\frac{1-v/c}{1+v/c}}\,\tau \qquad (19)$$

where

$$v = c\frac{k^2-1}{k^2+1} \qquad (20)$$

is the velocity of the moving observer.

So, we see that the delayed synchronization of clocks do not disturb the known form of the Lorentz transformations.

4.   Concluding remarks

The obtained above results convince in the impossibility to link any singled out reference frame with vacuum. In its uniformity there is no, neither kinematic nor dynamic, peculiarity which can be used as an "anchor" for such a frame. Nevertheless, the concept of vacuum cannot be completely waived, as it was proposed by Einstein. The both experiment and quantum theory prove that it is a specific material medium though all attempts to describe it in modern notions encounter a great number of contradictions. Construction of adequate theory of vacuum is now the main problem of physics.

We would like to stress also that although the Einsteinian synchronization of time does not contradict any experimental data, it is a macroscopic procedure. At small spacetime intervals the concept of vacuum is incomprehensible. It is not clear also in what sense one can speak about lengths inside elementary particles where the modern interpretation of form–factors describing the internal structure of the particles encounters difficulties and the usual image of an extended particle for which all the points have the same time becomes relativistic non invariant [2].

References

[1]   A. V. Sokolov, *Optical properties of metals,,* p. 240, (F-M., Moscow 1961).
[2]   V. S. Barashenkov, *Problems of microscopic space and time,* (Atomizdat, Moscow, 1978.)

On a Relativistic Magnetic Top

Mirjana Božić
Institute of Physics
P. O. Box 57, Belgrade, Yugoslavia

The development of theories of elementary particles as extended objects, from Thomson to Dehmelt and MacGregor, is reviewed. Most of the theories are inspired by the existence of particle spin. The arguments of their authors are contrasted to Pauli's assertion that spin is "a classically non–explainable two–valuedness," as well as to Dirac's theory of a point electron.

1.   Introduction

In the development of models and theories of elementary particles, two main lines of thinking and research have evolved: According to view (α) which goes back to Thomson's classical model of the electron [1,2], elementary particles are extended bodies (spheres) with internal dynamics and certain distribution of physical quantities (mass, charge, *etc*...). According to view (β), which has its roots in Pauli's concept of electron spin [3,4] (as a "classically nondescribable two–valuedness") and Dirac's equation for the electron [5,6], elementary particles are points with *abstract* internal quantum numbers. Somewhere in between those two views is a model of an electron as a soft quasi-orbital structure with a radius of about one Compton wavelength/2π formed by the circular *Zitterbewegung* of the hard point electron [7] of dimensions $< 10^{-16}$ cm. In this paper we suggest this intermediate picture as a starting point for a unification of the two views (α) and (β).

2.   On studies of a rotating charged sphere

In the modern physics of elementary particles, view (α) dominates over the view (β), and has been built into various classification schemes of elementary particles. Nevertheless, theoretical investigations of quantum states and quantum numbers, as well as of the relativistic dynamics of the rotating sphere models, have been pursued continuously since Thomson's evaluation of the classical radius of the electron ($R_0 = ke^2/mc^2 = 2,82 \times 10^{-15}$ m)

[1] and Thomson's subsequent discovery of the electron [8]. Lorentz investigated the relativistic dynamics of the electron [9]. In 1925, Uhlenbeck and Goudsmith proposed the theory [10] of spin based on the spinning sphere. Rasetti and Fermi [11] took into account the magnetic field associated with spin and determined the magnetic radius of the electron ($R_H \geq 4.09 \times 10^{-14}$ m). Casimir determined [12] angular momentum quantum numbers of the rotating sphere and showed that they take integer and half integer values. Bopp and Haag [13] solved Schrödinger's equation of a rotating sphere and determined its quantum states—wave functions of variables which determine the orientation of a sphere. Dahl wrote the relativistic equation [14] by generalizing the non–relativistic Bopp and Haag approach. Barut, Božić and Marić introduced [15] a notion of a magnetic top—a spherical top with magnetic moment proportional to its angular velocity—and investigated its classical and quantum dynamics. MacGregor proposed [16] the theory of electron spin based on the model of a relativistic rotating sphere with point charge on its surface.

3.   The electron radius

The existence of an *anomalous* magnetic moment of the electron has been taken by some authors as an indication of electron structure. The dynamics of an electron in an electron trap, from which the electron g–factor was determined by Van Dyck, Schwinberg and Dehmelt [7] inspired Dehmelt [17] to make the following observation about the structure of an electron:

> *"Today everybody 'knows' the electron is an indivisible atomon, a Dirac point particle with radius R = 0 and g = 2,00...But is it? Like the proton, it could be a composite object. History may well repeat itself once more. This puts a very high premium on precise measurements of the g factor of the electron.*

From known g and R values of other near–Dirac particles (proton, triton and helium3) and the g value of electron (measured by Van Dyck, Schwinberg and Dehmelt [7]), Dehmelt attempted to extrapolate [17] a value of electron radius. In this extrapolation Dehmelt used the relation, proposed by Brodsky and Drell [18]

$$|g - g_{\mathrm{Dirac}}| = (R - R_{\mathrm{Dirac}})/\lambda_C \tag{1}$$

where $\lambda_C = \lambda_C/2\pi$, and λ_C is a Compton wavelength of a particle

$$\lambda_C = h/m\,c \tag{2}$$

and $R_{\mathrm{Dirac}} = 0$ for all Dirac particles. For an electron $g_{\mathrm{Dirac}} = 2$ and $\lambda_C = 2.426 \times 10^{-12}$ m. In this way Dehmelt found an electron radius $R_{\mathrm{Dehmelt}} \approx 10^{-22}$ m.

Dehmelt's value of an electron radius is much smaller than the value estimated from electron–electron scattering $R_{e\text{-}e} < 10^{-18}$ m. But, the latter value is much smaller than the value of the classical electron radius, the magnetic electron radius [11] and MacGregor's electron radius [16], which is equal to Compton's wavelength/2π.

It is important to note that according to the relation (1), on which Dehmelt's extrapolation is based, only a small fraction of g value is associated with the extension of the electron in space. The largest part of the g value ($g = 2$) is associated with the point electron ($R = 0$). This is not the case in methods based on a rotating sphere. In those methods R is associated with g as a whole.

4.   *Zitterbewegung* of the hard point electron versus a model of a rotating sphere with point charge on its surface

We would like to point out that: the *size* of a soft quasi-orbital structure associated with *Zitterbewegung* of Dirac's electron and the radius of a sphere with point charge on its surface (having spin and magnetic moment of an electron) are both equal to the electron Compton radius. It seems to us that the equality of these two sizes is not accidental, but that this fact reveals eventual equivalence of the two pictures (models) of the electron.

In order to verify this guess, it would be necessary to make a synthesis of various existing results and works. In fact, despite many studies devoted to various properties of electron, a consistent quantum relativistic theory of electron as an extended object, which would explain all the properties of the electron, still does not exist.

For example, the Bopp and Hagg approach [13] is quantum–mechanical, but not relativistic. The same is true of the Barut *et al.* [15] magnetic top. Dahl's electron is quantum mechanical and relativistic, but Dahl does not treat an interaction between the electron and the electromagnetic field. Rivas started from classical Lagrangians of nonrelativistic and relativistic particles with internal degrees of freedom. He determined [19] the corresponding quantum systems, but did not treat their interaction with the electromagnetic field. On the other hand, MacGregor's internal dynamics is relativistic [16], but his theory is not quantum mechanical in the strict sense of the word.

We need a relativistic quantum mechanical theory of a spinning particle which would describe its free motion as well as its external and internal dynamics when subjected to a field (electromagnetic). This theory should have correct classical and nonrelativistic limit.

With this aim in mind we are trying to develop a quantum relativistic theory of the magnetic top with a moving center of mass, by taking into account the results of the various approaches mentioned above.

5.   A magnetic top with a moving center of mass

We assume that the total Hamilton operator is a sum of two parts:

$$H_{tot} = H_e + H_s \tag{3}$$

where: H_e is associated with a translational (external) motion of a classical charged (q) particle in an electromagnetic field ($\mathbf{A}, \Phi$)

$$H_e = \left[\frac{1}{2m}(\hat{\mathbf{p}} - q\mathbf{A})^2 + q\Phi\right] \tag{4}$$

and

$$H_s = \frac{(s - \gamma I \mathbf{B})^2}{2I} = \frac{\mathbf{s}^2}{2I} - \gamma \mathbf{s} \cdot \mathbf{B} + \frac{\gamma^2 I \mathbf{B}^2}{2} \tag{5}$$

describes [15, 20, 21] internal (rotational) motion, as well as the interaction of magnetic moment μ (proportional to the angular velocity of a top) with the (homogeneous) magnetic field $\mathbf{B}$, such that

$$\mathbf{A} = \frac{1}{2}\mathbf{B} \times \mathbf{r} \tag{6}$$

Here, I is moment of inertia of a top and $\mathbf{s}$ denotes canonical angular momentum-spin, which is related [15] to the angular momentum of a top

$$\mathbf{s} = \Sigma + \gamma I \mathbf{B} \tag{7}$$

With the aid of the relation (6) and neglecting nonlinear terms in $\mathbf{B}$, the total Hamiltonian is transformed into the form

$$H_{tot} = \frac{\mathbf{p}^2}{2m} - \frac{q}{2m}\mathbf{L} \cdot \mathbf{B} + \frac{\mathbf{s}^2}{2I} - \gamma \mathbf{s} \cdot \mathbf{B}. \tag{8}$$

The corresponding quantum Hamilton operator reads

$$\hat{H}_{tot} = \frac{\hat{\mathbf{p}}^2}{2m} - \frac{q}{2m}\hat{\mathbf{L}} \cdot \mathbf{B} + \frac{\hat{\mathbf{s}}^2}{2I} - \gamma \hat{\mathbf{s}} \cdot \mathbf{B}. \tag{9}$$

where $\hat{p}$ and $\hat{s}$ are operators associated with momentum $\mathbf{p}$ and spin $\mathbf{s}$:

$$\hat{p}_x = -i\hbar\frac{\partial}{\partial x}, \quad \hat{p}_y = -i\hbar\frac{\partial}{\partial y}, \quad \hat{p}_z = -i\hbar\frac{\partial}{\partial z} \tag{10}$$

$$\hat{s}_x \equiv i\hbar\left[-\cos\varphi\frac{\partial}{\partial\vartheta} + \sin\varphi\frac{\cos\vartheta}{\sin\vartheta}\frac{\partial}{\partial\varphi} - \frac{\sin\varphi}{\sin\vartheta}\frac{\partial}{\partial\chi}\right]$$

$$\hat{s}_y \equiv -i\hbar\left[\sin\varphi\frac{\partial}{\partial\vartheta} + \cos\varphi\frac{\cos\vartheta}{\sin\vartheta}\frac{\partial}{\partial\varphi} - \frac{\cos\varphi}{\sin\vartheta}\frac{\partial}{\partial\chi}\right]. \tag{11}$$

$$\hat{s}_z \equiv -i\hbar\frac{\partial}{\partial\varphi}$$

The square of the spin operator reads

$$\hat{s}^2 \equiv -\hbar^2\left[\frac{\partial^2}{\partial\vartheta^2} + \cot\vartheta\frac{\partial}{\partial\vartheta} + \frac{1}{\sin^2\vartheta}\left(\frac{\partial^2}{\partial\varphi^2} + \frac{\partial^2}{\partial\chi^2}\right) - 2\frac{\cot\vartheta}{\sin\vartheta}\frac{\partial^2}{\partial\chi\partial\varphi}\right]. \tag{12}$$

These operators form the algebra of the $SU(2)$ group:

$$[\hat{s}_x, \hat{s}_y] = i\hbar\hat{s}_z, \quad [\hat{s}_y, \hat{s}_z] = i\hbar\hat{s}_x, \quad [\hat{s}_z, \hat{s}_x] = i\hbar\hat{s}_y. \tag{13}$$

Components of the operator $\hat{s}$ along the axes of a system attached to the top

$$\hat{S}_x \equiv i\hbar\left[-\cos\chi\frac{\partial}{\partial\vartheta}+\sin\chi\frac{\cos\vartheta}{\sin\vartheta}\frac{\partial}{\partial\chi}-\frac{\sin\chi}{\sin\vartheta}\frac{\partial}{\partial\varphi}\right]$$

$$\hat{S}_y \equiv -i\hbar\left[\sin\chi\frac{\partial}{\partial\vartheta}+\cos\chi\frac{\cos\vartheta}{\sin\vartheta}\frac{\partial}{\partial\varphi}-\frac{\cos\chi}{\sin\vartheta}\frac{\partial}{\partial\varphi}\right] \qquad (14)$$

$$\hat{S}_z \equiv -i\hbar\frac{\partial}{\partial\chi}$$

also form the algebra of the $SU(2)$ group. Operators $\hat{S}_z$, $\hat{s}^2$ and $\hat{s}_z$ form the complete set of commuting observables in the internal space of a top.

As demonstrated by Bopp and Haag [13], eigenvalues of the operator $\hat{s}^2$ are $s(s+1)\hbar^2$, where s take integer and half–integer values. The spectrum of the operator $\hat{s}_z/\hbar$ for given s consists of the values: $-s, -s+1,...s$. The same spectrum is possessed by the operator $\hat{S}_z/\hbar$. Common eigenfunctions $U_{smn}(\varphi,\vartheta,\chi)$ of these operators have the general form

$$U_{smn}(\varphi,\vartheta,\chi) = \exp(im\varphi)\exp(in\chi)F_{smn}(\vartheta) \qquad (15)$$

where $s = \frac{1}{2}, 1, \frac{3}{2}, 2,...; m \in (-s, -s+1,..., s), n \in (-s, -s+1,..., s)$.

By comparing the Hamiltonian (9) with Pauli Hamiltonian, one concludes that in $\hat{H}_{tot}$ there exists the term $\hat{s}^2/2I$ which does not exist in the Pauli Hamiltonian, and that the term $-\gamma\,\hat{s}\times\mathbf{B}$ contains unknown parameter γ, instead of the gyromagnetic ratio (q/m) in the Pauli Hamiltonian.

These differences do not mean that a magnetic top is not an appropriate model of a spinning particle. On the contrary, they indicate avenues of research which could lead to a theory of spinning particles that would consistently describe all their properties.

6.   Spinning and nonspinning rest energies (masses)

Energy $s(s+1)\hbar^2/2$, associated with the term $\hat{s}^2/2I$, is the same in all states with given s, so that the presence of this term does not present any difficulty. Moreover, if one takes into account that Dirac's equation in the nonrelativistic limit [23] reduces to the equation

$$i\hbar\frac{\partial\varphi}{\partial t} = \left[mc^2 + \frac{\hat{\mathbf{P}}^2}{2m} - \frac{q}{2m}\left(\hat{\mathbf{L}}+\hbar\overline{\sigma}\right)\cdot\mathbf{B}\right]\varphi \qquad (16)$$

which, apart of the terms which exist in the Pauli equation [22], contains the additional term—the particle rest energy, one is lead to the idea to consider the energy $s(s+1)\hbar^2/2$ together with particle rest energy, where rest refers to absence of the center of mass motion. (φ in (16) denotes a spinor and $\sigma = \sigma_1\mathbf{I} + \sigma_2\mathbf{j}+\sigma_3\mathbf{k}$, where σ_i are Pauli matrices.) Elements of this idea may be found in MacGregor's distinction between spinning and non–spinning rest masses [16].

MacGregor evaluated relativistic mass increase of a rapidly spinning sphere by dividing it up into mass elements $dm_0(r)$ which are equidistant from the axis of rotation – rings centered on the rotations axis. The relativistic mass of this element (due to rotation with angular velocity ω) is:

$$dm(r) = \frac{dm_o(r)}{\sqrt{1 - \omega^2 r^2 / c^2}} \qquad (17)$$

The total mass of the spinning sphere (having nonspinning rest mass m_0) is given by the integral

$$m_s = \frac{3m_o}{R^3} \int_0^R \sqrt{\frac{R^2 - r^2}{1 - \omega^2 r^2 / c^2}} \, r \, dr \qquad (18)$$

The largest value ω can have is

$$\omega = c/R \qquad (19)$$

which represents the angular velocity at which the equator of the spinning sphere is moving at the velocity of light, c. By substituting this limiting value into the equation (18), MacGregor obtained the following relation between nonspinning and spinning rest mass:

$$m_s = \frac{3}{2} m_o . \qquad (20)$$

In an analogous way MacGregor evaluated a relativistic moment of inertia of a spinning sphere

$$I = \frac{3}{4} m_o R^2 = \frac{1}{2} m_s R^2 . \qquad (21)$$

This value is then substituted into the expression for spin angular momentum of a free rotating sphere

$$s = I\omega = \frac{1}{2} m_s R c . \qquad (22)$$

For an electron, spin is equal to $\hbar/2$, implying that radius of an electron should be equal to the Compton wavelength/2π

$$R \equiv \lambda_C = \frac{\hbar}{m_s c} \qquad (23)$$

where the electron mass $m = m_s$.

7.   Conclusion

As mentioned above, we feel that this picture of a rotating sphere with point charge on its surface bears a close resemblance to the Dirac point electron which executes a spontaneous periodic quasi-orbital motion at the speed of light [6], the *Zitterbewegung* of Schrödinger [24]. In particular, on the one hand, Huang showed [25] this motion (*Zitterbewegung*) to be circular and accompanied by the (orbital) angular momentum $\hbar/2$ and the magnetic moment of one Bohr magneton. On the other hand, Dahl showed [14] that the equation of motion of a relativistic rotor with moving center of mass becomes identical with the Dirac equation when transformed to a matrix representation.

However, further research, which is under way, is necessary to put all these pieces into a complete and consistent theory.

References

[1]   J. J. Thomson, *Phil. Mag.* **11** (1881) 229.

[2]   F. Rohrlich, *Classical Charged Particles*, Addison–Wesley, MA, 1965.

[3]   W. Pauli, *Z. Phys.*, **16** (1923) 155.

[4]   W. Pauli, *Z. Phys.*, **31** (1925) 765.

[5]   P. A. M. Dirac, *Proc. Roy. Soc]* (London) **A117**, (1928) 610; **A118** (1928) 351

[6]   P. A. M. Dirac, *The Principles of Quantum Mechanics*, 4th ed., Clarendon Press, Oxford, 1958.

[7]   R. S. Van Dyck, Jr., P. B. Schwinberg, H. G. Dehmelt, *Phys. Rev.* **D 34** (1986) 722.

[8]   J. J. Thomson, *Phil. Mag.*, **44** (1897) 298.

[9]   H. A. Lorentz, *The Theory of Electrons*, Dover, New York, 1952.

[10] G. E. Uhlenbeck, S. Goudsmit, *Naturwiss.*, **13** (1925) 953.

[11] F, Rasetti, E. Fermi, *Nuovo Cimento* **3** (1926) 226.

[12] H. Casimir, *Rotation of a Rigid Body in Quantum Mechanics*, Thesis, Leyden, Croningen, 1931; *Proc. Roy. Acad. Amsterdam* **34** (1931) 844.

[13] V. F. Bopp and R. Haag, *Proc. Roy. Acad. Amsterdam* . **5a** (1950) 644.

[14] J. P. Dahl, *Det Kongelige Danske Selskab Mat-fys. Medd.* **39** (1977) 12.

[15] A. O. Barut, M. Božić , Z. Mari ć , *Annals of Physics* **214** (1992) 53.

[16] M. H. MacGregor, *The Enigmatic Electron*, Kluwer Academic Publ. Dordrecht, 1992.

[17] H. Dehmelt, *Physica Scripta*, **T34** (1991) 47.

[18] S. J. Brodsky, S. D. Drell, *Phys. Rev.* **D 22** (1980) 2236.

[19] H. Rivas, *J. Math. Phys]* **35** (1994) 3380

[20] M. Božić , D. Arsenovi ć , in *Quantization and Infinite-Dimensional Systems.* 1994, Ed. J. P. Antoine, Plenum, Press, New York.

[21] A. O. Barut, M. Božić . in *Proc. II Int. Wigner Symposium*, 1993. Ed. H. D. Doebner, World Scientific, Singapore.,

[22] W. Pauli, Quantentheorie, in: *Handbuch der Physik*, **Bd. 23**, 1926, Eds. H. Geiger, K. Scheel, Springer-Verlag, Berlin.

[23] J. D. Bjorken, S. D. Drell, *Relativistic Quantum Mechanics*, McGraw–Hill Book Company, 1964

[24] E. Schrödinger, *Sitzungsber. Preuss. Akad. Wiss.*, **24** (1930) 418.

[25] K. Huang, *Am. J. Phys.* **20** (1952) 479.

The Interface between Matter and Time: a Key to Gravitation

Henrik Broberg
51, rue Ernest André
78110 Le Vesinet, France
email: henrik.broberg@wanadoo.fr

Introduction

We introduce the concept of an interface between matter and the dimension of time. This interface will have the form of a surface. Of course, this is not the first time surface has served as a significant measure for concepts in physics. Other examples are the probability density of the Schrödinger equation or the cross-sections of nuclear interactions.

We first make the comparison with the way an architect draws a house and its implementation by a builder. The architect may not bother with the thickness of his lines when making the drawing. But the builder must also consider the fine resolution of the lines, which will be enlarged to centimetres or decimetres in practise. In the same way, Special Relativity (SR) will need to be complemented by the introduction of a fine-structure given by the thickness of particle time-sheets, which replace the notion of geometrical time in the form of infinitely thin light-rays used heretofore. Only by making this extension to SR will it be possible to extend General Relativity (GR) from a static picture of the curved space-time to a dynamic theory which includes cause and effect.

The analysis will show that surface areas are significant for energy (or mass) which is confined in the vacuum fields. In particular, the cross-sections of time sheets are treated as surfaces in the complex plane of numbers. The absolute values of these surfaces are proportional to the masses of the particles. The actual surface values, which are negative for the particles, describe absolute sinks for energy in the vacuum space. It is these sinks which are the cause of gravitation. Mathematically, the gravitational effect can be calculated from the energy flows occurring in the vacuum as it tries to heal the ruptures caused by particles in its space-time web.

While the calculations of General Relativity are normally based on approximations of differential distances between points on surfaces in curved space-time in the form of metric functions, the present analysis focuses on the

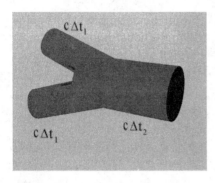

Figure 1 Particle strings and time-
sheets.

relation between the areas of the
surfaces and the energy confinement in
the vacuum of fields and particles.

The Mass-Time Interface

In the theory of strings, a particle
is associated with a string, which can be
an open loop, or rather for a stable
particle, a closed. When time is
included in the picture, the concept of a
time-sheet occurs. For example, the
fusing of two particles will be described
by the joining of two time-sheets into a
"pair of trousers."

The traditional light-lines used in SR (Figure 2) can easily be replaced by
tube-like space-sheets emanating from physical particles (Figure 3). This will
require the introduction of a plane perpendicular to the space-sheet of a
particle at rest *vis-à-vis* an observer.

The plane of Figure 3 is lined up with the velocity vector of the particle
in relation to an observer. It is therefore perpendicular to the time-sheet of the
particle in its own time-system, while it cuts through the time-sheet at a
different angle to the observer. The interface-surface so defined between the
plane and the time-sheet is therefore larger in the latter case.

Geometrically from the figure, the relation between the surfaces is equal
to the relativistic factor $\gamma = 1 \big/ \sqrt{1 - (v/c)^2}$ in SR, applicable to the relation
between time-intervals in the system of the observer and the system of the
moving particle. The relation between the surfaces is therefore equal to the
relation between the mass of the moving particle and the particle at rest.

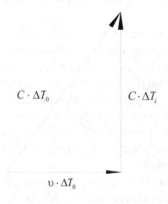

Figure 2 The traditional SR
picture

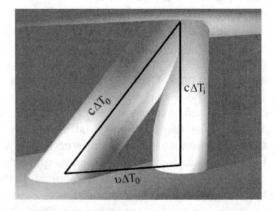

Figure 3 Time-sheets introduced into SR

Figure 4. A string rotating in its plane

It should be noted that if the above introduced plane (Figure 3) moves with the particle rather than being stationary in the observer system, the interface surface in that plane will become Lorenz-contracted relative to the surface of the particle at rest and as such invariant with respect to all observers. These surfaces are here defined as the mass-time interfaces of the particle.

In this essay, the latter, invariant surface will be called the particle MTI for convenience.

The geometry of the rest-mass string-system

We now consider a string which is rotating in its plane, as in Figure 4. The rest-mass/energy of the string is proportional to its MTI-surface, defined above. This proportionality is given by the parameter A:

$$MTI = A \cdot M_0$$

Initially, we will assume that the string has a radius r, angular velocity ω and rotates with velocity v:

$$v = r \cdot \omega$$

We will first study the case when all the mass of the string originates from the kinetic energy of its rotation while the rotational velocity approaches c and the rest-mass of the string itself becomes zero. The "surface-to-mass" relation then becomes:

$$\left[\Phi \middle/ m \right] = \frac{\pi \cdot r^2}{m_0 \cdot (\gamma - 1)} \underset{\substack{v \to c \\ m_0 \gamma \to M_k}}{\Rightarrow} \frac{\pi \cdot \left(\dfrac{c}{\omega_k} \right)^2}{M_k}$$

The MTI is defined to be that of a single surface:

$$[\Phi] = Am,$$

or

$$[\Phi] = \pi \cdot \left(\frac{c}{\omega_k} \right)^2$$

In the next step, we will study the relation between the surface across the string and the kinetic mass (energy) when the radius and, accordingly, the rotational velocity approaches zero while the angular velocity is constant:

$$\left[\Phi/m\right] = \frac{\pi \cdot r^2}{M_0 \cdot (\gamma - 1)} \underset{\substack{r \to 0 \\ v \to 0}}{\Rightarrow} \frac{\pi \cdot r^2}{M_0 \cdot \frac{1}{2}\left(\frac{r \cdot \omega_0}{c}\right)^2} = 2 \cdot \frac{\pi \cdot \left(\frac{c}{\omega_0}\right)^2}{M_0}$$

While preserving the definition of the MTI from the preceding, it becomes now:

$$[\Phi] = 2\pi \cdot \left(\frac{c}{\omega_0}\right)^2$$

which represents a double surface.

Therefore, in this case

$$[\Phi] = 2 \cdot Am$$

Geometrically, the two expressions for the MTI compare as a single surface system to that of a double surface. In the latter case, the mass/is that of a rest mass. It will be attributed to the contents of the encapsulated fields which constitute the mass.

The string will serve as a singularity in the sense that, in its own system, the surface covered by its radius is equivalent to the surface within the event horizon of a Schwarzschild singularity. The system of the string will be defined as a system which depends only on its own mass, and therefore is singular in the sense that it is independent of the rest of the Universe for its internal physical properties. It would be a black hole if Newton's constant applied with its numerical value as measured in our Universe. However, this is not the case for the string systems, except for a string with the same mass/energy content as our Universe, which would embrace the event horizon of the Universe.

When the transition is made from a physical radius, which approaches zero, to a radius in the time dimension, we also need to express the time dimension by an imaginary number. This can be done by substituting

$$\omega \to i\omega \,.$$

As a result, the surface becomes negative. However, for the time being we will only regard the absolute value of the surface, although the negative value will symbolize a sink for energy in the vacuum space.

The relation between mass, radius and surface in a "black hole"

The Schwarzschild radius, or the radius to the event horizon of a black hole is $R = 2GM/c^2$. Hence, with Newton's constant as a fixed parameter, the radius is proportional to the mass of the black hole. With the relation between particle mass and surface established earlier, we want to find a geometric configuration which also takes care of the proportionality between mass and surface. The answer is found in the geometry of a sphere, where the surface segments are of equal size when projected from equal distances along the diameter. See Figure 5, illustrating the well known Mercator projection of the surface of the Earth onto a cylinder, which maps the Earth with the correct areas of the surfaces.

In this model, we will treat the particle as an element of the Universe, while its mass is a part of the mass of the Universe. In accordance with the preceeding analysis of the limit of the surface to mass relation for the restmass of a string, the particle MTI will be represented by a surface equal to

$$\Delta\Phi = 2\pi R_U \cdot \Delta R \, ,$$

where R_U is the radius of the Universal sphere.

Obviously, if $\Delta R = R_U$ the above expression becomes that of the surface of half the Universal sphere. We need therefore two such delta surfaces to make the complete mapping on the sphere in proportion to the mass.

Each such delta element of the radius will therefore be defined as:

$$\Delta R = \tfrac{1}{2} r_g (\Delta M_U) \text{ , or } \Delta R = r_g \left(\frac{1}{2} \Delta M_U \right)$$

or with other words such that the sum of the two delta elements of the mass will correspond to the delta surfaces on the Universal sphere.

The small delta surfaces representing particles in the large-scale universe can be treated as the "polar tops" of opposing spheres overlapping by the depth of half the particle gravitational radius from each side. For illustration, regard Figure 11 as being rotated around the vertical diameter, which would generate a "polar top" as described.

The rest mass of a singular particle (a quark) is here defined as

$$m = \Delta M_U$$

The delta surface on *one* side of the particle plane (compare Figure 3 or Figure 4) will become:

$$\Delta\Phi = 2\pi \cdot R_U \cdot \frac{Gm}{c^2} \text{ , or } \Delta\Phi = \pi \cdot R_U \cdot r_g (m)$$

This is defined as the MTI (mass-time interface), which should be proportional to the particle mass:

$$Am = \Delta\Phi$$

Therefore, identifying the above two expressions will eliminate m and leave A as a parameter directly linked only to G and universal radius:

$$A = 2\pi \cdot R_U \cdot \frac{G}{c^2}$$

To evaluate the parameter A, we will use Hubble's constant in order to get a measure for the Universal radius:

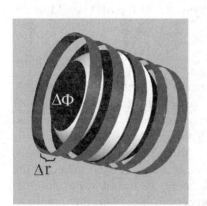

Figure 5 Mercator's projection

$$R = c/H .$$

Hence, $A = 2\pi \cdot \dfrac{G}{cH} \approx 0.7\left[\dfrac{M^2}{Kg}\right]h$

On the other hand, we are treating the Universe as one large gravitating system, or a "black hole" with an event horizon given by its Schwarzschild radius:

$$R_U = \frac{2G \cdot M_U}{c^2}$$

The MTI of the Universe is expressed as:

$$AM_U = \pi \cdot R_U{}^2 ,$$

which in combination with the preceding expression will give:

$$G = \sqrt{\frac{A \cdot c^4}{4\pi M_U}}$$

and

$$G = \frac{A \cdot c^2}{2\pi R_U}$$

These two expressions for the gravitational constant, with A as a parameter, give its dependence of the mass and the radius of the system.

We will assume, as our leading hypothesis, that A is a time-independent Universal constant, while G depends on the Universal Mass (or accordingly its radius) which may vary in time. Other autonomous singular systems, such as quarks, will have a similar dependance of the parameter A.

If we compare with Figure 4 or 11, the surface subscribed to a particle will be close to flat in the large-scale Universe, while it would be absolutely flat in an infinite Universe. If the particle was alone and the only mass in the Universe, its MTI would curl up from the two sides to form the two halves of a spherical surface. However, the total surface value of the MTI would be the same in all cases.

Particles as singularities

It is easy to verify that the relations found here apply to the large scale Universe as well as to particles, by simply comparing the observed masses with the surfaces and forces involved. As the analyses are made on the basis of Schwarzschild geometry, normally applicable to "black holes," it can be concluded that particles themselves, or their quarks, will serve as "singularities" in the same sense as black holes, and therefore be subject to the same kind of geometry and ability to absorb energy as if they were "absolute sinks." The mechanism of the absorption process may explain the quantized redshifts of galaxies and quasars, as in the Arp model (Arp, 1993).

The MTI and the mass of the electron

The electron can be used as an example. Suppose a string is charged by the quantum charge of the electron. The length of the string is equal to the classical radius of the electron curled up around the event horizon, the surface of which is equal to the Mass-Time Interface. The charged string is subject to an expansion force because it is rejecting its own charge. A counteracting force arises from the requirement that energy must be added if the MTI expands. Hence we find an equilibrium as follows:

The tension of the string has the following energy, because the charge sees itself one turn away around the horizon:

$$E_T = \frac{\mu \cdot e^2}{4\pi R_{cl}} \cdot c^2$$

The energy related to the surface inside the horizon is:

$$E_\Phi = \frac{\pi}{A} \cdot \left(\frac{R_{cl}}{2\pi}\right)^2 \cdot c^2$$

The equality of these expressions gives

$$R_{cl}^3 = A\mu e^2$$

and the mass is

$$M_e = \frac{1}{4\pi} \sqrt[3]{\frac{\left(\mu e^2\right)^2}{A}}$$

As a check, this expression is satisfied by A=0.7 M²/Kg in agreement with above.

In view of Quantum Dynamics and the importance of the fine structure constant α, an algebraic approach is introduced in the Appendix, where α is established as a member of a family of related numbers. Among other things, this algebra will explain why all the mass of the electron can be of electromagnetic origin.

The dynamic space

Einstein suggested that the force of gravitation is not any different from the force we would experience from standing in a lift which accelerates upwards. However, it is obvious that, for example, the surface of the Earth is not accelerating upwards. The explanation given by GR is that the space is curved in four-dimensional space-time, while a mass induces the curvature in its neighbourhood. Although GR gives an accurate description of the gravitational effect, it does not explain why space is curved. Hence it does not give a cause.

However, a cause can be found in the particles' Mass Time Interface, which causes them to act as absolute sinks of energy, thus forcing an energy exchange with the vacuum space, which creates the gravitational effect. The

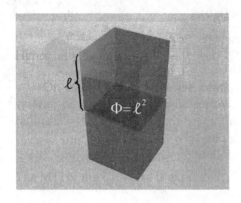

Figure 6 Density of "static space"

reason for the curvature of space in four-dimensional space-time would then be nothing other than the effect of the vacuum space being absorbed into the particles. To evaluate this 'dynamic space' theory the following parameters need to be calculated as functions of the distance from the particle MTI sink:

- the density of the space in the process of absorption,
- the velocity of the falling space.
- the energy flow to and from the particles.

It will then be possible to verify the theory with observed and predicted effects.

The density of vacuum space

"Static homogenous space"

In the case of a homogenous large volume of space, the MTI can be used to calculate a characteristic distance for the space. Consider a lattice of cubic volumes which fill up the vacuum space (Figure 6). Each cubic volume has the mass-equivalent of

$$m = l^3 \cdot \rho$$

The common surface of two cubes is l^2.

The relation between their common surface and the mass of the two cubes, which shall be equal to A is:

$$A = \frac{l^2}{2 \cdot l^3 \rho}$$

or

$$\rho = \frac{1}{2Al}$$

Inserting the Hubble length and the observed mass density in the Universe satisfies this relation as well as possible within the uncertainties of the observations.

"Spherical dynamic space"

Suppose that the vacuum space is a fluid that is streaming into a sink (Figure 7). The mass-equivalent of the flow towards the sink in a shell at distance R is:

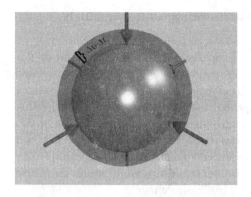

Figure 7. Flow towards the particle sink

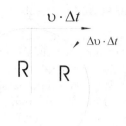

Figure 8. Velocity of "falling" space

$$\Delta M = 4\pi R \cdot \rho \cdot \Delta R$$

The surface of the MTI is

$$\Delta \Phi = 2\pi R \cdot \Delta R$$

The relation between the surface and the mass of the flow into the shell shall be:

$$A = \frac{\Delta \Phi}{\Delta M}$$

which gives

$$\rho = \frac{A}{2R}$$

or

$$A R \rho = \frac{1}{2},$$

in equivalence with the static case above.

It is interesting to note that the density as a function of distance from the absorbing sink depends only on A and R. Typically, a particle radius and therefore also its mass are truncated by the spin of the string, such that distinct quanta of angular momentum are set up.

Integrating the mass corresponding to a sphere with the above density function gives the typical mass-content of an elementary particle when integrated over a particle radius, while it gives an additional mass of Universal magnitude when integrated onwards over the Hubble length. Hence, the identified density function, in all its simplicity, fulfills the basic requirements for acceptance.

The space velocity field in the surrounding of a singularity

So far the way to progress has been found in the quantizing of properties. This has proven to be the case also for the establishment of the function of the falling space (Figure 8).

We make the analogy with an object which is in circular orbit around a central mass with such orbital velocity that the centrifugal force compensates the gravitational force, like a geostationary satellite. In a situation when the object is not rotating around the central force, it will still be subject to the same acceleration, or the corresponding force, towards the central mass.

Figure 8 gives the geometry of the vector components involved in the evaluation of the centrifugal force. From the figure we get:

$$R + (v \cdot \Delta t) = (R + \Delta v \cdot \Delta t)$$

This gives

$$\Delta v = \frac{1}{2} \cdot v^2 \cdot \frac{\Delta t}{R} \cdot \left[1 - \left(\frac{\Delta v}{v} \right)^2 \right] \approx \frac{1}{2} \cdot v^2 \cdot \frac{\Delta t}{R} .$$

Rather than letting $\Delta v \rightarrow 0$, which is usually done when calculating the centrifugal force, we will give a non-zero value to Δv, which corresponds to the flow of the vacuum space towards the central mass, which flow will be the cause of the gravitational effect. This will also leave finite values for the other differentials. As a starting point, the differential (quantum) of time in the system is chosen to be the time for a signal to go the distance R with the velocity c.

This quantizing of time leaves the classical formulas intact. The classical acceleration of the gravitational field is:

$$\frac{\Delta v}{\Delta t} = \frac{GM}{R^2}$$

or

$$\Delta v = \frac{GM}{R^2} \cdot \Delta t$$

Hence, in a reasonable neighbourhood of the particle the following will apply:

$$\Delta v = c \cdot \frac{GM}{c^2 R} \equiv c \cdot \frac{\frac{1}{2} r_G (M)}{R}$$

As a result we have created a ΔT and a Δv field in the surrounding of the central mass. The gravitational acceleration is given by these fields as:

$$\frac{\Delta v}{\Delta T} = \frac{GM}{R^2}$$

It may be noted that the same approach can be applied to the electric field leading to the development of Coulomb's law.

Particle energy absorption

The flow of energy, or rather its mass-equivalent, to the particle becomes:

$$\frac{\Delta M}{\Delta t} = 4\pi R^2 \cdot \Delta v(R) \cdot \rho(R)$$

By inserting the values of $\Delta v(R)$ and $\rho(R)$ the absorption rate becomes:

$$\frac{\Delta M}{\Delta t} = 4\pi R^2 \cdot \frac{GM}{CR} \cdot \frac{1}{2AR} = 2\pi \frac{GM}{AC}$$

With the expression established earlier

$$G_{M_U} = \frac{Ac^2}{2\pi R_U}$$

we get:

$$\begin{cases} R_U = \dfrac{c}{H_\circ} \\ \dfrac{\Delta M}{\Delta t} = \dfrac{Mc}{R_U} \end{cases}$$

from which it follows that

$$\frac{\Delta M}{\Delta t} = H_\circ M$$

where $H_\circ$ is Hubble's constant.

The latter relation will also be found if we study the region close to the particle MTI, in which case we have a double disc absorbing energy on both sides. We then return to the case of a "static homogenous space" (Figure 9). The absorption by each side of the disk becomes:

$$\frac{\Delta M}{\Delta t} = AM \cdot \frac{1}{2AR_H} \cdot c = \tfrac{1}{2} H_\circ M$$

Hence the total inflow becomes twice as large.

Alternatively, the absorption rate for the particle can be expressed in terms of the Hubble time:

$$\frac{\Delta M}{\Delta t} = \frac{M}{T_H}$$

The absorption flow depends only on the mass and the astronomical time parameter, whatever the significance of the latter may be.

The gravitational field

The mass equivalent of the vacuum energy in the interval ΔR becomes:

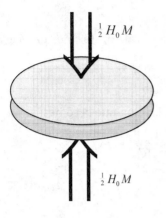

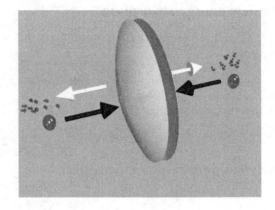

Figure 9 The absorption disk of   Figure 10. The negative gravitational flow
a particle

$$\Delta M = 4\pi R^2 \cdot \rho \cdot \Delta R = \frac{2\pi R}{A} \cdot \Delta R$$

This can also be calculated from the rate of the energy absorption:

$$\Delta M = H_\circ M \cdot \Delta t = H_\circ M \cdot \frac{R}{c} = M \cdot \frac{R}{R_U}$$

By identifying the two expressions we find that

$$\Delta R = \frac{AM}{2\pi R_U} = \frac{GM}{c^2} = \tfrac{1}{2} R_G(M) .$$

Hence ΔR is an invariant property in the field.

    We have also:

$$c \cdot \Delta R = R \cdot \Delta v .$$

The gravitational energy corresponding to ΔM is:

$$\Delta E_G = -\frac{GM\Delta M}{R} = -\frac{GM^2}{R_U} = -Mc^2 \cdot \frac{\Delta R}{R_U}$$

The corresponding differential mass equivalents contained within the differential ΔR for the fields, at the distance R from the absorption surface, are:

$$\begin{cases} \Delta M_G = -M \cdot \dfrac{\Delta R}{R_U} \\[2mm] \Delta M = +M \cdot \dfrac{R}{R_U} \end{cases}$$

    Assuming a velocity $-v_G$ for an energy flow corresponding to the negative gravitational energy, the mass-flows to and from the particle are:

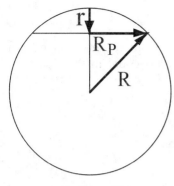

$$\begin{cases} \dfrac{dM_G}{dt} = \dfrac{\Delta M_G}{\Delta R} \cdot v_G = -M \cdot \dfrac{v_G}{R_U} = -H_\circ M \cdot \dfrac{v_G}{c} \\[2ex] \dfrac{dM}{dt} = \dfrac{\Delta M}{\Delta R} \cdot \Delta v = +M \cdot \dfrac{R}{R_U} \cdot \dfrac{\Delta v}{\Delta R} = +H_\circ M \end{cases}$$

If $v_G = c$, the two flows become equivalent, although with opposite signs.

Hence, the energy absorption by the particle causes a negative energy flow away from the particle, which creates the gravitational field (Figure 10). The process can be compared with the

Figure 11

transport of holes in a semiconductor. The curvature of the space in GR can be understood by the successive absorption of the relativistic bubbles of negative energy, which forces the vacuum space to "fall" towards the mass particle, thus creating a sink in the four-dimensional geometry.

Equilibrium in the process is reached if the particles emit energy back into the vacuum with the same rate as their energy absorption. An emission of electromagnetic energy with velocity c can compensate the outflow of negative energy in the gravitational field. Over the distance set by the Hubble time the energy flows will cancel out. For a similar treatment of gravitation see (Jaakkola, 1996).

Energy radiation from particles

While particles force an absorption flow of energy from the vacuum, they may return energy to the vacuum in one form or another. Possible processes for this are:

- decay because the particles increase their masses beyond an equilibrium state,
- Hawking's radiation near the MTI surface,
- electromagnetic waves emanating from vibrating charged strings.

Suppose, for example that the particle temperature is increased until an equilibrium temperature is reached, at which the particle radiates energy in the form of a Planck spectrum. For this case, the temperature can be calculated with the Stefan-Boltzmann law:

$$Temp = \sqrt[4]{\frac{\frac{dm}{dt} \cdot c^2}{\sigma \cdot \Phi_{rad}}} ,$$

where σ is Stefan-Boltzmann's constant:

$$\sigma = 5{,}670 \cdot 10^{-8} \left[\frac{W}{m^2 \cdot s \cdot k^4} \right]$$

The radiation from an electron

We will estimate the equilibrium radiation from the electron over a spherical shell, or a surface with the circumference equal to the Compton wavelength. The radiating surface becomes:

$$\Phi_{rad} = 4\pi \cdot \left(\frac{\lambda_c}{2\pi}\right)^2 = \frac{R_{cl}^2}{\pi \cdot \alpha^2}$$

The energy absorption is related to the electron mass:

$$\frac{dM_e}{dt} = H_\circ M_e$$

The radiation temperature becomes:

$$Temp = \sqrt[4]{\frac{\pi \cdot H_\circ M_e \cdot c^2 \cdot \alpha^2}{\sigma \cdot R_{cl}^2}} \approx 2.8^\circ K$$

This temperature agrees with good approximation to that of the microwave background radiation from space.

This model may also apply to radiation from spontaneously created and annihilated electron-positron pairs emanating from the mass of imaginary particles in the vacuum, according to the Dirac theory. It is therefore possible that the background radiation is caused by free electrons in space, as well as by interactions of virtual vacuum particles. The relation between the level of the radiation and the age of the Universe is implicit in the presence of Hubble's "constant" in the formula for the temperature. Hence, in a dynamic Universe the temperature would gradually decrease.

The radiation from stars

Applying the above described process to the sun, for example, would give the following radiation temperature:

$$Temp = \sqrt[4]{\frac{H_\circ \cdot M_\otimes \cdot c^2}{\sigma \cdot \Phi_\otimes}} = 32000^\circ K \,,$$

This temperature is higher than the 6 000°K observed at the surface of the sun, at a distance of about 700 000 km from the centre. The difference from the surface temperature may be attributed to the fact that the sun is not solid, apart from in its core body, where it has a temperature of about 15 000 000 °K. At a distance of 500 000 km from the centre the estimated temperature is 1 000 000 °K.

However, the process described here involves a much larger energy flow than that needed to explain the solar radiation. It may therefore also account for energy storage in the fusion of heavy elements in the core of the sun and for an expansion of the sun.

This way it is possible to explain why the observed neutrino flow from the sun is much lower (only about one third) of what it is expected to be if all the energy radiated by the sun were to come from fusion, the reason being that the major part of the radiated energy instead originates in the absorption of energy from the vacuum in the gravitational process.

This can also explain how a sun can be fused into one element, a neutron star. In this scenario, the light of the stars appears as the visible indicator of gravitational sinks in dynamic vacuum space.

The expansion of the planets

Everyone who has studied global maps of our Earth has probably been struck by the pattern of continents and islands in the oceans, with their contours everywhere seeming to fit into each other as if they were part of a surface which has been cracking up due to a large scale expansion of the globe.

Considering the above described scenario, it seems indeed possible that the planets and the stars, and maybe even the particles and atoms, are expanding, which might be due to absorption of energy from the vacuum in the gravitational process.

This effect has been suggested by many scientists, although the theoretical framework for an explanation has not been present so far. However, the mechanism suggested here for the gravitational interaction may also explain planetary expansion, and perhaps also the accumulation of energy in the core of the Earth, manifesting itself in the well known effects of earthquakes and volcanic activities (*cf.* Myers, 1997).

Acknowledgment

The author wishes to express his gratitude to Dr. Franco Selleri for the invitation to present this material at the conference on "Relativistic Physics and Some of its Applications."

References

Arp, Halton, 1993. Fitting theory to observation—from stars to cosmology, in: *Progress in New Cosmologies*, ed. H. Arp *et al.*, (New York and London: Plenum Press).
Broberg H., 1993. On the kinetic origin of mass, *Apeiron* Vol. 1, Nr. 15.
Jaakkola, Toivo, 1996. Action at a distance and local action in gravitation, *Apeiron* Vol. 3, No. 3-4.
Myers, Lawrence, 1997. An unheralded giant of geology, *Apeiron* Vol. 4, Nr. 4.

Internal Structures of Electrons and Photons and some Consequences in Relativistic Physics

W. A. Hofer
Institut für Allgemeine Physik
Technische Universität Wien, A-1040 Vienna, Austria

The theoretical foundations of quantum mechanics and de Broglie-Bohm mechanics are analyzed and it is shown that both theories employ a formal approach to microphysics. By using a realistic approach it can be established that the internal structures of particles comply with a wave-equation. Including external potentials yields the Schrödinger equation, which, in this context, is arbitrary due to internal energy components. The uncertainty relations are an expression of this, fundamental, arbitrariness. Electrons and photons can be described by an identical formalism, providing formulations equivalent to the Maxwell equations. Electrostatic interactions justify the initial assumption of electron-wave stability: the stability of electron waves can be referred to vanishing intrinsic fields of interaction. Aspect's experimental proof of non-locality is rejected, because these measurements imply a violation of the uncertainty relations. The theory finally points out some fundamental difficulties for a fully covariant formulation of quantum electrodynamics, which seem to be related to the existing infinity problems in this field.
PACS numbers: 03. 65. Bz, 03. 70, 03. 75, 14. 60. Cd
Keywords: electrons, photons, EPR paradox, quantum electrodynamics

1. Introduction

It is commonly agreed upon that currently two fundamental frameworks provide theoretical bases for a treatment of microphysical processes. The standard quantum theory (QM) is accepted by the majority of the physical community to yield the to date most appropriate account of phenomena in this range. It can be seen as a consequence of the Copenhagen interpretation, and its essential features are the following: It (i) is probabilistic [1], (ii) does not treat fundamental processes [2],[3], (iii) is restricted by a limit of description (uncertainty relations [4]), (iv) essentially non-local [5],[6], and (v) related to classical mechanics [7].

Open Questions in Relativistic Physics
Edited by Franco Selleri (Apeiron, Montreal, 1998)

The drawbacks and logical inconsistencies of the theory have been attacked by many authors, most notably by Einstein [8], Schrödinger [9], de Broglie [10], Bohm [11], and Bell [12]. Its application to quantum electrodynamics (QED) provides an infinity problem, which has not yet been solved in any satisfying way [13], for this reason QED has been critized by Dirac as essentially inadequate [14].

These shortcomings have soon initiated the quest for an alternative theory, the most promising result of this quest being the de Broglie-Bohm approach to microphysics [15],[16],[3]. While the interpretations of these two authors differ in detail, both approaches are centered around the notion of "hidden variables" [3]. The de Broglie-Bohm mechanics of quantum phenomena (DBQM) is still highly controversial and rejected by most physicists, its main features are: It (i) is deterministic [3], (ii) has no limit of description (trajectories), (iii) is highly non-local [17], (iv) based on kinetic concepts, and (v) ascribes a double meaning to the wave function (causal origin of the quantum potentials and statistical measure if all initial conditions are considered) [18].

Analyzing the theoretical basis of these two theories, it is found that QM and DBQM both rely on what could be called a *formal approach* to microphysics. The Schrödinger equation [19]

$$\left(-\frac{\hbar^2}{2m}\Delta + V\right)\psi = i\hbar\frac{\partial\psi}{\partial t}$$ (1)

is accepted in both frameworks as a fundamental axiom; DBQM derives the trajectories of particles from a quantum potential subsequent to an interpretation of this equation, while the framework of standard QM is constructed by employing, in addition, the Heisenberg commutation relations [20]:

$$[X_i, P_j] = i\hbar\,\delta_j$$ (2)

Analogous forms of the commutation relations are used for second quantization constitutional for the treatment of electromagnetic fields in the extended framework of QED [13].

2. Realistic approach

Although these existing approaches have been highly successful in view of their predictive power, they do not treat—or only superficially—the question of the *physical* justification of their fundamental axioms. This gap can be bridged by reconstructing the framework of microphysics from a physical basis, and the most promising approach seems to be basing it on the experimentally observed wave features of single particles. For photons this point is trivial in view of wave optics, while for electrons diffraction experiments by Davisson and Germer [21] have established these wave features beyond doubt.

Using de Broglie's formulation of wave function properties [22] and adapting it for a non-relativistic frame of reference we get (particle velocity $|\mathbf{u}| \ll c_0$):

$$\psi(x_\mu) = \psi_0 \exp - i\,(k^\mu\, x_\mu) \mapsto \psi(\mathbf{r}, t) = \psi_0 \sin\,(\mathbf{k}\,\mathbf{r} - \omega t), \quad \psi_0 \in R \qquad (3)$$

The immediate consequence of the approach is a periodic and local density of mass within the region occupied by the particle and which is described by:

$$\rho(\mathbf{r}, t) = \rho_0 \sin^2 (\mathbf{k}\,\mathbf{r} - \omega t) \quad \rho_0 = C\,\psi_0^2 \qquad (4)$$

The main reason that this approach—leading to a local and realistic picture of internal structures—so far has remained unconsidered is the dispersion relation of matter waves, based on two independent statements [22],[23]. With the assumption that energy of a free particle is kinetic energy of its inertial mass m, the phase velocity of the matter wave is not equal to the mechanical velocity of the particle, an obvious contradiction with the energy principle:

$$\lambda(\mathbf{p}) = \frac{h}{|\mathbf{p}|} \quad E(\omega) = \hbar\omega \rightarrow c_{phase} = \left(\frac{|\mathbf{u}|}{2}\right)_{non-rel} \qquad (5)$$

In the case of real waves, however, the kinetic energy density is a periodic function, and the energy principle then requires the existence of an equally periodic intrinsic potential of particle propagation. Using the total energy density for a particle of finite dimensions and volume V_P then yields the result, that internal features described by a monochromatic plane wave of phase velocity c_{phase} comply with propagation of a particle with mechanical velocity $|\mathbf{u}|$:

$$\frac{1}{2}\int_{V_P} dV\,\rho(\mathbf{r}, t = 0)|\mathbf{u}|^2 = \frac{1}{2}\overline{\rho}\,V_P|\mathbf{u}|^2 = \frac{m}{2}|\mathbf{u}|^2$$

$$W_K = \frac{m}{2}|\mathbf{u}|^2 \quad W_P = \frac{m}{2}|\mathbf{u}|^2 \qquad (6)$$

$$W_T = W_K + W_P = m|\mathbf{u}|^2 =: \hbar\omega$$

$$c_{phase} = \lambda\nu = |\mathbf{u}|$$

where W_K denotes kinetic and W_P potential energy of particle propagation, the total energy is given by W_T. The significance of intrinsic field components will be shown presently, in this case the wave function and also the local density of mass comply with a wave equation:

$$\Delta\psi(\mathbf{r}, t) - \frac{1}{|\mathbf{u}|^2}\frac{\partial^2\psi(\mathbf{r}, t)}{\partial^2 t} = 0$$

$$\Delta\rho(\mathbf{r}, t) - \frac{1}{|\mathbf{u}|^2}\frac{\partial^2\rho(\mathbf{r}, t)}{\partial^2 t} = 0 \qquad (7)$$

3.   Quantum theory in a realistic approach

For a periodic wave function $\psi(\mathbf{r}, t)$ the kinetic component of particle energy can be expressed in terms of the Laplace operator acting on ψ, its value given by:

$$|\mathbf{u}|^2 \Delta \psi = \frac{\partial^2 \psi}{\partial t^2} = -\omega^2 \psi, \quad W_K \psi = \frac{m}{2}|\mathbf{u}|^2 \psi = \frac{\hbar}{2}\omega\psi$$

$$\left(W_k + \frac{\hbar^2}{2m}\Delta\right)\psi = 0 \rightarrow E_k = -\frac{\hbar^2}{2m}\Delta \tag{8}$$

If, furthermore, the total energy W_T of the particle is equal to kinetic energy and a local potential $V(\mathbf{r})$ the development of the wave function is described by a time-independent Schrödinger equation [19]:

$$\left[-\frac{\hbar^2}{2m}\Delta + V(\mathbf{r})\right]\psi(\mathbf{r},t) = W_T\psi(\mathbf{r},t) \tag{9}$$

Its interpretation in a realistic context is not trivial, though. If the volume of a particle is finite, then wavelengths and frequencies become *intrinsic* variables of motion, which implies, due to energy conservation, that the wave function itself is a measure for the potential at an arbitrary location $\mathbf{r}$. But in this case the wave function must have physical relevance, and the EPR dilemma [8] in this case will be fully confirmed. The result seems to support Einstein's view, that quantum theory in this case cannot be complete. The only alternative, which leaves QM intact, is the assumption of zero particle volume: in the context of electrodynamics this assumption leads to the, equally awkward, result of infinite particle energy.

The second major result, that the Schrödinger equation in this case is not an exact, but an essentially arbitrary equation, where the arbitrariness is described by the uncertainty relations, can be derived by transforming the equation into a moving reference frame $\mathbf{r}' = \mathbf{r} - \mathbf{u}t$:

$$\left[-\frac{\hbar^2}{2m}\Delta' + V(\mathbf{r}' + \mathbf{u}t)\right]\psi(\mathbf{r}') = W_T\psi(\mathbf{r}') \tag{10}$$

Since the time variable in this case is undefined we may consider the limits of the pertaining k-values of plane wave solutions in one dimension. Resulting from a variation of the potential $V(\mathbf{r}', t)$ they are given by:

$$V_1 = V(\mathbf{r}' + \mathbf{u}t_1) = V(\mathbf{r}') + \Delta V(t)$$
$$V_0 = V(\mathbf{r}' + \mathbf{u}t_0) = V(\mathbf{r}') - \Delta V(t) \tag{11}$$
$$\hbar\Delta k(t) = \frac{m\Delta V(t)}{\hbar k}$$

If the variation results from the intrinsic potentials, which are neglected in QM, then uncertainty of the applied potentials has as its minimum the amplitude ϕ_0, given by:

$$\phi_0 = m\,\mathbf{u}^2 = \Delta V \tag{12}$$

Together with the uncertainty $\Delta x = \lambda/2 = 2\,(x(\phi_0) - x(0))$ for the location, which is defined by the distance between two potential maxima (the value can deviate in two directions), we get for the product:

$$\Delta x\,\Delta k \geq k\,\lambda/2 \mapsto \Delta x\,\Delta p \geq h/2 \tag{13}$$

Apart from a factor of 2π, which can be seen as a correction arising from the Fourier transforms inherent to QM, the relation is equal to the canonical formulation in quantum theory [24]:

$$\Delta X_i\,\Delta P_i \geq \hbar/2 \tag{14}$$

The uncertainty therefore does not result from wave-features of particles, as Heisenberg's initial interpretation suggested [4], nor is it an expression of a fundamental principle, as the Copenhagen interpretation would have it [2]. It is, on the contrary, an *error margin* resulting from the fundamental assumption in QM, *i.e.* the interpretation of particles as inertial mass aggregations. Due to the somewhat wider frame of reference, this result also seems to settle the long-standing controversy between the *empirical* and the *axiomatic* interpretation of this important relation: although within the principles of QM the relation is an *axiom*, it is nonetheless a *result* of fundamental theoretical shortcomings and not a physical principle.

Since this feature is inherent to any evaluation of the Schrödinger equation, it also provides a reason for the statistical ensembles pertaining to its solutions: the structure of the ensembles treated in QM is based on a fundamental arbitrariness of Schrödinger's equation, which has to be considered in all calculations of measurement processes [25].

4. Classical electrodynamics

The fundamental relations of classical electrodynamics (ED) are commonly interpreted as axioms which cannot be derived from physical principles [26]. However, within the realistic approach they are an expression of the intrinsic features of particle propagation and related to the proposed intrinsic potentials, as can be shown as follows. We define the longitudinal and intrinsic momentum $\mathbf{p}$ of a particle by:

$$\mathbf{p}\,(\mathbf{r},t) := \rho\,(\mathbf{r},t)\,\mathbf{u} \qquad \mathbf{u} = \text{constant} \tag{15}$$

From the wave equation and the continuity equation for $\mathbf{p}$:

$$\nabla^2\mathbf{p} - \frac{1}{|\mathbf{u}|^2}\frac{\partial^2\mathbf{p}}{\partial t^2} = 0 \qquad \nabla\mathbf{p} + \frac{\partial\rho}{\partial t} = 0 \tag{16}$$

the following expressions can be derived:

$$\nabla^2\mathbf{p} = -\nabla\frac{\partial\mathbf{p}}{\partial t} - \nabla\times(\nabla\times\mathbf{p})$$

$$\frac{1}{|\mathbf{u}|^2}\frac{\partial^2\mathbf{p}}{\partial t^2} = \frac{\partial}{\partial t}\left(\frac{\overline{\sigma}}{|\mathbf{u}|^2}\frac{1}{\overline{\sigma}}\frac{\partial\mathbf{p}}{\partial t}\right) \tag{17}$$

where $\bar{\sigma}$ shall be a dimensional constant to guarantee compatibility with electromagnetic units. With the definition of electromagnetic **E** and **B** fields by:

$$\mathbf{E}(\mathbf{r},t) := -\nabla \frac{1}{\bar{\sigma}} \phi(\mathbf{r},t) + \frac{1}{\bar{\sigma}} \frac{\partial \mathbf{p}}{\partial t}$$

$$\mathbf{B}(\mathbf{r},t) := -\frac{1}{\bar{\sigma}} \nabla \times \mathbf{p} \tag{18}$$

where ϕ $(\mathbf{r},t)$ shall denote some electromagnetic potential, we derive the following expression:

$$\frac{\partial}{\partial} \nabla \left(\frac{1}{|\mathbf{u}|^2} \phi + \rho \right) + \bar{\sigma} \left(\frac{1}{|\mathbf{u}|^2} \frac{\partial \mathbf{E}}{\partial t} - \nabla \times \mathbf{B} \right) = 0 \tag{19}$$

Since the total intrinsic energy density $\phi_T = \rho \, |\mathbf{u}|^2 + \phi$ is, for a single particle in constant velocity, an intrinsic constant, the following equation is valid for every micro volume:

$$\frac{1}{|\mathbf{u}|^2} \frac{\partial \mathbf{E}}{\partial t} = \nabla \times \mathbf{B} \tag{20}$$

From the definition of electromagnetic fields we get, in addition, the following expression:

$$\nabla \times \mathbf{E} = -\frac{\partial \mathbf{B}}{\partial t} \tag{21}$$

which equals one of Maxwell's equations. Including the source equations of ED and the definitions of current density **J** as well as the magnetic **H** field:

$$\nabla(\varepsilon \mathbf{E}) = \nabla \mathbf{D} = \sigma \quad \nabla \mathbf{B} = 0 \tag{22}$$

$$\nabla \mathbf{J}(\mathbf{r},t) := -\dot{\sigma} \quad \mathbf{H} := \mu^{-1} \mathbf{B} \tag{23}$$

where μ and ε are the permeability and the dielectric constant, which are supposed invariant in the micro volume, and computing the source of (20) and the time derivative of (22), we get for the sum:

$$\nabla \left(-\frac{1}{|\mathbf{u}|^2} \frac{\mu^{-1} \partial \mathbf{E}}{\partial t} + \mathbf{J} + \frac{\partial \mathbf{D}}{\partial t} \right) = 0 \tag{24}$$

For a constant of integration equal to zero we then obtain the inhomogeneous Maxwell equation [26]:

$$\mathbf{J} + \frac{\partial \mathbf{D}}{\partial t} = \nabla \times \mathbf{H} \tag{25}$$

The energy principle of material waves can be identified as the classical Lorentz condition. To this aim we use the continuity equation (16) and the definition of electric fields (18). For linear $\nabla \times \nabla \times \mathbf{p} = 0$ and uniform motion $\bar{\sigma}\,\mathbf{E} = 0$ the equations lead to:

$$\frac{1}{|\mathbf{u}|^2}\frac{\partial \phi}{\partial t} - \nabla \mathbf{p} = 0 \qquad (26)$$

In case of $|\mathbf{u}| = c_0$, a comparison with the classical Lorentz condition [26]:

$$\nabla \mathbf{A} + \frac{1}{c_0}\frac{\partial \phi}{\partial t} = 0 \qquad (27)$$

yields the result, that the vector potential of ED is related to the intrinsic momentum of the particle:

$$\mathbf{A} = -c_0 \mathbf{p} = \frac{1}{\alpha}\mathbf{p} \qquad (28)$$

It can thus be said that the Maxwell equations follow from intrinsic properties of particles in constant motion. But the result equally means, that electrodynamics and quantum theory *must have* the same dimensional level: both theories can be seen as a *limited account* of intrinsic particle properties.

5. Internal structures of photons and electrons

On this basis a mathematical description of internal properties of photons and electrons can be given, which comprises, apart from the kinetic and longitudinal properties, also the transversal features of electrodynamics. The two aspects of single particles are related by the proposed intrinsic potential (see Section 2), in case of photons it leads to a total intrinsic energy density described by Einstein's energy relation [27]. To derive the intrinsic properties of photons we proceed from the wave equation (16), the energy relation (28) and the equation for the change of the vector potential in ED:

$$\frac{1}{c_0}\frac{\partial \mathbf{A}}{\partial t} = -\nabla \phi - \mathbf{E} \qquad (29)$$

A wave packet shall consist of an arbitrary number N of monochromatic plane waves:

$$(30)$$

where p_i^0 is the amplitude of longitudinal momentum, and ϕ_i^0 the corresponding intrinsic potential of a component i. For a single component of this wave packet, which shall define the *photon*, we get the equation:

$$p_i^0 k_i - \frac{\omega_i}{c_0^2}\phi_i^0 = 0 \qquad (31)$$

Together with the relation $p_i^0 = \rho_{ph,i}^0 c_0$ for the momentum and the dispersion of plane waves in a vacuum $c_0 = \omega/k_i$ it leads to the following expression:

$$\phi_i^0 = \rho_{ph,i}^0 c_0^2 \frac{c_0 k_i}{\omega_i} = \rho_{ph,i}^0 c_0^2 \qquad (32)$$

The intrinsic energy of photons thus complies with Einstein's energy relation. With the expression for the potential ϕ in classical electrodynamics:

$$\phi_{em} = \frac{1}{8\pi}\left(\mathbf{E}^2 + \mathbf{B}^2\right) \tag{33}$$

the electromagnetic and transversal fields for a photon are given by:

$$
\begin{aligned}
\mathbf{E}_i &= \mathbf{e}^t c_0 \sqrt{4\pi\rho^0_{ph,i}}\ \cos(\mathbf{k}_i\mathbf{r} - \omega_i t) \\
\mathbf{B}_i &= \left(\mathbf{e}^k \times \mathbf{e}^t\right) c_0 \sqrt{4\pi\rho^0_{ph,i}}\ \cos(\mathbf{k}_i\mathbf{r} - \omega_i t)
\end{aligned}
\tag{34}
$$

The units of electromagnetic fields are in this case dynamical units, they can be in principle determined from the energy relations of electromagnetic fields and their relation to the intrinsic momentum of a photon. These electromagnetic fields are, furthermore, *causal* and not statistical variables resulting from the intrinsic potentials of single photons. This result, essentially incompatible with the current framework of QED [13], has been derived by Renninger on the basis of a gedankenexperiment [28].

For electrons the same procedure and a velocity $\mathbf{u} < c_0$ leads to similar expressions of longitudinal and transversal properties.

$$
\begin{aligned}
\mathbf{p} &= \rho^0_{el}\mathbf{u}\sin^2(\mathbf{k}\mathbf{r} - \omega t) \\
\phi_{em} &= \rho^0_{el}\mathbf{u}\cos^2(\mathbf{k}\mathbf{r} - \omega t) \\
\phi^0_{el} &= \phi_{em} + \rho_{el}|\mathbf{u}|^2 = \rho^0_{el}|\mathbf{u}|^2 = \text{const} \\
\mathbf{E}_{el} &= \mathbf{e}^t|\mathbf{u}|\sqrt{4\pi\rho^0_{el}}\ \cos(\mathbf{k}\mathbf{r} - \omega t)
\end{aligned}
\tag{35}
$$

The intrinsic electromagnetic fields are of transversal polarization and solutions of the Maxwell equations for $|\mathbf{u}| < c_0$ (subluminal solutions). These theoretical results establish also, that electrodynamics and quantum theory are not only formally analogous, a result frequently used in interference models, but that they are complementary theories of single particle properties.

6. Spin in quantum theory

One of the most difficult concepts in quantum theory is the property of particle spin [29]. This is partly due to its abstract features, partly also to its relation with the magnetic moment, in electrodynamics a well defined vector of a defined orientation in space. To relate particle spin to the polarizations of intrinsic fields, we first have to consider the definition of the magnetic field (18) not only for the intrinsic but, in case of electrons, also for the external magnetic fields due to curvilinear motion. As can be derived from the solution for $\mathbf{u}$ in a homogeneous magnetic field $\mathbf{B} = B_0\ \mathbf{e}^z$, the definition in this case has to be modified for electrons by a factor of two, it then reads:

$$B_{el} = \frac{1}{2\bar{\sigma}}\nabla \times \mathbf{p}_{el} = \mathbf{B}_{ext} \tag{36}$$

where the subscript denotes its relation to external fields. The justification of this modification is the fact, that the intrinsic magnetic fields of electrons cannot be derived in quantum theory due to its fundamental assumptions. On this basis particle spin of photons and electrons can be deduced from the expression for energy of magnetic interactions in ED:

$$W = -\mu B \tag{37}$$

and by four assumptions: (i) The energy of electrons or photons is equal to the energy in quantum theory, (ii) the magnetic field referred to is the intrinsic (photon) or external (electron) magnetic field, (iii) the frequency ω can be interpreted as a frequency of rotation, and (iv) the magnetic moment is described by the relation in quantum theory:

$$\vec{\mu} = g_s \frac{e}{2mc_0} \mathbf{s} \tag{38}$$

For a photon the energy is the total energy, and with:

$$W_{ph} = \hbar\omega \quad \mathbf{B}_{ph} = -\frac{1}{\sigma}\nabla \times \mathbf{p} \approx \frac{2\overline{\rho}}{\sigma}\vec{\omega} \tag{39}$$

as well as the ratio $\overline{\rho}/\overline{\sigma} = m/e$, and considering, in addition, that the magnetic fields in the current framework are c_0 times the magnetic fields in classical electrodynamics, we obtain the relation:

$$W_{el} = \hbar\omega = g_{ph}\vec{\omega}\,\mathbf{s}_{ph} \tag{40}$$

For constant g_{ph} the relation can only hold, if:

$$\mathbf{s}_{ph} \quad || \quad \omega \mapsto \mathbf{s}_{ph}\,\omega = s_{ph}\,\omega \quad g_{ph}\,s_{ph} = \hbar \tag{41}$$

The energy relation of electrodynamics is only consistent with the energy of photons in quantum theory if they possess an intrinsic spin of $\hbar$, a gyromagnetic ratio of 1, and if the direction of spin-polarization is equal to the direction of the intrinsic magnetic fields.

The same calculation for electrons, where the energy and the magnetic fields are given by (intrinsic potentials of electrons remain unconsidered in QM):

$$W_{el} = \frac{1}{2}\hbar\omega \quad \mathbf{B}_{el} = -\frac{1}{2\sigma}\nabla \times \mathbf{p} \approx \frac{\overline{\rho}}{\sigma}\vec{\omega} \tag{42}$$

leads to the same result for the product of s_{el} and g_{el}, namely:

$$\mathbf{s}_{el} \quad || \quad \omega \mapsto \mathbf{s}_{el}\,\omega = s_{el}\,\omega \quad g_{el}\,s_{el} = \hbar \tag{43}$$

The difference to photons ($g_{el} = 2$ and $s_{el} = \hbar/2$) seems to originate from the assumptions of Goudsmit and Uhlenbeck [29], that the energy splitting due to the electron spin in hydrogen atoms should be symmetric to the original state. This general multiplicity of two for the spin states allows only for the given solution. The direction of spin polarizations is also for electrons equal to the polarization of intrinsic fields.

7. Bell's inequalities and Aspect's measurements

We may reconsider Aspect's experimental proof of non-locality [30], based on Bell's inequalities [5] from the viewpoint of the derived quality of photon spin. Adopting the view on spin-conservation of quantum theory, the measurement must account for the spin correlations of two photons with an arbitrary angle ϑ of polarization in the states +1 and −1, respectively.

Since the intrinsic magnetic fields oscillate with mbox$\mathbf{B} = \mathbf{B}_0 \cos (\mathbf{k}\, \mathbf{r} - \omega\, t)$, the spin variable $s(x)$ *cannot* remain constant, but must equally oscillate from −1 to +1. And in this case a valid measurement of spin polarizations $s(x)$ of the two particles is only possible, if the variable is measured in the interval $\Delta t < \tau/2$. The local resolution of the measuring device must therefore be equal to $\Delta x < \lambda/2$.

But as demonstrated in the deduction of the uncertainty relations, this interval is lower than the local uncertainty in quantum theory. A valid measurement of spin correlations thus violates the uncertainty relation and, for this reason, *cannot* be interpreted within the limits of quantum theory, which means, evidently, that it cannot be interpreted with Bell's inequalities. From this viewpoint Aspect's measurements are unsuitable for a proof of non-locality, they can therefore not contradict the current framework, which is essentially a local one.

8. Electron photon interactions

Defining the Lagrangian density by total energy density of an electron in an external field ϕ, including total energy of a presumed photon, we may state:

$$\mathcal{L} := T - V = \rho_{el}^0 \dot{x}_i^2 + \rho_{ph}^0 c^2 - \sigma_{el}^0 \phi \qquad (44)$$

$$\rho_{el}^0 = \text{constant} \quad \dot{x}_i^2 = \dot{x}_i^2(\dot{x}_i)$$
$$\rho_{ph}^0 = \rho_{ph}^0(\dot{x}_i) \quad \phi = \phi(x_i)$$

where ρ_{el}^0 is the amplitude of electron density, ρ_{ph}^0 the amplitude of photon density, and σ_0 the amplitude of electron charge. An infinitesimal variation with fixed endpoints yields the result:

$$\int_{t_1}^{t_2} dt \int d^3x \left\{ \sigma_{el}^0 \frac{\partial \phi}{\partial x_i} + \frac{d}{dt}\left(\rho_{el}^0 \frac{\partial x_i^2}{\partial \dot{x}_i} + c^2 \frac{\partial \rho_{ph}^0}{\partial \dot{x}_i} \right) \right\} \delta x_i = 0 \qquad (45)$$

Therefore the following expression is valid:

$$\rho_{ph}^0 c^2 = -\int dt \underbrace{\sigma_{el}^0 \dot{x}_i \nabla \phi}_{-j_0 \mathbf{E} = -\partial V_{em}/\partial t} - \rho_{el}^0 \dot{x}_i^2 = +V_{em} - \rho_{el}^0 \dot{x}_i^2 \qquad (46)$$

where we have used a relation of classical electrodynamics. From (46) and (45) it can be derived that the partial differential of $\mathcal{L}$ may be written:

$$\frac{\partial \mathcal{L}}{\partial \dot{x}_i} = \frac{\partial V_{em}}{\partial \dot{x}_i} \tag{47}$$

Since the dependency of photon density on the velocity of the electron is unknown, we may consider a small variation of electron velocity and evaluate, with 46), the Hamiltonian of the system in first order approximation of a Taylor series. In this case we get:

$$H = \frac{\partial \mathcal{L}}{\partial \dot{x}_i} \dot{x}_i - \mathcal{L} \approx V_{em} - \mathcal{L} = \sigma_{el}^0 \phi \tag{48}$$

The result seems paradoxical in view of kinetic energy of the moving electron, which does not enter into the Hamiltonian. Assuming, that an inertial particle is accelerated in an external field, its energy density after interaction with this field would only be altered according to its alteration of location. The contradiction with the energy principle is only superficial, though. Since the particle will have been accelerated, its energy density *must* be changed. If this change does not affect its Hamiltonian, the only possible conclusion is, that photon energy has equally been changed, and that the energy acquired by acceleration has simultaneously been emitted by photon emission. The initial system was therefore over-determined, and the simultaneous existence of an external field *and* interaction photons is no physical solution to the interaction problem.

The process of electron acceleration then has to be interpreted as a process of simultaneous photon emission: the acquired kinetic energy is balanced by photon radiation. A different way to describe the same result, would be saying that electrostatic interactions are accomplished by an exchange of photons: the potential of electrostatic fields then is not so much a function of location than a history of interactions. This can be shown by calculating the Hamiltonian of electron photon interaction:

$$\begin{aligned} H_0 &= \rho_{el}^0 \dot{x}_i^2 + \sigma_{el}^0 \qquad H = \sigma_{el}^0 \phi \\ H_w &= H - H_0 = -\rho_{el}^0 \dot{x}_i^2 = \rho_{ph}^0 c_0^2 \end{aligned} \tag{49}$$

But if electrostatic interactions can be referred to an exchange of photons, and if these interactions apply to accelerated electrons, then an electron in constant motion does not possess an intrinsic energy component due to its electric charge: electrons in constant motion are therefore stable structures.

It is evident that neither emission, nor absorption of photons at this stage does show discrete energy levels: the alteration of velocity can be chosen arbitrarily small. The interesting question seems to be, how quantization fits into this picture of continuous processes. To understand this paradoxon, we consider the transfer of energy due to an infinitesimal interaction process. The differential of energy is given by:

$$dW = A \cdot dt \rho_{ph}^0 c^2 \tag{50}$$

A denotes the cross section of interaction. Setting d*t* equal to one period τ, and considering, that the energy transfer during this interval will be αW, we get for the transfer rate:

$$\alpha\frac{dW}{dt}=\frac{A\lambda\rho_{ph}^{0}c^{2}}{dt}=\alpha\cdot\frac{V_{ph}\rho_{ph}^{0}c^{2}}{\tau}=\alpha\frac{h\nu}{\tau} \tag{51}$$

Since total energy density of the particle as well as the photon remains constant, the statement is generally valid. And therefore the transfer rate in interaction processes will be:

$$\frac{dW}{dt}=\frac{d}{dt}m\mathbf{u}^{2}=\hbar\cdot\frac{\omega}{\tau}=h\cdot\nu^{2} \tag{52}$$

The term energy quantum is, from this point of view, not quite appropriate. The total value of energy transferred depends, in this context, on the region subject to interaction processes, and equally on the duration of the emission process. The view taken in quantum theory is therefore only a good approximation: a thorougher concept, of which this theory in its present form is only the outline, will have to account for *every* possible variation in the interaction processes.

But even on this, limited, basis of understanding, Planck's constant, commonly considered the fundamental value of energy quantization is the fundamental constant not of energy values, but of transfer rates in dynamic processes. Constant transfer rates furthermore have the consequence, that volume and mass values of photons or electrons become irrelevant: the basic relations for energy and dispersion remain valid regardless of actual quantities. Quantization then is, in short, a *result of energy transfer* and its characteristics.

9. Consequences in relativistic physics

The framework developed has some interesting consequences in relativistic physics. So far all expressions derived are only valid in a non-relativistic inertial frame. To estimate its impact on Special Relativity we may consider the wave equation in a moving frame of reference, and estimate the result of energy measurments in the system in motion *S'* and the system at rest *S*. From the wave equation in $S' = S'(V)$:

$$\Delta'p_{x}'-\frac{1}{u_{x}'^{2}}\frac{\partial^{2}p_{x}'}{\partial t'^{2}}=0 \quad p_{x}'=\rho'u_{x}' \tag{53}$$

With the standard Lorentz transformation:

$$x'^{\mu}=\Lambda_{\nu}^{\mu}x^{\nu} \quad \Lambda_{\nu}^{\mu}=\begin{pmatrix} \gamma & -\beta\gamma & 0 & 0 \\ -\beta\gamma & \gamma & 0 & 0 \\ 0 & 0 & 1 & 0 \\ 0 & 0 & 0 & 1 \end{pmatrix} \tag{54}$$

The differentials in *S'* are given by:

$$\Delta' = \left(\frac{\partial x}{\partial x'}\right)^2 \Delta = \left(1 - \beta^2\right)\Delta$$

$$\frac{\partial^2}{\partial t'^2} = \left(\frac{\partial t}{\partial t'}\right)^2 \frac{\partial^2}{\partial t^2} = \left(1 - \beta^2\right)\frac{\partial^2}{\partial t^2}$$

(55)

Density of mass and velocity in S' are described by (transformation of velocities according to the Lorentz transformation, α presently undefined):

$$\rho' = \alpha \cdot \rho \qquad u'_x = \frac{u_x - V}{1 - u_x V / c^2}$$

(56)

Then the transformation of the wave equations yields the following wave equation for density ρ:

$$\Delta \rho - \frac{1}{u'^2_x} \frac{\partial^2 \rho}{\partial t^2} = 0$$

(57)

Energy measurements in the system S at rest must be measurements of the phase-velocity c_{ph} of the particle waves. Since, according to the transformation, the measurement in S yields:

$$c_{ph}^2 (S) = u_x'^2$$

(58)

the total potential ϕ_0, measured in S will be:

$$\frac{\phi_0(S)}{\rho_S} = u'^2_x \quad \Rightarrow \quad \phi_0(S) = \frac{\rho_S}{\rho_{S'}}\phi_0(S')$$

(59)

And if density $\rho_{S'}$ transforms according to a relativistic mass-effect with:

$$\rho_{S'} = \gamma \rho_S$$

(60)

then the total intrinsic potential measured in the system S' will be higher than the potential measured in S:

$$\phi_0 (S') = \gamma \phi_0 (S)$$

(61)

But the total intrinsic energy of a particle with volume V_P will not be affected, since the x-coordinate will be, again according to the Lorentz transformation, contracted:

$$V_P(S') = \frac{1}{\gamma} V_P(S)$$

$$E_0(S) = \phi_0(S)V_P(S) = \frac{1}{\gamma}\phi_0(S')V_P(S) =$$

$$= \phi_0(S')V_P(S') = E_0(S')$$

(62)

The interpretation of this result exhibits some interesting features of the Lorentz transformation applied to electrodynamic theory. It leaves the wave equations and their energy relations intact if, and only if energy *quantities* are evaluated, *i.e.* only in an integral evaluation of particle properties. In this case it seems therefore justified, to consider Special Relativity as a necessary adaptation of kinematic variables to the theory of electrodynamics. The same does not hold, though, for the intrinsic potentials of particle motion. Since the

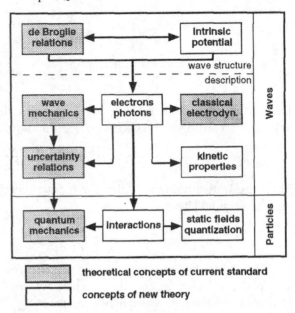

Figure 1. Logical structure of micro physics according to matter wave theory. Classical electrodynamics and wave mechanics describe intrinsic features of particles, quantization arises from interactions.

intrinsic potentials depend on the reference frame of evaluation and increase by a transformation into a moving coordinate system, the theory suggests that the infinity problems, inherent to relativistic quantum fields, may be related to the logical structure of QED.

10. Conclusion

We have shown that a new and *realistic* approach to wave properties of moving particles allows to reconstruct the framework of quantum theory and classical electrodynamics from a common basis.

As it turns out, retrospectively, the theoretical framework removes the three fundamental problems, which made a physical interpretation (as opposed to the probability-interpretation [1]) of intrinsic wave properties impossible.

- The contradiction between mechanical velocity of a particle and phase velocity of its wave is removed by the discovery of intrinsic potentials.
- The problem of intrinsic Coulomb interactions is removed, because electrostatic interactions can be referred to an exchange of photons. A particle in uniform motion is therefore stable.
- The experimentally verified non-locality of micro physical phenomena could be referred to measurements violating the fundamental principles of quantum theory: they are therefore not suitable to disprove any local and realistic framework. itemize
- From the viewpoint of electrodynamics the theory will reproduce any result the current framework provides, because the basic relations are

not changed, merely interpreted in terms of electron and photon propagation. Therefore an experiment, consistent with electrodynamics, will also be consistent with the new theory.

In quantum theory, the only statements additional to the original framework are beyond the level of experimental validity, defined by the uncertainty relations. Since quantum theory cannot, in principle, contain results beyond that level, every experiment within the framework of quantum theory must necessarily be reproduced. That applies to the results of wave mechanics, which yields the eigenvalues of physical processes, as well as to the results of operator calculations, if von Neumann's proof of the equivalence of Schrödinger's equation and the commutation relations is valid.

In quantum field theory, the same rule applies: the new theory only states, what is, in quantum theory, no part of a measurement result, therefore a contradiction cannot exist. If, on the other hand, results in quantum electrodynamics exist, which are not yet verified by the new theory, then it is rather a problem of further development, but not of a contradiction with existing measurements.

Experimentally, the new theory cannot be disproved by existing measurements, because its statements at once verify the existing formulations and extend the framework of theoretical calculations. The same does not hold, though, for the verification of existing theoretical schemes of calculation by the new framework. Since the theory extends far beyond the level of current concepts, established theoretical results may well be subject to revision. The theory requires that every valid solution of a microphysical problem can be referred to physical properties of particle waves. Essentially, this is not limiting nor diminishing the validity of current results, since it extends the framework of micro physics only to regions, which so far have remained unconsidered. The logical structure of the new framework and its relation to existing theories is displayed in Fig. 1.

Acknowledgements

Thanks are due to the *Österreichische Forschungsgemeinschaft* for generous financial support to attend the Athens conference.

References

[1]   Born M. Z. *Phys*ik, **37**, 863 (1926)
[2]   Bohr N. "Discussion with Einstein on Epistemological Problems in Atomic Physics", in *A. Einstein: Philosopher-Scientist*, Schilpp P. A. (ed.), New York (1959)
[3]   Bohm D. *Phys. Rev.*, **85**, 166; 180 (1952)
[4]   Heisenberg W. Z. *Phys*ik, **43**, 172 (1927)
[5]   Bell J. S. *Physics*, **1**, 195 (1964)
[6]   Aspect A., Dalibard J., and Roger G., *Phys. Rev. Lett.*, **49**, 1804 (1982)
[7]   George C., Prigogine I., and Rosenfeld L. *Mat. Fys. Medd. Dan. Vid. Selsk.*, **38** (12), 1 (1972)

[8]   Einstein A., Rosen N., and Podolsky B. *Phys. Rev.*, **47**, 180 (1935)

[9]   Röseberg U. in *Erwin Schrödinger's world view*, Götschl J. (ed.), Dordrecht (1992)

[10]   de Broglie L. in his foreword to [11]

[11]   Bohm D. *Causality and Chance in Modern Physics*, Princeton (1957)

[12]   Bell J. S. *Found. Phys.*, **12**, 989 (1982)

[13]   Schweber S. *QED and the Men Who Made It*, Princeton (1994)

[14]   Dirac P. A. M. *Eur. J. Phys.*, **5**, 65 (1984)

[15]   de Broglie L. *C. R. Acad. Sci. Paris*, **183**, 447 (1926); **185**, 580 (1927)

[16]   de Broglie L. *Nonlinear Wave Mechanics*, Amsterdam (1960)

[17]   Bell J. S. *Rev. Mod. Phys.*, **38**, 447 (1966)

[18]   Holland P. R. *The Quantum Theory of Motion*, Cambridge (1993)

[19]   Schrödinger E. *Ann. Physik*, **79**, 361; 489 (1926)

[20]   Heisenberg W. *Z. Physik*, **33**, 879 (1925); Born M., Heisenberg W., and Jordan P. *Z. Physik*, **35**, 357 (1926)

[21]   Davisson C. and Germer L. H. *Phys. Rev.*, **30**, 705 (1927)

[22]   de Broglie L. *Ann. Phys.*, **3**, 22 (1925)

[23]   Planck M. *Ann. Physik*, **4**, 553 (1901)

[24]   Cohen-Tannoudji C., Diu B., and Laloe F. *Quantum Mechanics*, New York (1977)

[25]   Hofer W. A. *Measurement processes in quantum physics: a new theory of measurements in terms of statistical ensembles*, to be published

[26]   Jackson J. *Classical Electrodynamics*, New York (1984)

[27]   Einstein A. *Ann Physik*, **20**, 627 (1906)

[28]   Renninger M. *Z. Physik*, **136**, 251 (1953)

[29]   Uhlenbeck G. E. and Goudsmit S. A. *Naturwiss.*, **13**, 953 (1925); *Nature*, **117**, 264 (1925)

[30]   Aspect A. PhD thesis, Universite de Paris-Sud, Centre d'Orsay (1983)

Generally Covariant Electrodynamics in Arbitrary Media

Edward Kapuscik
Institute of Physics and Informatics
Cracow Pedagogical University
30-084 Kraków, ul. Podchorazych 2, Poland
   and
Henryk Niewodniczanski
Department of Theoretical Physics
Institute of Nuclear Physics
ul. Radzikowskiego 152, 31 342 Kraków, Poland

The general covariance of the new form of Maxwell electrodynamics is established. The physical meaning of the new source term present in field equations is clarified. The Faraday induction law in arbitrary media is discussed.

It has been shown in [1] that classical Maxwell electrodynamics in arbitrary medium may be formulated as a theory of two pairs of electromagnetic fields, namely $\left[\vec{D}(\vec{x},t),\vec{H}(\vec{x},t)\right]$ and $\left[\vec{P}(\vec{x},t),\vec{M}(\vec{x},t)\right]$ which satisfy two sets of Maxwell equations

$$\operatorname{rot}\vec{D}(\vec{x},t) = -\frac{1}{c^2}\frac{\partial\vec{H}(\vec{x},t)}{\partial t} - \frac{1}{c^2}\vec{j}_M(\vec{x},t) \tag{1}$$

$$\operatorname{div}\vec{H}(\vec{x},t) = \rho_M(\vec{x},t) \tag{2}$$

$$\operatorname{rot}\vec{H}(\vec{x},t) = \frac{\partial\vec{D}(\vec{x},t)}{\partial t} + \vec{j}(\vec{x},t) \tag{3}$$

$$\operatorname{div}\vec{D}(\vec{x},t) = \rho(\vec{x},t) \tag{4}$$

and

$$\operatorname{rot}\vec{P}(\vec{x},t) = -\frac{1}{c^2}\frac{\partial\vec{M}(\vec{x},t)}{\partial t} - \frac{1}{c^2}\vec{j}_M(\vec{x},t) \tag{5}$$

$$\operatorname{div}\vec{M}(\vec{x},t) = \rho_M(\vec{x},t) \tag{6}$$

$$\text{rot}\,\vec{M}(\vec{x},t) = \frac{\partial \vec{P}(\vec{x},t)}{\partial t} + \vec{j}_P(\vec{x},t) \tag{7}$$

$$\text{div}\,\vec{P}(\vec{x},t) = \rho_P(\vec{x},t) \tag{8}$$

where $\rho(\vec{x},t)$ and $\vec{j}(\vec{x},t)$ are the usual densities of external charge and current, respectively, while $\rho_P(\vec{x},t)$ and $\vec{j}_P(\vec{x},t)$ are the usual densities of polarized charge and current, respectively. The nature of the new scalar and vector densities $\rho_M(\vec{x},t)$ and $\vec{j}_M(\vec{x},t)$ was not clarified in Ref. 1.

The electromagnetic fields $\vec{D}(\vec{x},t)$, $\vec{H}(\vec{x},t)$, $\vec{P}(\vec{x},t)$ and $\vec{M}(\vec{x},t)$ have the same meaning as in the standard Maxwell equations, and may be used to define the standard vacuum electromagnetic fields

$$\vec{E}(\vec{x},t) = \frac{1}{\varepsilon_0}\left[\vec{D}(\vec{x},t) - \vec{P}(\vec{x},t)\right] \tag{9}$$

and

$$\vec{B}(\vec{x},t) = \mu_0\left[\vec{H}(\vec{x},t) + \vec{M}(\vec{x},t)\right], \tag{10}$$

where ε_0 and μ_0 are the standard electromagnetic constants of the vacuum.

The aim of the present paper is to describe the generally covariant form of the new Maxwell equations.

Generally covariant vacuum electrodynamics

As is well known, the generally covariant form of macroscopic electrodynamics was found only for the vacuum case. In this formulation two objects are utilized: the antisymmetric tensor field $F_{\mu\nu}(x)$ and the antisymmetric tensor density $\mathscr{H}^{\mu\nu}(x)$ which satisfy the generally covariant Maxwell's equations

$$\partial_\mu F_{\nu\lambda} + \partial_\nu F_{\lambda\mu} + \partial_\lambda F_{\mu\nu} = 0 \tag{11}$$

$$\partial_\mu \mathscr{H}^{\mu\nu} = -\mathscr{I}^\nu \tag{12}$$

where $\mathscr{I}^\nu$ is the vector density of external charge and current. In the linear approximation the basic electromagnetic fields may be related by the generally covariant constitutive relation

$$\mathscr{H}^{\mu\nu}(x) = \eta^{\mu\nu\lambda\rho}(x)F_{\lambda\rho}(x) \tag{13}$$

where the tensor density $\eta^{\mu\nu\lambda\rho}(x)$ describes the electromagnetic properties of the medium. Due to several symmetry properties, the tensor density $\eta^{\mu\nu\lambda\rho}$ contains only 20 independent components. These components in the case of the vacuum are usually expressed by the metric tensor $g_{\mu\nu}(x)$. As a result the relation (13) takes the form

$$\mathcal{H}^{\mu\nu}(x) = \mu_o \sqrt{-g}\, g^{\mu\lambda}(x) g^{\nu\rho}(x) F_{\lambda\rho}(x) \tag{14}$$

or conversely

$$F_{\mu\nu}(x) = \frac{1}{\mu_o \sqrt{-g}}\, g_{\mu\lambda} g_{\nu\rho} \mathcal{H}^{\lambda\rho}(x) \tag{14a}$$

where as usual

$$g = \det\left(g_{\mu\nu}\right). \tag{14b}$$

Generally covariant new Maxwell equations

Our result rests on the fact that the left hand sides of the new Maxwell equations present in the set (1) – (8) have exactly the same form as in the standard vacuum Maxwell equations. From this it follows that the only possible version of their generally covariant form is

$$\partial_\mu F_{\nu\lambda}^{(D,H)} + \partial_\nu F_{\lambda\mu}^{(D,H)} + \partial_\lambda F_{\mu\nu}^{(D,H)} = j_{\mu\nu\lambda} \tag{16}$$

$$\partial_\mu H_{(D,H)}^{\mu\nu} = -\mathcal{q}^\nu \tag{17}$$

$$\partial_\mu F_{\nu\lambda}^{(P,M)} + \partial_\nu F_{\lambda\mu}^{(P,M)} + \partial_\lambda F_{\mu\nu}^{(P,M)} = j_{\mu\nu\lambda} \tag{18}$$

$$\partial_\mu \mathcal{H}_{(P,M)}^{\mu\nu} = -\mathcal{q}_P^\nu \tag{19}$$

where the corresponding tensor fields and tensor densities are constructed form the pairs of fields (D,H) and (P,M) according to the content in parenthesis. Clearly, each tensor field is related to the corresponding tensor density by the same relations as in (14) and (14a).

For linear media the polarized charge and current densities are linearly related to the external charges and currents. In our case this means that all the polarization and magnetization properties of such media are described by the relation

$$\mathcal{q}_P^\mu(x) = \varepsilon_\nu^\mu(x) \mathcal{q}^\nu(x) \tag{20}$$

where $\varepsilon_\nu^\mu(x)$ is the polarization and magnetization mixed tensor of the medium. It contains 16 independent components. The remaining 4 functions allowed by the general form of the material tensor density $\eta^{\mu\nu\lambda\rho}$ are provided by the 4 independent components of the totally antisymmetric tensor $j_{\mu\nu\lambda}(x)$ formed from $\rho_M(x)$ and $\bar{j}_M(x)$. It is erroneous to try to relate this antisymmetric tensor to the external current like in (19) because this will introduce an additional 16 new functions. It is also erroneous to treat this tensor as any kind of current, because it is not a tensor density. It is therefore impossible to obtain global covariant quantities from it by the process of integration. The only physically correct interpretation is to treat the source terms in eqs. (15) and (17) as terms describing the influence of the medium on the Faraday induction law. Such terms are absent in the case of the vacuum.

Electromagnetic potentials

The relations between electromagnetic tensors $F_{\mu\nu}^{(D,H)}(x)$ and $F_{\mu\nu}^{(P,M)}(x)$ and tensor densities $\mathscr{H}_{(D,H)}^{\mu\nu}$ and $\mathscr{H}_{(P,M)}^{\mu\nu}$ of the type (14)–(14a) create serious problems in the case when the metric tensor $g_{\mu\nu}(x)$ is treated as the gravitational field. The problem is in the choice which electromagnetic field is the primary field and which one is the composite field composed from the electromagnetic and gravitational fields. Both choices are unsatisfactory because both basic electromagnetic fields should be equally fundamental as the gravitational field. The solution of this problem lies in the introduction of the electromagnetic potentials. Due to the fact that in our case all Maxwell equations are inhomogeneous, we must introduce more potentials than usual. Taking into account that only part of the electromagnetic fields are tensors and the others are tensor densities, we must introduce the customary vector potentials $A_{\mu}^{(D,H)}(x)$ and $A_{\mu}^{(P,M)}(x)$ and new potentials $\mathscr{A}_{(D,H)}^{\mu\nu\lambda}(x)$ and $\mathscr{A}_{(P,M)}^{\mu\nu\lambda}$, which are antisymmetric tensor densities of third rank. Treating these quantities as primary electromagnetic fields we arrive to the following representations of the customary electromagnetic tensor and tensor density fields

$$F_{\mu\nu} = \partial_{\mu}A_{\nu} - \partial_{\nu}A_{\mu} + \left(-g\right)^{-\frac{1}{2}} g_{\mu\lambda}g_{\nu\rho}\partial_{\sigma}\mathscr{A}^{\sigma\lambda\rho} \tag{20}$$

$$\mathscr{H}^{\mu\nu} = \partial_{\lambda}\mathscr{A}^{\lambda\mu\nu} + \left(-g\right)^{-\frac{1}{2}} g^{\mu\lambda}g^{\nu\rho}\left(\partial_{\lambda}A_{\rho} - \partial_{\rho}A_{\lambda}\right) \tag{21}$$

where we omitted the corresponding sub- and superscripts because the relations of the electromagnetic fields to the corresponding potentials are the same in the case of fields of the types (D,H) and (P,M).

The old and new potentials are not unique. They undergo the following gauge transformations

$$A_{\mu} \rightarrow A_{\mu} + \partial_{\mu}\Phi \tag{22}$$

$$\mathscr{A}^{\lambda\mu\nu} \rightarrow \mathscr{A}^{\lambda\mu\nu} + \partial_{\sigma}\mathscr{A}^{\sigma\lambda\mu\nu} \tag{23}$$

where $\Phi(x)$ and $\mathscr{A}^{\sigma\lambda\mu\nu}(x)$ are the scalar and tensor density gauging quantities. Here $\mathscr{A}^{\sigma\lambda\mu\nu}(x)$ is a totally antisymmetric tensor density of the fourth rank.

The basic electromagnetic fields $A_{\mu}(x)$ and $\mathscr{A}^{\lambda\mu\nu}(x)$ satisfy the following field equations

$$\partial_{\mu}\left[\left(-g\right)^{-\frac{1}{2}} g^{\mu\nu}g^{\nu\rho}\left(\partial_{\lambda}A_{\rho} - \partial_{\rho}A_{\lambda}\right)\right] = -\mathscr{q}^{\nu} \tag{24}$$

$$\partial_{\lambda}\left[\left(-g\right)^{-\frac{1}{2}} g_{\mu\alpha}g_{\nu\beta}\partial_{\sigma}\mathscr{A}^{\sigma\alpha\beta}\right] + \text{cycle. in }\left(\lambda,\mu,\nu\right) = j_{\lambda\mu\nu}. \tag{25}$$

These equations show that the basic and independent electromagnetic fields $A_\mu(x)$ and $\mathcal{A}^{\mu\nu\lambda}(x)$ propagate in spacetime only as aggregates of composite fields formed together with the gravitational field. This determines a specific interaction between gravity and electromagnetism.

As a final remark we want to stress that in the approach presented here we use all mathematical quantities provided by the mathematics of arbitrary four-dimensional manifolds without using the notion of covariant derivative [2]. This is an advantage because any use of the covariant derivative introduces additional interactions of electromagnetism with gravity.

References

[1]   E. Kapuscik; *Comm. JINR* E2-91-272, Dubna, 1991.
[2]   E. Schrödinger; *Spacetime Structure* (Cambridge University Press, 1950).

On Weyl's Extension of the Relativity Principle as a Tool to Unify Fundamental Interactions

Marek Pawlowski[*****]
Soltan Institute for Nuclear Studies
Warsaw, Poland

The principle of democracy between reference systems which was formulated by Galileo, modified by the special relativity theory and generalised by general relativity could be extended and applied also to reference scales. Basing on this Weyl's idea we review a model that unifies fundamental interactions of elementary particles described by the Standard Model and gravity. Local conformal symmetry is a pivotal feature of this model. The most important phenomenological consequence of this approach is the absence of the dynamical scalar field in the predicted particle spectrum.

1.   Introduction

The principle of democracy of description languages in physics is one of the most fruitful principles in science. Its trivial part says that different languages can be used to describe physical reality. The deep part of this principle says that "good" description languages belonging to the same class are not only equivalent but they also express general physical rules in formally identical way. This principle was one of the basis of classical mechanics, it led to special relativity and found its geometrized version in general relativity. The next step on this way was originated by Weyl [1]. He attempted to extend the democracy originally ascribed to the class of reference systems. In his opinion also the scale that is used to measure dimensional quantities could be considered as something subject to the principle of democracy. Weyl's attempts were criticised and were finally abandoned by the author but they originated further investigations in many fields of physics where conformal symmetry could be identified and the methods based on this notion could be applied.

*****   Supported by the Committee for Scientific Researches grant nb. 603/P03/96.
Open Questions in Relativistic Physics
Edited by Franco Selleri (Apeiron, Montreal, 1998)   247

In this talk I would like to show that the original Weyl's idea may be realised in a sense leading to a unification of gravity and other interactions. The essential point in the reasoning is the notion of conformal symmetry and its physical interpretation. Thus in the first part (Section 2) I will discuss shortly this subject. Then (in Section 3) I shall review a model of electro-weak, strong and gravitational interactions which was proposed and developed by R. Raczka in collaboration with the present author. Local conformal symmetry is the essential feature of this model.

2.   Local conformal symmetry

Local conformal transformation (LCT) of a metric space $M, \hat{g}$ is a "stretching" by a factor Ω of all lengths measured with the metric $\hat{g}$:

$$LCT \ni \quad c:(M,\hat{g}) \to (M,\hat{g}^c) = (M,\Omega\hat{g}). \tag{1}$$

where the conformal factor Ω depends only on the location of the objects in question—Ω is a scalar field on M, positive and smooth.

Local conformal transformations form the Local Conformal Group (LCG). It is, in a sense, a generalisation of the 15 parameter conformal group of angle conserving transformations of a flat four-dimensional space-time.

Generic field defined on M—a purely mathematical object—is in principle independent on the metric structure defined on M. Consequently LCT need not to affect a generic field on M.

A physical, dynamical field on the space-time M is defined by dynamical relations it has to fulfil. The relations are usually given in the form of equations of motion or a variational principle. Dynamical relations involve the metric and this is the point where LCT can affect a physical field.

Let us denote symbolically by

$$DR\big[\{\Lambda\},\hat{g}\big] \tag{2}$$

the dynamical relations that define a set of physical fields Λ.

It sometimes happens that dynamical relations are invariant with respect to LCT of metric and simultaneous redefinition of physical fields

$$\{\Lambda\} \to \{\Lambda^c(\Lambda,\Omega)\}. \tag{3}$$

where, because of group properties of LCT, the inverse relation is also defined:

$$\Lambda = \Lambda(\Lambda^c,\Omega).$$

The invariance of dynamical relations means that

$$DR\big[\{\Lambda\},\hat{g}\big] = DR\Big[\big\{\Lambda(\Lambda^c,\Omega)\big\},\Omega^{-1}\hat{g}^c\Big] \equiv DR^c\big[\{\Lambda^c\},\Omega,\hat{g}^c\big] = DR\big[\{\Lambda^c\},\hat{g}^c\big] \tag{5}$$

If there exists such field transformation (3) that leads to (5) the transformation (3) is called local conformal transformation of the field Λ. Such transformations can be defined for example for the massless Dirac system and the Maxwell and Yang-Miels systems in four dimensions. For those systems the appropriate conformal transformations for the fields (3) take the simple multiplicative form

$$\Lambda \to \Lambda^c(\Lambda,\Omega) = \Omega^s \Lambda \qquad (6)$$

where s is a real number—the conformal weight of the field of a given kind.

It can be easily shown that there is no scalar field transformation of the form (6) leading to the conformal invariance of massless Laplace equation. This situation will change if we add the Penrose coupling of the scalar field and the curvature [2]. Such a system is conformally invariant if one transforms the scalar field according to (6) with $s = 1$.

Observe that in the above considerations we do not admit for the transformations of coupling constants (or alternatively, for example, to [3]).

The explicit form and content of (2) specify whether

i)   the metric is nondynamical, externally specified by the investigator of the system or

ii) the metric itself is dynamical (is defined by appropriate part of dynamical relations DR and has to be calculated from those relations).

We draw attention to the elementary fact that if a conformally invariant dynamical system for the material fields (DR of the case i)) is supplemented by the Einstein equation defining dynamical relations for the metric tensor (so we get a set of DR of the case ii)) then the obtained gravito-material system is no longer conformally invariant because the Einstein dynamic with material sources is not conformally invariant itself.

The cases i) and ii) are essentially different from practical and interpretative point of view. Let us dwell for a moment on the interpretation of conformal transformations in these two distinct dynamical cases.

In the case i) the observer describe a dynamical system of fields $\{\Lambda\}$ propagating in a space-time with a metric. The dynamic of $\{\Lambda\}$ is given by dynamical relations involving the fields and a generic metric. The metric enters to the dynamical relations DR but the metric itself is assumed to be independent on the system of fields in question. It could be measured by the observer with a specified system of reference apparatus or phenomena (e.g., reference rods and clocks) that are not a subject of description of the constructed theory or it is given to the observer in a definite form being not a subject of the considered dynamics.

The crucial "conformal" question is: what would happen with the dynamics DR of $\{\Lambda\}$ if the metric involved in DR is conformally changed. In the case i) the change of a metric in question physically means a change of the considered space-time region into the conformally equivalent one. (Alternatively a change of the set of reference apparatus can be also considered in this case as a physical interpretation of the act of conformal transformation, however the last interpretation has to be considered with a caution.) As was said for the conformally invariant system it is possible to redefine the fields according to (3) in a way that the dynamical relations expressed with the new fields and metric will be formally identical with the relations at the origin according to (5). Physically it means that having found a solution $\{\Lambda_0\}$ valid in a region with metric $\hat{g}$ we can formally construct a solution $\{\Lambda_0^c\}$ valid in the conformally equivalent region with metric $\hat{g}^c$. The lack of conformal invariance means that solutions in a region with a given metric cannot be constructed from solutions found in conformally equivalent regions—they have to be searched from the beginning.

The possibility of constructing new solutions valid in conformally equivalent regions is of only conditional practical importance: It is interesting only if the conformally equivalent regions really exist in our externally given manifold. However there are interesting examples of metric spaces with conformally equivalent regions. The most important case is the case of the flat space-time which is (locally) a self-equivalent manifold with respect to the mentioned above global conformal transformations.

In the case ii) the dynamics of metric tensor is included into the system DR of dynamical relations. Now the metric is not given to the observer from the outside of the system but has to be calculated from the dynamics and supplementary conditions. We are not restricted to the externally given metric space with a given metric and consequently all conformally transformed configurations of metric can be potentially considered. If the system of dynamical relations has the property of conformal invariance then all conformally transformed solutions $\left(\{\Lambda_0^c\}, \hat{g}^c\right)$ of the system DR are physically acceptable. But here the crucial point is that all those conformally equivalent solutions have to be considered as physically equivalent if there is no other physical system which could serve as a conformal reference tool—an independent reference scale. In this sense the local conformal symmetry of the dynamical system of the case ii) is a trivial symmetry of the description language of a physical system but not the symmetry of the system itself. This fact manifests in the absence of a dynamical term for the field degree of freedom connected with the conformal factor Ω.

Despite the above mentioned triviality the conformally symmetric models with dynamical metric can be fruitfully considered. The freedom of change of the description language could be useful because it could be easier to analyse some properties of the models when a suitable language of analysis is used.

Let us mention that the freedom of choice of the length scale is a natural feature which we use in practice changing the scale standards depending on the problem in consideration (astrophysics or microphysics for example) and depending on the country (as different length and mass standards are used in different countries).

An additional qualitative difference between the case i) and ii) appears on the quantum level. In the case i) the anomaly connected with LCT could appear [4]. There is no room for such an anomaly in the case ii) because only the physical degrees of freedom are the subject of quantization [4].

3.   The model

In a series of papers [5] we have proposed a model of type ii) in which local conformal symmetry serves as a tool which allows for a unification of electro-weak and strong interactions described by the Standard Model (SM) and gravity.

The proposed lagrangian of the model reads:

$$L = \left[L_{SM^c} + L_{\Phi/g} + L_{grav} \right] \sqrt{-g} \tag{7}$$

where L_{SM^c} is an ordinary Standard Model lagrangian with the Higgs doublet Φ without the Higgs mass term, however. Instead there is a Higgs-gravity interaction term

$$L_{\Phi/g} = \beta \partial_\mu |\Phi| \partial^\mu |\Phi| - \frac{1}{6} (1+\beta) R \Phi^\dagger \Phi, \tag{8}$$

which contains a Penrose inspired coupling and a β proportional conformally invariant part that assures the proper ratio of electro-weak and gravitational couplings at the classical level already. A conformally invariant pure gravitational part is given by

$$L_{grav} = -\rho C^2, \qquad \rho \geq 0, \tag{9}$$

where $C^\delta_{\alpha\beta\gamma}$ is the Weyl tensor.

The metric tensor is one of the dynamical degrees of freedom so the model is of the type ii) of the classification of Section 2. The local conformal symmetry changes all dimensional quantities (lengths, masses, energy levels, *etc.*) in every point of the space--time but it leaves theirs ratios unchanged. It is a symmetry of the language of description but it reflexes the deep truth of the nature that nothing except the numbers has an independent physical meaning. This symmetry is of the same nature as gauge symmetries in gauge theories with compensating vector potentials. In fact the idea of gauge symmetry introduced by Weyl [1] was connected with the conformal symmetry. Later it was shown by Padmanabhan [6] that the conformally invariant theory with Penrose term is a special case of a more general class of models with compensating potential.

The gauge fixing choice freedom—or in the present case rather a scale fixing choice freedom—allows us to settle a more convenient language of description. In the conventional approach we define the length scale in such a way that elementary particle masses are the same for all times and in all places. This will be the case when we rescale all fields in (7) with the x-dependent conformal factor $\Omega(x)$ in such a way that the length of the rescaled scalar field doublet is fixed

$$\widetilde{\Phi}^{\dagger}\widetilde{\Phi} = \frac{v^2}{2} = const. \tag{10}$$

In this scale fixing the Higgs field disappears from the theory but simultaneously we obtain the Einstein like term in the gravity sector

$$-\frac{1}{6}(1+\beta)\frac{v^2}{2}R. \tag{11}$$

If we take β big enough and negative then the term (11) can be fixed to the ordinary Einstein form.

Obviously we can choose other conditions and obtain other equivalent descriptions that could be more convenient for other specific considerations. For example we can keep the scalar field and instead reduce the metric to the unimodular case.

Since in our formalism we do not use the spontaneous symmetry breaking mechanism (SSB) for mass generation (we will come to this point below) the self-interaction term $\lambda\left(\widetilde{\Phi}^{\dagger}\Phi\right)^2$ can be set to zero by setting $\lambda = 0$. In this case the cosmological constant obtained in our formalism is also zero in agreement with experiments and the conviction of Einstein and many others.

The condition (10) together with the unitary gauge fixing of $SU(2)_L \times U(1)$ gauge group, reduce the Higgs doublet to the form

$$\Phi^{scaled} = \frac{1}{\sqrt{2}}\begin{pmatrix} 0 \\ v \end{pmatrix}, \quad v > 0. \tag{12}$$

Inserting this condition into (7) we produce the tree level mass terms for leptons, quarks and vector bosons associated with $SU(2)_L$ gauge group.

The fermion--vector boson interactions in our model are the same as in SM. Hence analogously as in the case of conventional formulation of SM one can deduce the tree level relation between v and G_F—the four-fermion coupling constant of β-decay:

$$v^2 = (2G_F)^{-1} \rightarrow v = 246 \text{ GeV}.$$

The resulting expressions for masses of physical particles are identical as in the conventional SM and we see that the Higgs mechanism and SSB is not indispensable for the fermion and vector mesons mass generation!.

Several things need to be said about the renormalizability. The lagrangian (7) contains three parts. Two of them—the Standard Model and the R^2 gravity—seem to be separately renormalizable. The nonpolynomial term of the Higgs-gravity interaction part (8) spoils this property. The β proportional terms were introduced in order to square the strengths of (quantum) weak interactions and the (classical) gravity. If there would be found a more subtle mechanism of coordination these quantum and classical effects then the nonpolynomial term will be redundant.

We were able to derive definite dynamical predictions of our model. These are electro-weak predictions for the present accelerator experiments.

It is natural and reasonable for this purpose to ignore gravity and consider a flat space-time approximation of the model. In the limit of flat space-time our model represents a massive vector boson (MVB) theory for electro-weak and strong interactions, which is perturbatively nonrenormalizable. The perturbative nonrenormalizability means that we are not able to improve the accuracy of predictions by inclusion of more and more perturbation orders but it does not mean that we are not able to derive predictions at all. We can obtain definite predictions at one loop if we introduce some ultraviolet (UV) cut-off Λ and if we consider processes with energy scale E below this cut-off. We have shown that the cut-off Λ is closely connected with the Higgs mass m_H appearing in the Standard Model. From this point of view Higgs mass is nothing else as the UV cut-off which assures that the truncated perturbation series is meaningful.

The introduction of Λ makes all predictions Λ-dependent. In order to obtain the independent predictions we have to select some reference observable $R_0(\Lambda)$ which is measured with the best accuracy in the present EW experiments. It will replace the unknown variable Λ in the expressions for the other physical quantities R_i. Choosing the total width of Z-meson Γ_Z as the reference quantity we have calculated one loop predictions for all interesting observables measured at Z_0 peak [7]. The results almost coincide with SM predictions and equally well (or poorly) describes the present data. One can expect that even more accurate measurements taken at the single energy point would not be able to discriminate between the Standard Model and a class of effective models with massive vector mesons—as I said the Higgs mass plays the role of UV cut-off and *vice versa*.

One can distinguish these models finding the Higgs boson directly (of course!). But we have shown also another way of independent comparison [8]. Predictions of an effective model can be in principle calculated for experiments performed in various energy regions. These predictions would depend on the cut-off. And inversely: the value of the appropriate cut-off derived from experimental data collected in different energy regions can be energy dependent in principle. The cut-off is an artificial element that we introduce in order to cover an incompleteness of the model or imperfection of our calculational methods. We try to hide our ignorance in a simplest way: we introduce one additional parameter Λ. We hope that this parameter can be the same for a class of similar phenomena. It would be nice to have a universal cut-off valid for all phenomena below some energy scale but in principle it needs not to be the case. Thus we have to admit that the cut-off is energy dependent.

We have already mentioned, that the UV cut-off Λ is closely connected with the Higgs mass of SM. But of course the value of physical Higgs mass derived from various sets of experiments should be the same—in contrast to the supposed energy dependence of the cut-off Λ. This is the difference which makes a room for comparison.

In practice we need predictions for the Higgs mass (or the cut-off Λ) derived independently from two separate energy regions. One of them can be of course the Z_0 peak. The second can be provided for example by 10-20GeV e^+e^- colliders of luminosity high enough ($\sim 10^{34}$ cm^2 s^{-1}} [8]). We have estimated that the necessary sample of produced tau pair should be of order of $\sim 10^8$. Then the observed sensitivity to the value of Higgs mass will be of order of 100GeV and the predicted Higgs mass could be compared with the value of m_H derived from Z_0-peak data. We have to stress that one can relax from the particular modified model inspiration and regard the proposed test as a self-consistency check of the SM itself.

4.   Conclusions

In conclusion, we want to stress that at the level of material part the character of our modification of SM differs essentially from the usual extensions like supersymmetry, extended internal symmetries, strings and so on. We have not introduced any new fields; on the contrary the number of physical particle modes has been reduced on expense of yet unobserved scalar particle. It was the result of the conformal symmetry which in the present model is a symmetry of whole dynamical system of "interacting" metric and matter fields. At the level of coupling of SM with gravity the presented approach is minimal and in a sense "economical." It turns out that the dynamical scalar field is not necessary now and the same is true for the interaction which it transmits in the standard approach.

The new essential feature of the proposed modification is the local conformal symmetry. Although the new symmetry is trivial in a sense (its constrains can be easily resolved leading to a kind of massive vector meson model or sigma model coupled to more or less modified gravity) it seams to be a useful tool which allows for manipulation with degrees of freedom and the language of description. Special Relativity told us that all inertial frames are equally good reference systems of description and are indistinguishable at the level of dynamical equations. The General Relativity theory extended this democracy to all reference systems. The present approach shows that not only different reference systems but also different reference scales could be treated on an equivalent foot. The choice of the particular scale is a subject of a free will of the investigator and could be modified according to his actual convenience.

Acknowledgments

I am indebted to Professor Franco Selleri and members of the Organising Committee of the International Conference "Relativistic Physics and Some of its Applications," Athens, Greece, 25-28 June 1997 for this marvellous meeting and for creating the scientific atmosphere.

References

[1] H. Weyl, *Space-Time-Matter* Berlin 1918.

[2] R. Penrose, *Ann. of Phys.* **10**, 171 (1960).

[3] J.D. Bekenstein and A. Meisels, *Phys. Rev.* **D22**, 1313 (1980).

[4] M.J. Duff, *Twenty Years of Weyl Anomaly*, preprint CTP-TAMU-06/93, hep-th/9308075

[5] M. Pawlowski and R. Raczka, *Found. of Phys.* **24**, 1305 (1994); M. Pawlowski and R. Raczka, in: *Modern Group Theoretical Methods in Physics*, eds. J. Bertrand *et al.*, (Kluwer Acad. Publishers 1995); M. Pawlowski and R. Raczka, *A Higgs-Free Model for Fundamental Interactions, Part I: Formulation of the Model*, preprint, ILAS/EP-3-1995, Trieste; e-Print Archive: hep-ph/9503269.

[6] T. Padmanabhan, *Class. Quant. Grav.* **2** (1985) L105.

[7] M. Pawlowski and R. Raczka, *A Higgs-Free Model for Fundamental Interactions, Part II: Predictions for Electroweak Observables*, preprint, ILAS/EP-4-1995, Trieste; e-Print Archive: hep-ph/9503270.

[8] M. Pawlowski and R. Raczka, *Consistency test of the Standard Model*, preprint JINR-Dubna E2-97-17; e-Print Archive: hep-ph/9610435.

Evidence for Newtonian Absolute Space and Time

J. P. Wesley
Weiherdammstrasse 24
78176 Blumberg, Germany

The permanency of the celestial sphere is a result of a cosmological limit velocity for all bodies relative to absolute space. Neomechanics, prescribing momentum as $mv\big/\sqrt{1 - v^2/c^2}$, where v is the absolute velocity, accounts for the cosmological limit velocity as c. Local laws of physics must then involve absolute velocities. The Monstein-Wesley experiment confirms neomechanics. Absolute accelerations imply absolute space. Newtonian mechanics requires absolute space. Relativity theories can only be approximately valid for slowly varying effects. The oneway energy velocity of light, having the velocity c with respect to absolute space, has the observed velocity $c^* = c - v$, where v is the absolute velocity of the observer, as confirmed by many experiments. The Voigt-Doppler effect yields the null Michelson-Morley result. Absolute time is established by induction from corrected synchronized clock rates.

1. The Celestial Sphere Defines a Cosmological Limit Velocity and Absolute Space

The distant stars and galaxies that constitute the celestial sphere appear to be permanently fixed in position relative to each other. This permanency can only arise if the velocity of every body in the universe does not exceed some common finite limit velocity. If no limit existed and all velocities were equally likely, then the celestial sphere would appear like a swarm of gnats rushing wildly about at random. Since the limit velocity applies to every body in the universe; the limit velocity for each body must be measured with respect to a single unique zero velocity frame, that then defines absolute space.

Open Questions in Relativistic Physics
Edited by Franco Selleri (Apeiron, Montreal, 1998)

2.   Physical Laws Must Depend upon the Cosmological Limit Velocity and Absolute Space

Physical laws governing the motion of bodies must be such that no individual body can ever exceed the absolute finite cosmological limit velocity. Local laboratory physics must thus include explicitly this cosmological limit velocity. This means that the precise laws of physics will depend upon absolute space. Local laboratory physics must thus depend upon the absolute velocity of the laboratory. This further means that the laws of physics cannot be the same in different inertial frames with different constant uniform absolute velocities.

3.   Neomechanics Yields the Cosmological Limit Velocity as c

Since the time of Newton (1730) there has been speculation that matter with mass m can be converted into radiant energy E; thus,

$$E = mk, \tag{1}$$

where k is some constant. Toward the end of the last century this coefficient was generally speculated to be c^2; so

$$E = mc^2. \tag{2}$$

By the 1930's nuclear and particle physics established the correctness of mass-energy equivalence with the coefficient c^2 (to about a 2 place accuracy).

"Neomechanics" is defined here as mechanics hased upon mass-energy equivalence in absolute space and time that reduces to Newtonian mechanics for small velocities, $v/c \ll 1$. In particular, the mass to be associated with a kinetic energy K is K/c^2. This means that a change in momentum $\mathbf{p}$ of a particle of rest mass m is given by

$$d\mathbf{p} = d[(m + K/c^2)\mathbf{v}]. \tag{3}$$

The corresponding change in kinetic energy dK is then given by

$$\mathbf{v} \bullet d\mathbf{p} = dK = \mathbf{v} \bullet d[(m + K/c^2)\mathbf{v}]. \tag{4}$$

Integrating Eq.(4), choosing the constant of integration such that $K = 0$ when $\mathbf{v} = 0$, yields

$$K = mc^2(\gamma - 1), \tag{5}$$

where

$$\gamma = \frac{1}{\sqrt{1 - v^2/c^2}}. \tag{6}$$

The neomechanical momentum then becomes

$$\mathbf{p} = m\gamma\mathbf{v} = \frac{m\mathbf{v}}{\sqrt{1 - v^2/c^2}} \tag{7}$$

And Newton's second law becomes

$$F = d\mathbf{p}/dt = d(m\gamma\mathbf{v})/dt. \tag{8}$$

The velocity $\mathbf{v}$ appearing in these expressions (3)-(8) is the absolute velocity of the mass m.

These neomechanical results for the kinetic energy and momentum, Eqs.(5) and (7) go to infinity as v approaches the absolute velocity c. Since there is no infinite source of energy to give a single body infinite kinetic energy; and since no infinite forces exist in nature that can accelerate a single body to an infinite momentum; the absolute velocity of any body with rest mass m must always remain less than the limit velocity c. This result then says that neomechanics yields the absolute cosmological limit velocity as c.

4.   The Monstein-Wesley Experiment Confirms Neomechanics

Monstein and Wesley (1996) measured the anisotropy of the cosmic-ray muon flux using a cosmic-ray telescope to obtain the absolute velocity of the solar system. In particular, taking into account the statistical neomechanics for the decay rate of moving radioactive particles, the half-life is given by

$$\tau = \tau_o \gamma = \frac{\tau_o}{\sqrt{1 - v^2/c^2}}, \tag{9}$$

where τ_o is the half-life for a stationary particle and v is the absolute velocity of the particle (Wesley 1991).

In terms of the muon velocity relative to the Earth $\mathbf{v}'$ and the absolute velocity of the Earth $\mathbf{v}_e$ the absolute velocity squared to be used in Eq.(9) becomes

$$v^2/c^2 = (\mathbf{v}' + \mathbf{v}_e)^2/c^2 \approx v'^2/c^2 + 2\mathbf{v}' \bullet \mathbf{v}_e/c^2, \tag{10}$$

where v_e^2/c^2 has been neglected compared with the terms retained and where the absolute velocity of the Earth is approximately equal to the velocity of the solar system, $\mathbf{v}_e \approx \mathbf{v}_o$, the absolute solar system velocity $\mathbf{v}_o$ being of the order of 300 km/s (A slight improvement can be expected by also including the known orbital velocity of the Earth about the Sun.).

From the theory for the expected sea-level muon flux as a function of τ and thus of $2\mathbf{v}' \bullet \mathbf{v}_o/c^2$ and from 32,400 coincident counts registered by the telescope collected over 18 years the magnitude and direction of the absolute velocity of the solar system $\mathbf{v}_o$ was found to be that presented in Table 1.

This result is in reasonable agreement with the values of $\mathbf{v}_o$ obtained by other methods. This means that the velocity to be used in the gamma factor γ, Eq.(6), is, in fact, the absolute velocity of the body or particle, thereby confirming neomechanics. In addition, the Monstein-Wesley experiment demonstrates the fact that the absolute velocity of the local inertial frame or laboratory is explicitly involved in the local laws of physics.

5.   Absolute Accelerations Imply Absolute Space

Newton's second law reveals absolute accelerations. This permits one to speak of "inertial frames" that are not accelerating. For each point in the

Table 1. Observed values of the absolute velocity of the solar system

method	observer	v_o km/s	α_o hr	δ_o deg
Galactic red shifts anisotropy	De Vaucouleurs & Peters (1968)	300 ± 50	7 ± 1	50 ± 10
	Rubin et al. (1976)	600 ± 100	2 ± 2	50 ± 20
2.7 K cosmic background anisotropy	Conklin (1969) from ground	200 ± 100	13 ± 2	30 ± 30
	Henry (1971) from balloon	320 ± 80	10 ± 4	-30 ± 25
	Smoot et al. (1977) from airplane	390 ± 60	11.0 ± 0.5	5 ± 10
oneway light velocity anisotropy	Marinov (1974) coupled mirrors	300 ± 20	13.3 ± 0.3	-20 ± 4
	Marinov (1984) toothed wheels	360 ± 40	12 ± 1	-24 ± 7
	Müller (1994) Geostationary satellite time signals	250 ± 50	6 ± 1	-14 ± 3
muon flux anisotropy	Monstein& Wesley (1996)	360 ± 180	9 ± 4	-1 ± 10

universe a nonaccelerating inertial frame may be chosen with some arbitrary constant uniform velocity $\mathbf{v}_{0i}$ Since all points in the universe are equivalent in a uniform isotropic universe; there can be no priviledged point with some priviledged velocity with some priviledged direction. Thus, the constant uniform velocity to be associated with each point in space must be the same and must be zero in a universe with no preferred direction; thus,

$$\mathbf{v}_{0i} = \mathbf{v}_0 = 0. \tag{11}$$

The inertial frame with this unique zero velocity, that can be assigned to every point in the universe, defines absolute space.

6. Newtonian Mechanics Requires Absolute Space

Newtonian mechanics involves integrals of the motion that involve velocities of bodies. The values of these velocities are valid only for a particular inertial frame with a particular constant uniform velocity. These integrals of the motion involving the velocities of bodies must take on different values in different inertial frames with different constant uniform velocities. For Newtonian mechanics to be valid for the universe as a whole a single unique inertial frame for the whole universe must be chosen, which to preserve isotropy, must be a zero velocity frame or absolute space.

In neomechanics the absolute velocity is involved in Newton's second law, Eq.(8), itself; so the need for absolute space is not limited to merely the integrals of the motion.

7.   Relativity Theories Are Insufficient

Relativity theories assume fundamental laws of physics can be deduced from the interaction between two bodies, which involve only their relative position, their relative velocity, and their relative acceleration.

Such theories obviously cannot account for the absolute cosmological limit velocity; since absolute velocities and absolute space are not involved.

Such theories involve action at a distance, such as Coulomb's law or Newton's law of universal gravitation. From symmetry involving two bodies the action of one body on the other must be assumed to occur instantaneously. Relativity theories cannot incorporate radiation or action transmitted with a finite velocity. Such theories require no propagating fields nor any fields at all.

Relativity, such as classical celestial mechanics, works as an approximation, where the time for the transmission of action between two bodies may be neglected as small. In particular, the finite velocity of action compatible with the cosmological limit velocity may be taken as c; so the relativity approximation works when

$$R/c << \Delta t, \tag{12}$$

where R is the separation distance between the two bodies and Δt is the shortest time interval observed or that is of interest. Thus, relativity theories are approximately valid for slowly varying effects.

Generally only the velocity of a particle relative to the laboratory is used in the γ factor, Eq.(6), instead of the absolute velocity required by neomechanics. This does not produce any large error; since the absolute velocity of the laboratory is approximately that of the solar system which is small compared with c, or $v_0/c \sim 10^{-3}$. However, opprotunities to measure the absolute velocity of the solar system by using a little care are thereby overlooked.

8.   The Oneway Energy Velocity of Light Reveals Absolute Space

The oneway energy velocity of light is empirically found to be c with respect to absolute space. If an observer has the absolute velocity v, then the apparent oneway energy velocity of light is observed to be

$$c^* = c - v, \tag{13}$$

where c and v are chosen positive in the same direction. This formula (13) accounts trivially for the observations of Roemer (1677), Halley (1694), Bradley (1728), Sagnac (1913), Michelson-Gale (1925), Conklin (1969), who first observed the 2.7 K cosmic background anisotropy, Marinov (1974) with his coupled mirrors experiment, Marinov (1984) with his toothed wheels experiment, and Müller-Dale (1994) with the use of geostationary satellite time signals.

The failure of the Michelson-Morley experiment to detect the absolute velocity of the set-up was predicted by Voigt (1887) as a Doppler effect for light in absolute space and time before the experiment had even been performed. Michelson expected a positive result; because he erroneously considered the oneway velocity of energy propagation instead of the phase velocity. The setup is only sensitive to the phase velocity; and in Doppler effects the phase velocity need not have the same magnitude nor direction as the oneway velocity of energy propagation. A thorough analysis of the Michelson-Morley experiment has been made by Wesley (1984). (Voigt's unfortunate mathematical representation of his Doppler effect in space and time variables, yielding the so-called "Lorentz transformation," instead of in the propagation constant and frequency, gave rise to the strange idea that space and time could themselves somehow change in a moving system, as is now assumed in "special relativity").

9. Absolute Time

The concept of time is an abstraction by induction from the observation of many different processes occurring in nature. It is found that periodically reproducible processes can be compared and synchronized with each other to establish a universal time. It is thus possible to construct clocks that have steady reproducible periodic behavior. The fact that standard clocks can be constructed (at least in principle) to run at constant rates independent of their location in the universe, the date, the gravitational field, the velocity, the acceleration, and other such extraneous conditions defines by induction the concept of absolute time.

The peak frequency of the 2.7 K cosmic background radiation provides in principle a single unique clock that can be seen and used by all observers anywhere in the whole universe to establish an absolute unit of time.

The naturally occurring frequency of a spectral line of Hydrogen can also be used to define a universal unit of time valid on any distant star or galaxy. Such clocks on distant stars and galaxies when viewed on the Earth must be corrected for the cosmological redshift, the gravitational redshift, and Doppler shifts. The reliability of such atomic clocks may also be used as possible evidence for the distance, the gravitational field, or the velocity of the distant star or galaxy.

The most accurate practical standard unit of time today on the Earth is given by cesium beam clocks that have practical fractional accuracies of about 10^{-13}.

A clock based on the half-life of a radioactive element, would have to be corrected for its absolute velocity using Eq.(9). On the other hand, a clock based upon the resonating frequency of an electrodynamic standing wave in a cavity is not sensitive to its absolute velocity and does not need the correction given by Eq.(9).

There is no known or conceiveable natural phenomenon that necessitates the idea that time itself must run faster or slower under certain circumstances; because appropriate corrections can always be introduced to yield synchrony with ordinary accurate clocks.

References

J. Bradley, *Lond. Phil. Trans.* 35, #406 (1728).

E.K. Conklin, *Nature* **222**, 971 (1969).

G. de Vaucouleurs and W.I. Peters, *Nature* **220**, 868 (1968).

E. Halley, *Phil. Trans. 18*, 237 (1694).

P.S. Henry,*Nature* 231, 516 (1971).

S. Marinov, *Czechosl. J. Phys. 4*, 965 (1974): and *Gen. Rel. Grav.* **12**, 57 (1980).

S. Marinov, *The Thorny Way of Truth* (East-West, Graz, Austria, 1984).

A.A. Michelson and H. G. Gale, *Astrophys. J. 61*, 137 (1925).

C. Monstein and J.P. Wesley, *Apeiron 3*, 33 (1996).

F.J. Müller and D. Means, *Galilean Electro. 5*, 90 (1994).

Sir Isaac Newton, *Optiks,* 4th ed. London 1730, reprint (Dover, New York, 1952) Book 3, Part 1, Question 30.

O. Roemer, *Phil. Trans.* **12**, 895 (1677).

V.C. Rubin, W. K. Ford, N. Thonnard, M. S. Roberts, and J. A. Graham, *Astrophys. J. 81, 687,* 719 (1976)-.

G. Sagnac, *Comptes Rendus 157*, 708 (1913).

G. F. Smoot, M. V. Gorenstein, and R. A. Muller, *Phys. Rev. Lett. 39*, 898 (1977).

W. Voigt, *Gött. Nachr. Math.-Phys. Kl,* p. 41 (1887).

J. P. Wesley, *Advanced Fundamental Physics* (Benjamin Wesley, 78176 Blumberg, Germany, 1991) pp. 27-29.

J. P. Wesley, *Found. Phys. 14*, 155 (1984); & *Classical Quantum Theory* (Benjamin Wesley, 78176 Blumberg, Germany, 1996) pp. 210-217.

Cosmology and
Astrophysics

Evolution of Quasars into Galaxies and its Implications for the Birth and Evolution of Matter

Halton Arp
Max–Planck–Institut für Astrophysik
85740 Garching, Germany

In recent years satellite observations have recorded a number of point X–ray sources in the sky. Such sources are overwhelmingly identified with medium to high redshift quasars. These quasars can now be shown to originate in low redshift, active galaxies which eject them along their minor (rotation) axis. It can be empirically demonstrated that this is how galaxies are born and evolve. The observations invalidate the assumptions of Friedmann and Einstein in General Relativity and require a more general, Machian solution of the field equations in flat space-time.

Introduction

Quasars are stellar appearing objects which have large redshifts. They were discovered in 1963 and interpreted as having high recessional velocities which required them to be out near the limits of the visible universe. Already in 1966, however, evidence started to accumulate that quasars were in fact ejected from the nuclei of nearby galaxies and that their redshifts were due to some intrinsic cause, *not* translational velocity.

Evidence for nearby quasars continued over the years to come from many different kinds of observations. The observations also included young galaxies which had physical continuity with quasars and which also showed non–velocity redshifts. This data, up to about 1987, is discussed in *Quasars, Redshifts and Controversies'* (Interstellar Media). In the following decade X–ray observations have picked out quasars with increasing efficiency and massively confirmed the previous association of young higher redshift objects with nearby, low redshift parent galaxies.

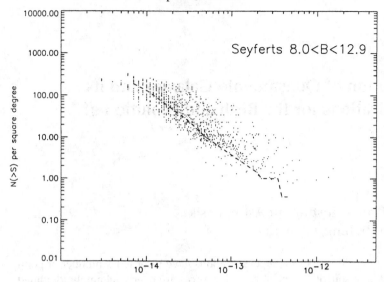

Figure 1. The density of X–ray sources greater than a given brightness S, for X–ray fields around 24 bright Seyfert galaxies. The dashed line represents the average relation from 14 high latitude control fields.

X-ray Sources around Seyfert Galaxies

In 1996 H. Arp and H.-D. Radecke analyzed archival X–ray fields around a nearby complete sample of Seyfert Galaxies. These galaxies had been investigated by many different observers because their nuclei were strong sources of high energy emission, including X-rays. In the surrounding areas, however, the results showed an excess of point–like X-ray sources around these Seyferts (Radecke 1997). The results for 24 fields are shown here in Fig. 1. The X-ray sources were obviously physically associated with the central galaxy and almost all of those identified turned out to be quasars (Arp 1997).

The associated X-ray quasars confirmed strongly the pairing and alignment across the ejecting galaxy which had first been observed in 1966. (Radio sources, of which quasars are a sub class, had been accepted since the 1950's as being ejected from the nuclei of active galaxies.) Fig. 2 here shows one example of a pair of bright X-ray quasars across a bright, active Seyfert Galaxy, NGC 4235.

This example illustrates a number of properties of the configurations which cannot be due to chance arrangement of background objects. The chance of accidental arrangement for *each one* of the many examples runs from between one in a million to one in ten million. The configuration aspects which are repeated in all the examples however, furnish the data which leads to understanding the physical processes involved.

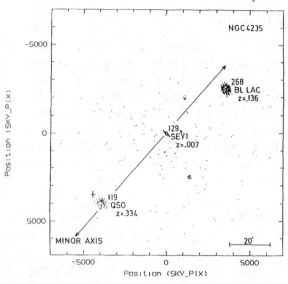

Figure 2. An X-ray map of the area around the Seyfert galaxy NGC 4235. The very bright quasar–like objects of z = .334 and .136 are aligned closely along the minor axis.

The Charasteristic Configuration of X-ray Sources and Parent Galaxy

Fig. 3 shows the average behaviour of physical companions around an active galaxy. The salient characteristics are:

- Quasars emerge as high redshift, low luminosity objects.
- As they travel outward their ejection velocity slows, their luminosities increase and their intrinsic redshifts decrease.
- When they reach maximum extension from the parent galaxy (about 400 kpc) they increasingly behave as BL Lac objects—that is they experience a burst of synchrotron continuum emission which drastically decreases the visibility of emission lines in the spectrum. They readily break up at this stage and exhibit secondary ejection.
- The BL Lac phase begins to show spectral evidence of an underlying stellar population. The quasar is beginning to evolve into a galaxy.
- Subsequent evolution into compact galaxies, blue galaxies and galaxies which are themselves ejecting new generations of objects continues until normal companion galaxies are reached. These companions can be less closely distributed along the line back to the original parent galaxy, as if they had experienced some perturbations during their aging. But they never lose all of their excess redshift.
- All galaxies and quasars have quantized redshifts. The quasars are quantized in large steps: $z = 1.96, 1.41, .96, .60, .30$ and $.06$. The low redshift galaxies into which they evolve have redshifts quantized in steps of $z = .0002$ and $.0001$ ($cz = 72$ and 37.5 km/sec).

Confirmation by a Single Example

In April 1997 Prof. Yaoquan Chu communicated to me the sensational result shown in Fig. 4. He had measured the bright X-ray sources around the very active Seyfert NGC 3516. They turned out to be quasars ordered in redshift, culminating in the most distant, a bright optical and X-ray BL Lac type object. All six of these quasars fell within ± 20 degrees of the minor axis of NGC 3516 (a chance, just in this one property, of only 10^{-4} of being accidental). As the bottom of the Figure shows, the redshifts of these six ejecta fit very closely the periodicity which has been known for quasars for more than 20 years.

Evolution of an Intrinsic Property

If high redshift quasars are physically associated with low redshift galaxies ($z \leq .01$), then the redshifts cannot be interpreted as recession velocities. Observations of the high redshift objects rule out gravitational redshifting or "tired light" effects. If the electrons making the orbital transitions in atoms of the high redshift object are less massive than those of the observer, however, the emitted photon will be redshifted. Can this be reconciled with physics as we know it?

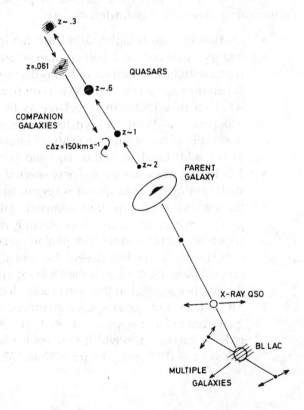

Figure 3. Summarizing the empirical data for low redshift ejecting galaxies and their associated quasars and companion galaxies since 1966. The high redshift quasars lose their intrinsic redshifts as they travel out and eventually evolve into companion galaxies of only slightly higher redshift.

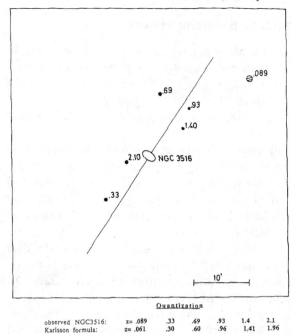

Figure 4. All bright X-ray
objects arround the very
active Seyfert galaxy
NGC 3516 have had
their redshifts measured
by Y. Chu. Redshifts
decline as they separate
from the ejecting galaxy.

	Quantization					
observed NGC3516:	z= .089	.33	.69	.93	1.4	2.1
Karlsson formula:	z= .061	.30	.60	.96	1.41	1.96

Actually the 1922 Friedmann solution of the Einstein field equations of General Relativity *assume* the particle masses comprising matter are constant. A more *general* solution of the field equation yields:

$$m = at^2$$

where m is particle mass, t is cosmic time and a is a constant.

With this one simple solution particle masses grow with time, young objects start with high intrinsic redshifts and evolve to lower redshift as they age. This is exactly what 30 years of empirical evidence has required (Arp 1991, Narlikar and Arp 1993).

For example, Erik Holmberg showed in 1969 that companion galaxies concentrated primarily within ±35 degrees of the minor axis of disk galaxies. Now as Table 1 shows, *quasars* concentrate within about ±20 degrees of the minor axis and reach a 400 kpc extension from the ejecting galaxy *exactly the same extent and alignment as companion galaxies*. As quasars age into companion galaxies they clearly spread more from the line of initial ejection either by axial precession or gravitational perturbations. Whatever causes the quantized steps in their redshifts, they must become smaller, however, as their total redshifts become smaller. So they are also continuous in the property of quantization with age. There should no longer be any reasonable denial that the variable particle mass theory uniquely fits the observations.

Continuous Creation Replaces Big Bang Theory

The variable mass theory is Machian, not a local theory as Einstein ruefully conceded about conventional General Relativity. When a new particle is created it sees a very small universe. As time goes on it exchanges signals within a light sphere which is growing at c. As the particle mass increases electron transitions emit higher energy quanta and the intrinsic redshift decreases.

New matter is created with near zero mass, therefore it is travelling with nearly the signal speed of the medium, namely the velocity of light. As the particle mass grows, however, its translational velocity drops in order to conserve momentum. It finally comes to rest near the observed 400 kpc maximum extension shown in Table 1. This is the same maximum extension quantitatively predicted by Narlikar and Das (1980).

Moreover, the random velocities within a young plasma must diminish as the particles gain mass and therefore its temperature drops. We have the ideal explanation for how a hot plasmoid ejected from an active galaxy can cool and condense into a quasar and eventually evolve into a normal galaxy. The synchrotron "knots" observed leaving the nucleus of M87, for example, could never make the transition into the older galaxies observed to be aligned along the direction of the jet. In a conventional plasma the large particle masses would contain too much energy and consequently blow the plasmoid apart before it could cool. What we now have with the low particle mass plasma is Viktor Ambartsumian's "superfluid" which he intuitively described just from looking at a photograph of new galaxies being formed in ejections from other galaxies in 1957.

Where the Friedmann/Einstein Expanding Universe Fails

Fig. 5 shows that the conventional solution of the field equations leads to expanding spatial coordinates and redshifts caused by recessional velocity. This violates the observations previously summarized in this paper. The variable mass solution outlined in the right column, however, explains nicely the intrinsic redshifts as due to the creation age of the matter.

Table 1 Companion Objects around Spiral Galaxies

No.	Companions	$\Delta\Theta_1$	$\Delta\Theta_2$	$r_1 \sim r_2$	Reference
2	quasars across NGC 4258	13°	17°	25-30 kpc	Pietsch et al. 1994
2 + (4)	quasars across NGC 2639	0°	13°(31°)	10-400	Fig. 3
2	quasars across NGC 4235	2°	12°	500-600	Fig. 4
4	quasars nearest NGC 1097	~ 20°		100-500	Arp 1987
6	quasars nearest NGC 3516	±20°		100-400	Chu et al. 1997
218	comp^{n}s around 174 spirals	~35°		40 kpc	Holmberg 1969
96	distbd. comp^{n}s around 99 sp.	~60°		150	Sulentic et al. 1978
115	comp^{n}s around 69 spirals	~35°		500	Zaritsky et al. 1997
12	comp^{n}s of M31	~0°		(700)	Arp 1987

Friedmann (1922)	Narlikar (1977)
Special solution	**General solution**
• $m = $ constant	• $m = m(t)$
$\dfrac{S(\tau_o)}{S(\tau)} = 1 + z$	$\dfrac{m_o}{m} = \dfrac{t_o^2}{t^2} = 1 + z$
$H_o = \left.\dfrac{\dot{S}}{S}\right|_{\tau=\tau_o}$	$H_o = \dfrac{2}{t_o}$
• Expanding coordinates	• Non expanding Universe (Euclidean)
• Singularities at $m = 0$ $\tau = 0$	• Creation points at $m = 0$
• $z \equiv$ velocity	• Quantum $\Leftrightarrow$ classical physics
• distance $\equiv \dfrac{z}{H_o}$	• Merging time scales t, τ
	• Cascading, episodic creation
	• Indefinitely large, old Universe
$z = z(t)$	↑

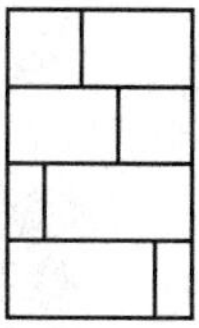

Figure 5. A schematic summary of the Big Bang (left hand side) versus the more general, variable mass solution (right hand side) of the General Relativistic field equations. The conventional assumption that particle mass, m, is constant leads to an expanding universe and collision with the brick wall of observation that redshifts are not primarily velocity but intrinsically age related. The Machian solution on the right gives redshift (z) as a function of age (t), predicts the correct Hubble constant, turns conventional singularities into creation points of "new" matter and permits connection with non-local theories such as quantum mechanics.

As we have mentioned, the Narlikar (1977) solution is a general solution of what are essentially just energy conservation equations. The general solution, however, is non-local, Machian. It does not require the unrealistic assumption of homogeneity. It avoids the embarassing singularities where physics simply breaks down in the current theory. Instead these singularities become the necessary creation points of "new" matter. By conformally transforming to our galaxy's time scale (τ) we recover all standard local physics. The continual creation solution takes place in flat (Minkowski) space-time. There is no need for the semantic contradictions of "curved space-time".

Since the large redshifts are predominantly due to young age there is no evidence, and indeed no place for, recessional velocities and no evidence for an expanding universe. It can be argued that the cosmic microwave background is simply the temperature of the intergalactic medium, averaged through the line of sight to obtain the extreme smoothness which is observed. The CMB is therefore likely to be the primary reference frame which is such an anathema to conventional relativity.

References

Arp, H. 1987, *Quasars, Redshifts and Controversies,'* Interstellar Media, Berkeley.
Arp, H. 1991, *Apeiron*, Winter–Spring, 18.
Arp, H. 1997, *Astron. Astrophys.* **319**, 33.
Chu, Y., Wu, J., Hu, J., Zhu, X. and Arp, H. 1997, *Ap. J.* in press.
Holmberg, E. 1969, *Arkiv. of Astron.*, Band 5, 305.
Narlikar, J. V. 1977, *Annals of Physics* **107**, 325.
Narlikar, J. V. and Das, P. K. 1980, *Astrophys. J.* **240**, 401.
Narlikar, J. V. and Arp, H. 1993, *Astrophys. J.* **405**, 50.
Pietsch, W. *et al.* 1994, *Astron. Astrophys.* **284**, 386.
Radecke, H. –D. 1997, *Astron. Astrophys.* **319**, 18.
Sulentic, J. W., Arp, H. and di Tullio, G. A. 1978, *Ap. J.* **220**, 47.
Zaritsky, D., Smith, R., Frenk, C. S. and White, S. D. M. 1997, *Ap. J.* **478**, L53.

A Lorentzian Approach to General Relativity: Einstein's Closed Universe Reinterpreted

J. Brandes
Danziger Str. 65
D-76307 Karlsbad, Germany

Within Lorentzian interpretation of general relativity (GR) curvilinear space is not reality itself, and has to be projected to Euclidean space. A finite, closed universe is even more complicated. When these difficulties are resolved, black holes disappear. This explains another point: An expanding universe should stay at its beginning within a black hole but later on leave it.

Key words: special and general relativity, finite, closed universe, black holes, white holes, big bang

1. Lorentzian Interpretation (LI) of Special Relativity (SR)

The Lorentzian interpretation of special relativity (LI of SR) demands a priviledged inertial system; even if there is no such system it is important for SR that there could be one; the relativity principle (RP) holds if you look only at the results of measurements. While Einstein's interpretation of SR is closely connected to the physical space of Minkowski, LI prefers the physical space of inertial transformations which results from F. Selleri (and before him Tangherlini) [1]. The Minkowski space describes most excellent measuring situations when the whole laboratory is transferred to another inertial system but the physical object remains where it is. The physical space of inertial transformations gives more evidence to Sagnac effect, Ehrenfest paradox, twin paradox, *etc.*, which rely on considerations relative to accelerated systems. Details and further literature are in textbooks [6], [7]. Both physical spaces belong together and possess their own mathematical elegance since for both of them a RP holds.

LI of SR rests on space and time and not on spacetime. Therefore general relativity (GR) has to be reinterpreted—especially Einstein's finite, closed universe being locally but not globally understandable within LI. Giving a reasonable explanation black holes become suspicious, too.

2. Lorentzian Interpretation of GR— Contraction of Standards in Gravitational Fields

observer

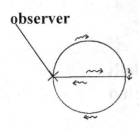

Fig. 1 Light rays in a closed
universe return from the
opposite direction

The LI of GR is above all an application of Poincaré's thesis stating that curvilinear space can be reinterpreted as shrinking of measuring sticks in Euclidean space caused by gravitational fields. This is a well accepted idea and even part of more tolerant textbooks [2]. The finite closed universe gives a fundamental objection since a light ray transmitted in one direction will return to the observer from the opposite direction, Fig. 1—contrary to reasonable projections to Euclidean space similiar to Fig. 2 ff. Therefore some arguments against a finite, closed universe are discussed with the following result: Thought experiments with the Friedmann model and its application to the expanding universe state that our universe may be interpreted similiar to exploding or collapsing dust stars (3) with a static SM outside and a RWM inside of it, Fig. 2. During big bang and later up to $z = 1000$ the expanding universe then stayed within a black hole – and left it. This is an experimental

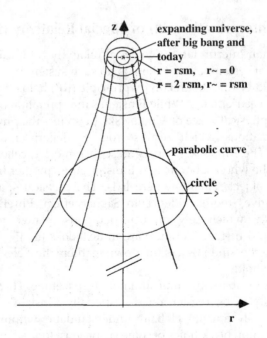

Fig. 2 Static Schwarzschild metric (SM) – gravitational
field of stars and expanding universe (and the
universe just after big bang and today)

evidence against black holes and is easily understandable using LI of GR. In Ch. 3 it is shown what this means for SM and RWM, Ch. 4 belongs to H. Weyl's objection [4], Ch. 5 – 6 to more detailed arguments against a finite, closed universe and black holes.

3. Lorentzian Interpretation of Schwarzschild Metric (SM) and Robertson-Walker Metric (RWM)

The SM is described by

$$ds^2 = \frac{1}{A}dr^2 + r^2\left[d\theta^2 + \sin^2(\theta)d\phi^2\right] - Ac^2dt^2 \tag{1}$$

with $A = 1 - r_{sm}/r$, r, θ, ϕ the spherical coordinates of a point P, t the time, G the gravitational constant, M the mass of spherical body (dust star or universe), ds spacetime distance of two narrow points, and r_{sm} the Schwarzschild radius $2GM/c^2$, and its LI means: Within this system of coordinates a measuring stick ds will have the components dr, $r\sin(\theta)$ dϕ and $rd\theta$ at position r,θ,ϕ in the three-dimensional Euclidean space. This is illustrated in the special case with ds in radial direction and within a plane through the center of a star (or universe) in Fig. 3. Here SM becomes the curvilinear surface of Fig. 2 and ds is outside the dust star (or "outside" the universe).

Inside the dust star (or expanding universe) the RWM holds with its most important version (finite space, known as $k = 1$):

$$ds^2 = R^2[d\alpha^2 + \sin^2(\alpha)(d\theta^2 + \sin^2(\theta)d\phi^2)] - c^2dt^2 \tag{2}$$

(In the other cases $\sin(\alpha)$ is replaced by α (normal space, $k = 0$) and $\sinh(\alpha)$ (hyperbolic space, $k = -1$), in normal space r is described by $R\alpha$), while α, θ, ϕ, ds are similiar as before, R is the radius.

A measuring stick ds now becomes shrunk as belongs to some projection into three dimensional Euclidean space. In a plane through the center of the star the RWM becomes a sphere (or a part of it). ds lies on the sphere and the Euclidean components are e.g. identical to vertical projection of ds to the equatorial plane ($R\cos(\theta)$ dθ, $R\sin(\theta)$ dϕ illustrated by the spaceships in Fig.5).

In the sense of LI the SM as well as the RWM describe only certain curvilinear physical spaces and not the real space itself which is Euclidean and is represented by one

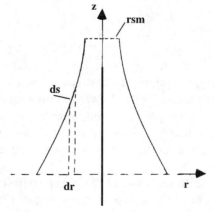

Fig. 3 Part of Fig. 2. Illustrates how measuring sticks become contracted

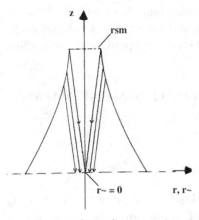

Fig. 4 System of coordinates r~

of the several possible coordinate spaces.

4. H. Weyl's Objection to Lorentzian Interpretation of GR

In his famous textbook [4] H. Weyl states: *"Im Gravitationsfeld koennen wir... die Euklidische Geometrie... retten... So gibt es hier offenbar unendlich viele gleichmoegliche und gleichberechtigte Vorschriften... deren jede zur Euklidischen als der wahren Geometrie fuehrt; kein Anhaltspunkt ist da, um eine von ihnen im Gegensatz zu allen anderen als die allein richtige auszuwaehlen."* (In gravitational fields the Euclidean geometry can be saved. But there are infinitely many possible and equivalent rules, all of them leading to the Euclidean geometry as the true one without any indication which will be the right one.) This means: The curvilinear physical space (*e.g.* the two dimensional surface in Fig. 2) is not changed itself by choosing another system of coordinates but every new one presents another projection of ds to Euclidean space. As far as none of these can be preferred H. Weyl concludes that there is none.

The contrary position: Since GR doesn't know the correct one, GR is *incomplete. E.g.* replacing in SM

$$r \to r\sim + \frac{r_{sm}}{\exp(r\sim/r_{sm})} \tag{3}$$

with θ, ϕ remaining unchanged. Near r_{sm} $r \to r\sim + r_{sm}$, but

$$\lim_{r \to \infty} r = r\sim \tag{4}$$

SM becomes:

$$ds^2 = 1/A \, (dr/dr\sim)^2 \, dr\sim^2 + r^2(d\theta^2 + \sin^2(\theta) \, d\phi^2) - Ac^2dt^2 \tag{5}$$

with $dr/dr\sim = 1 - 1/\exp(r\sim/r_{sm})$ and r from (3) part of which is illustrated in Fig. 4. Excluding negative $r\sim$, the black hole vanishes. There is no infinity of SM for $r\sim \to 0$. The well known facts that radially falling clocks reach r_{sm} in finite proper time τ but infinite coordinate t time become reasonable. Severe gravitational fields prevent falling clocks from reaching the center and stop them running. There is no crossing into another world within finite proper time τ.

In the sense of LI GR is incomplete since it cannot be decided which one of both coordinate systems—(1) or (5), physically quite different—is the true projection to Euclidean space. With evidence given in Ch. 5 – 6 equation (5) will be nearer to reality excluding a finite, closed space and black holes, too.

The well known experimental validations of GR and those formulas remain valid since they belong to the unchanged part of curvilinear space.

5. Einstein's Closed Universe—a Fundamental and Valid Objection to Lorentzian Interpretation?

H. Stephani [3] states: "*Ein expandierender Staubstern, also ein Ausschnitt eines Friedmann-Kosmos, der nach aussen von einer statischen Schwarzschild-Metrik umgeben ist,... ist vielleicht auch ein recht gutes Weltmodell.*" (An expanding dust star being a section of a Friedmann cosmos surrounded by a static SM may be a reasonable model of universe, too.) In more detail:

Starting with the following assumptions:

a) cosmological principle (universe is homogenous and isotropic, analog for the interior of dust stars.)

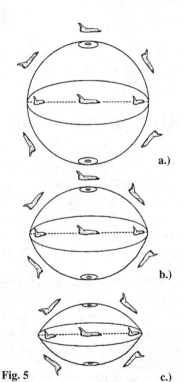

Fig. 5

b) Universe has a finite mass (reasonable if big bang)

c) $p = O$, pressure is zero. (galaxies are dust)

d) $g = g(t)$, the density is only a function of time.

e) cosmological constant is zero (Einstein: 'The biggest blunder of my life.")

GR proves RWM and $R = R(t)$ *via* Friedmann's differential equation (6) for collapsing or expanding dust stars as well as for an expanding universe. In the latter case dust consists of galaxies but normally its interpretation being different: now one postulates a finite, closed universe without further confirming arguments.

The SM in the outside of stars or the expanding universe is static because there are only radial motions. Therefore:

Fig. 5 The finite closed universe consists of two cups put together. One cup and its projection to circular plane are the rationale counterparts. Two-dimensional "planes" only. a.) – c.): The same density of the universe but different masses.

A) collapsing dust has the same formulas as the expanding universe but is not a closed space.

B) look at the universe at $R'(t) = 0$ *i.e.* $R = R_{max}$. Then the whole universe becomes a static star with SM in its outside but a closed universe shouldn't have something like this.

C) in GR density $g(t)$ of the universe and its mass M are accidental values. More likely a closed universe should look like cups of a sphere (see Fig. 5 and below).

D) following the principle that all physical laws on earth and nearby are the same as far away it is more likely our universe being open like stars of dust leaving a finite, closed universe only a theoretical possibility.

From C). From the Friedmann differential equation

$$R'(t)^2 = 8\pi GM_{const}/(c^4 3R(t)^3) - 1 \qquad (6)$$

with $M_{const} = g(t)c^2R(t)^3 = $ const one gets in the case $R'(t) = 0$ (maximum extension, $R = R_{max}$):

$$g(t_O) = 3c^2 R_{max}/[8\pi GR(t_O)^3] \qquad (7)$$

where t_O is the present time.

Formula (7) means: Knowing $R_{max,}$ $R(t_O)$ one gets $g(t_O)$. With

$$M_{univ} = g(t_O) 2\pi^2 R(t_O)^3 \qquad (8)$$

one gets the mass of the universe.

With known $R(t_O)$, R_{max} (ca. 1.8×10^{10} Ly or 3.6×10^{10} Ly, resp.) both, M_{univ} and $g(t_O)$ are fixed values. A *spherical* closed universe becomes unlikely and normally a closed universe will have a center s as in Fig 5.

6. Expanding Universe: Experimental Evidence against Black Holes

In an open expanding universe inside there is the RWM together with $R = R(t)$ outside there is the static SM. *The SM is static because there are only radial velocities.* At the beginning (and later up to $z > = 1000$) the radius of the universe is smaller than r_{sm}. This means: our universe stayed within a black hole and then left it which is theoretically impossible since "nothing" can escape a black hole. Black holes become suspicious since the RWM and $R = R(t)$ have no singularities at $r = r_{sm}$ but only at $r = 0$.

Given a closed expanding universe there might be no outside and no SM but similiar arguments are applicable to suited chosen inner parts of the universe which have a SM. A further possible objection: Our universe is not a Friedmann one. But, taking all these considerations as a thought experiment within GR the contradiction remains between RWM and $R = R(t)$ on one side allowing $r < = r_{sm}$ and SM on the other which forbids it.

A *possible* solution is due to Lorentzian interpretation (LI) concerning coordinate transformations similiar to $r\sim$ in Fig. 4: $r\sim$ *allows* us to select a projection to Euclidean space without a black hole (only a "black point" remains). If the standard model is correct (which relativists should believe) such a selection seems to be necessary.

Some reasonable now not proven ideas: black holes (points) turn into white holes (particles under great gravitational stresses become photons, the white hole explodes); there is not only one big bang but there are smaller ones everywhere, H. Arp (5); the infinite universe on the whole is stable since gravitational collapses big enough turn into expansion; the very high energy background radiation and its particles are relicts of exploding white holes; further ideas s. (6), (7).

Conclusion

While LI of SR gives evidence for an ether and more reasonable philosophical principles LI of GR has provable physical consequences *e.g.* gravitational collapse (black hole) may turn into expansion (white hole) if its kinetic energy increases.

Acknowledgments

My thanks are due to F. Selleri, H. Arp, E. Kapuscik and all the other participants in the "Relativistic Physics and some of its Applications" International Conference Athens 1997.

References

[1]   Selleri, F.: Noninvariant One-Way Velocity of Light. *Found. of Phys*. 26, 641 (1996)

[2]   Sexl, R. U.; Sexl, H.: *Weisse Zwerge – Schwarze Loecher*. 2. Aufl. Braunschweig, Wiesbaden: Friedr. Vieweg u. Sohn 1979.

[3]   Stephani, H.: *Allgemeine Relativitaetstheorie*. 4. Aufl. Berlin: Deutscher Verlag der Wissenschaften 1991.

[4]   Weyl, H.: *Raum, Zeit, Materie*. Berlin: Springer Verlag 1920.

[5]   Arp, H.: *Quasar Creation and Evolution into Galaxies*, Max-Planck-Institute fuer Astrophysics Jan. 1997, private communication and Arp, H.; Narlikar, J.: Flat spacetime cosmology: a unified framework for extragalactic redshifts, *Astrophysical J*. 405, 51 (1993).

[6]   Brandes, J.: *Die relativistischen Paradoxien und Thesen zu Raum und Zeit. Interpretationen der speziellen und allgemeinen Relativitätstheorie*. 2. Aufl. Karlsbad: VRI 1995

[7]   Brandes, J.; Czerniawski, J.; Hoyer, U.; Selleri, F.; Wohlrabe, K.: *Die Einstein'sche und lorentzianische Interpretation der speziellen und allgemeinen Relativitaetstheorie*. Karlsbad: VRI 1997.

The Simplest Inflationary Scenario in Relativistic Quantum Cosmology

Zbigniew Jacyna-Onyszkiewicz
Faculty of Physics
A.Mickiewicz University
61-614 Poznań, Poland

Bogdan Lange
Institute of Philosophy and Sociology
Gdańsk University
80-851 Gdańsk, Poland

A proposition of the simplest quantum relativistic model of the universe is analysed with an emphasis on the inflationary scenario. Although it is a much simplified model of the closed universe, it permits presentation of the most important features of the relativistic quantum cosmology.

I. Introduction

According to the theory of relativity, the universe homogeneity on a large scale implies a homogeneity of the curvature of the three-dimensional space of the universe [1]. Should we wish to construct a sensible model of the universe consistent with the quantum theory principles, we have to assume that the space of the universe is a three-dimensional hypersphere as only for such a space the total energy of the universe is finite and equal to zero [1]. In the other two possible kinds of homogeneous space the total energy of the universe is infinite. Moreover, the three-dimensional hypersphere has a finite volume and includes the greater distance from the observer. Denoting such a distance by l_0 at the present moment, we can calculate it at the moment t from the formula:

$$l(t) = a(t)l_o \quad [a(t) \geq 0],$$ (1)

where $a(t)$ is the scale factor.

If $\hat{H}_g + \hat{H}$ stands for the operator of the energy of the whole universe, assuming that the universe space is a three-dimensional hypersphere, we arrive at the eigenequation of the Wheeler-De Witt type, in the form:

$$\left(\hat{H}_g + \hat{H}\right)|\Phi\rangle = 0,$$ (2)

where $\hat{H}_g$ is the operator of the gravitational energy and $\hat{H}$ is the operator of the energy of the universe matter. The question arises for which the quantum effects in the description of the universe become important. The greatest distance in the universe can be treated as an observable endowed with a hermitian operator $\hat{l} = \hat{l}^+ = l$. The expected value of this observable is given by the expression:

$$\langle \hat{l} \rangle = \langle \Phi(l) | \hat{l} | \Phi(l) \rangle, \tag{3}$$

and uncertainty of this observable is given by:

$$\Delta l = \sqrt{\langle \Phi(l) | \left(\hat{l} - \langle \hat{l} \rangle \right)^2 | \Phi(l) \rangle}. \tag{4}$$

Quantum effects are important in the universe in which

$$\frac{\Delta l}{|\langle \hat{l} \rangle|} \approx 1. \tag{5}$$

The three universal constants: the gravitational Newton constant G which is a universal constant in gravitational interactions, the Planck constant $\hbar$ which is a universal constant appearing in the microscopic description and c the light velocity in vacuum, can be combined into a universal constant with the characteristics of length l_p known as the Planck length: $l_p = \sqrt{G\hbar c^{-3}} \cong 1.6 \cdot 10^{-35} m$. In a similar way we can define a universal constant of time characteristics: $t_p = l_p c^{-1} = \sqrt{G\hbar c^{-5}} \cong 5.4 \cdot 10^{-44} s$ known as the Planck time, and a universal constant mass with characteristics: $m_p = \sqrt{c\hbar G^{-1}} \cong 2.2 \cdot 10^{-8} kg$ known as the Planck mass.

Thus we can expect that eq. (5) holds in the universe in which the maximum distance l is comparable with the Planck length l_p, so in which $l \approx l_p$. It also means that the classical description of the universe is possible only if $l \gg l_p$.

As follows from the form of eq. (2) the total energy of the universe is equal exactly to zero. The positive energy of matter contained in the universe is exactly compensated by the negative gravitational energy [1]. In order to present eq. (2) in the explicit form, we shall write it in the base of the eigenvectors of the operator $\hat{l}$ meeting the eigenequation $(l - l') | l' \rangle = 0$, analogous to the eigenequation of the position operator. The conditions for the completeness and orthonormality of these vectors take the form:

$$\int_0^\infty dl' \, | l' \rangle \langle l' | = 1, \quad \langle l | l' \rangle = \delta(l - l'). \tag{6}$$

Multiplying eq. (2) from the left by $\langle l |$ and making use of the conditions (6), we get;

$$\langle l|(\hat{H}_g + \hat{H})l|\Phi\rangle = \int_0^\infty dl'\,\langle l|(\hat{H}_g + \hat{H})|l'\rangle\langle l'|\Phi\rangle = 0.$$

Then, assuming that $\hat{H}$ does not act on the eigenvectors $|l\rangle$, this equation can be rewritten as:

$$\left(\hat{H}_g(l) + \hat{H}\right)|\Phi(l)\rangle = 0, \tag{7}$$

where: $\langle l|\hat{H}_g|l'\rangle = \hat{H}_g(l)\delta(l-l')$ and $|\Phi(l)\rangle = \langle l|\Phi\rangle$ is the state vector in the space of the hamiltonian $\hat{H}$ states, where the hamiltonian describes the matter of the universe. The state vector $|\Phi(l)\rangle$ is a function of the greatest distance in the universe $0 \le 1 < \infty$.

II. Schrödinger equation from Quantum Cosmology

According to the results obtained within the general theory of relativity [2] the gravitational energy can be formally presented as a sum of the kinetic and potential ones. Therefore we postulate that $\hat{H}_g(l)$ can be formally presented as (the parameter representing the ambiguity in the ordering of non-commuting operators $\hat{l}$ and $\hat{p}$ is equal to 0):

$$\hat{H}_g(l) = \frac{\hat{p}^2}{2m_p} + V(\hat{l}), \tag{8}$$

where $\hat{p}$ is the operator of the momentum defined as:

$$\hat{p} = -i\hbar\frac{\partial}{\partial l}, \tag{9}$$

and $V(\hat{l})$ is the operator of the potential energy. We shall assume the following simple expression for the potential energy:

$$V(\hat{l}) = -\frac{1}{2}m_p c^2 \left(\alpha + \frac{l}{l_p}\right)^2, \tag{10}$$

where α is a constant defined by the initial conditions.

The eigenequation (2) takes the form:

$$\left(\frac{\hat{p}^2}{2m_p} - \frac{1}{2}m_p c^2 \left(\alpha + \frac{l}{l_p}\right)^2 + \hat{H}\right)|\Phi(l)\rangle = 0. \tag{11}$$

Let us note that this equation does not depend on time, however we will show that for a universe of $l > l_p$ there is a parameter numbering states which can be interpreted as time t.

We assume that a solution of eq. (11) can be written in the form used in the so-called quasi-classical approximation (WKB solution):

Open Questions

$$\left|\Phi(l)\right\rangle = e^{i\frac{S(l)}{\hbar}}\left|\Psi\right\rangle, \tag{12}$$

where the terms $e^{i\frac{S}{\hbar}}$ and $\left|\Psi\right\rangle = \left|\Psi(l)\right\rangle$ describe the rapidly and slowly changing parts of the state vector, respectively, and $S(l)$ is the gravitational function of the universe action.

Introducing (9) and (12) into the eigenequation (11) we get:

$$\left(-\frac{\hbar^2}{2m_p}\frac{\partial^2}{\partial l^2} - \frac{1}{2}m_p c^2\left(\alpha + \frac{l}{l_p}\right)^2 + \hat{H}\right)e^{i\frac{S(l)}{\hbar}}\left|\Psi\right\rangle = 0. \tag{13}$$

Assuming that $\hat{H}$ acts only on the state vector $\left|\Psi\right\rangle$, and having calculated the second derivative with respect to l and divided equation (13) by $e^{iS(l)/\hbar}$, we arrive at

$$\left(\frac{1}{2m_p}\left(\frac{\partial S}{\partial l}\right)^2 - \frac{1}{2}m_p c^2\left(\alpha + \frac{l}{l_p}\right)^2\right)\left|\Psi\right\rangle$$
$$-\frac{i\hbar}{2m_p}\left(\frac{\partial^2 S}{\partial l^2}\left|\Psi\right\rangle + 2\frac{\partial S}{\partial l}\frac{\partial\left|\Psi\right\rangle}{\partial l}\right) + \hat{H}\left|\Psi\right\rangle \tag{14}$$
$$-\frac{\hbar^2}{2m_p}\frac{\partial^2\left|\Psi\right\rangle}{\partial l^2} = 0.$$

A comparison of the terms of the same powers of $\hbar$ brings about a set of three equations:

$$\left(\frac{\partial S}{\partial l}\right)^2 - m_p^2 c^2\left(\alpha + \frac{l}{l_p}\right)^2 = 0, \tag{15}$$

$$-\frac{i\hbar}{2m_p}\left(\frac{\partial^2 S}{\partial l^2}\left|\Psi\right\rangle + 2\frac{\partial S}{\partial l}\frac{\partial\left|\Psi\right\rangle}{\partial l}\right) + \hat{H}\left|\Psi\right\rangle = 0, \tag{16}$$

$$-\frac{\hbar^2}{2m_p}\frac{\partial^2\left|\Psi\right\rangle}{\partial l^2} = 0. \tag{17}$$

The first of them is the Hamilton -Jacobi equation for gravitational interactions, and in this equation the momentum is defined as the derivative:

$$p = \frac{\partial S}{\partial l}. \tag{18}$$

On the other hand, the momentum is also defined as

$$p = m_p\frac{\partial l}{\partial t}. \tag{19}$$

A comparison of these equations gives:

$$\frac{\partial S}{\partial l} = m_p \frac{\partial l}{\partial t}.$$

(20)

Putting this equation into the Hamilton-Jacobi equation assuming that $\frac{\partial l}{\partial t} \geq 0$, we get:

$$t_p \frac{\partial l}{\partial t} = \alpha l_p + l.$$

(21)

whose solution is:

$$l = Ae^{\frac{t}{t_p}} - \alpha l_p.$$

(22)

The integration constant A and the constant α are determined by the initial conditions:

$$l(t = 0) = 0 \quad and \quad l(t = t_p) = l_p$$

(23)

and having found them we arrive at the exponential time dependence of the greatest distance in the universe:

$$l = l_p (e-1)^{-1} \left(e^{\frac{t}{t_p}} - 1 \right).$$

(24)

The period of the exponential expansion of the universe is known as the inflation process [3-8].

Equations (15) and (24) imply:

$$\frac{\partial S}{\partial l} = \pm \frac{m_p}{t_p} \left[(e-1)^{-1} l_p + l \right]$$

(25)

and thus:

$$S = \pm \frac{m_p}{t_p} \left((e-1)^{-1} l_p l + \frac{1}{2} l^2 \right) + S_\pm,$$

(26)

where S_+ and S_- are the integration constants.

Introducing (20) and (25) into (17) we obtain:

$$-i\hbar \frac{\partial |\Psi\rangle}{\partial t} - \frac{i\hbar}{t_p} |\Psi\rangle + \hat{H} |\Psi\rangle = 0.$$

(27)

The second term of this equation can be omitted as it is a constant, and finally we arrive at:

$$i\hbar \frac{\partial |\Psi\rangle}{\partial t} = \hat{H} |\Psi\rangle,$$

(28)

which is the Schrödinger equation. This equation describes the time changes of the slowly changing part of the state vector depending on the hamiltonian describing the matter of the universe.

The Schrödinger equation is usually treated as one of the main postulates of the quantum theory, however, in the line of our reasoning it is an approximate equation following from the exact eigenequation (2). Thus, the Schrödinger equation can be treated as a semiclassical approximation of the exact equation (2) which does not depend on time. Time dependence appears only in the semiclassical approximation when the uncertainty Δl can be neglected. This is the reason why a time operator does not appear in the quantum theory. In the latter time is not an observable but a parameter. The above is a very important result of the relativistic quantum cosmology.

Substituting (26) into (12) under the assumption that $\left|\Phi(l=0)\right\rangle = \left|0\right\rangle$, we get the following expression for the universe state vector:

$$\left|\Phi(l)\right\rangle = \sin\left((e-1)^{-1}\frac{l}{l_p}+\frac{1}{2}\left(\frac{l}{l_p}\right)^2\right)\left|\Psi\right\rangle. \tag{29}$$

The state when l=0 can be called nothingness. In this state there is no space, time or matter, and the state vector $\left|\Phi(l=0)\right\rangle$ is the zero vector of the Hilbert space. However, this is not nothingness in the ontic sense (see ref. [9] for more details). After the creation for $t \geq t_p$ the universe undergoes exponential expansion according to eq. (24).

The general theory of relativity (Friedman-Robertson-Walker metric) which is a classical theory valid for $t > t_p$, for the proposed model of the universe leads to the following equation [1]:

$$\left(\frac{dl}{dt}\right)^2 + \pi^2c^2 = \frac{8\pi G}{3}\rho l^2, \tag{30}$$

where ρ is the mass density in the universe.

Introducing eq. (24) into eq. (30), on condition that $t >> t_p$ we get:

$$\rho \cong \frac{3}{8\pi G t_p^2} = \text{constant(t)} \tag{31}$$

Taking into regard eq. (1), for $t >> t_p$, that is when :

$$\exp\left(-\frac{t}{t_p}\right) << 1, \tag{32}$$

we can rewrite eq. (30) as:

$$\left(\frac{dl}{dt}\right)^2 = \frac{8\pi G}{3}\rho l^2. \tag{33}$$

The exponential dependence $l = l(t)$ follows from the formula assumed for the potential (10). When $l >> l_p$, also other terms of the equation for $V(\hat{l})$ can be important, therefore we assume that the exponential expansion of the universe takes place only to a certain moment t_1, and later the universe evolution can be described by the classical equation (33). The latter equation corresponds to the universe of the flat space [1], so for $t > t_1$ with the accuracy determined by expression (32), the universe has non-curved space.

Taking into regard the requirement of the constant mass density of the universe in the inflation period, see eq. (31), the mass of the universe must grow in this period. If at the Planck time t_p the universe mass was equal to the Planck mass m_p, then after the end of the inflation period at t_1, the universe mass, as follows from eq. (24) is:

$$m(t_1) = m_p(e-1)^{-3}\left(e^{\frac{t_1}{t_p}} -1\right)^3. \tag{34}$$

According to some more complex models [4,5], at the end of the inflation period oscillations of the field describing the matter appear and cause a rapid increase of the universe temperature. Moreover, a state of strong thermodynamical nonequilibrium in the inflation period seems to be a necessary condition for production of a certain excess of matter over antimatter. In the present Universe there are $2 \cdot 10^{10}$ photons produced as a result of matter and antimatter annihilation per a single nucleon. After the inflation period which is estimated to have ended in $t_1 < 10^{-35}$ s, the universe was hot, of a temperature of 10^{28} K and a huge but finite volume, much greater than that of the observable part of the universe. The appearance of such a universe is frequently referred to as the big bang. If in the Planck time $t_p \approx 10^{-44}$ s, the greatest distance in the universe was equal to the Planck distance $l_p \approx 10^{-35}$ m, then after the inflation period, at $t_1 \approx 10^{-35}$ s it grows enormously to reach, according to eq. (24), $l(t_1) \approx 10^{10^9}$ m. This is the distance much greater than the greatest possible for us to observe in the universe estimated as 10^{26} m. The universe has become so huge that we are able to observe only its very small, insignificant, part. That is why the part of the universe observable by us seems flat . At present the universe is smooth and homogeneous as all inhomogeneities have been extended 10^{10^9} times. A large density of the original monopoles and other exotic topological defects whose existence follows from of the great unification theory (GUT), has been reduced by a factor of the order of $10^{-3 \cdot 10^9}$. A single quantum fluctuation might have expanded in a split of a second into a huge universe which we shall never be able to fully recognize.

After the inflation period, the further global evolution of the universe is governed by the classical equation (33) supplemented with the condition following from the energy conservation principle [1]:

$$\rho(t)l^4(t) = \rho(t_1)l^4(t_1). \tag{35}$$

Having put this condition into eq. (33) we get:

$$\left(\frac{dl}{dt}\right)^2 = \frac{8\pi G}{3}\frac{\rho(t_1)l^4(t_1)}{l^2}. \tag{36}$$

whose solution gives:

$$l(t) = l(t_1)\left(\frac{t}{t_1}\right)^{1/2} \qquad dla \quad t \geq t_1. \tag{37}$$

III. Conclusion

In the above presented considerations the most important point was to show the possibility of derivation of the Schrödinger equation within the inflationary scenario of the quantum cosmology, so we have been considering the extremely simplified model of the universe. Rich literature on the subject proposes much more sophisticated versions of quantum cosmology [6-8], however, they are all more or less speculative because a generally accepted quantum theory of gravitation has not been developed yet [2]. In the quantum cosmology we do not know a full form of eq. (2) and the initial condition for the state vector $|\Psi\rangle$, nevertheless, even under substantial simplifications of the quantum cosmology formulations we are able to reproduce in a general aspect the origin and advent of the universe as a whole.

References

[1]   L.D. Landau and E.M. Lifshitz, *Field Theory*, Pergamon, Oxford 1964.

[2]   E. Alvarez, *Rev. of Mod. Physics*, **61** (1989) 561.

[3]   A.H. Guth, *Phys.Rev.* D **23** (1981) 347.

[4]   A.D. Linde, *Phys.Lett.* **129B** (1983) 177.

[5]   A.D. Linde, "Inflation and Quantum Cosmology," in: *300 Years of Gravitation*, edited by S.W. Hawking and W. Israel, Cambridge University Press, Cambridge 1987, pp.604-630.

[6]   A. Vilenkin, *Phys.Lett.* **117B** (1982) 25.

[7]   A. Vilenkin, *Nuclear Phys.* B **252** (1985) 141.

[8]   J.B. Hartle and S.W. Hawking, *Phys.Rev.* D **28** (1983) 2960.

[9]   Z. Jacyna-Onyszkiewicz, *Phys. Essays* **10** (1997).

An Analysis of 900 Rotation Curves of Southern Sky Spiral Galaxies: Are the Dynamics Constrained to Discrete States?

D. F. Roscoe
School of Mathematics
Sheffield University, Sheffield, S3 7RH, UK.
Email: D.Roscoe@ac.shef.uk
Tel: 0114-2223791, Fax: 114-2223739

One of the largest rotation curve data bases of spiral galaxies currently available is that provided by Persic and Salucci (1995; hereafter, PS) which has been derived by them from unreduced rotation curve data of 965 southern sky spirals obtained by Mathewson, Ford and Buchhorn (1992; hereafter, MFB). Of the original sample of 965 galaxies, the observations on 900 were considered by PS to be good enough for rotation curve studies, and the present analysis concerns itself with these 900 rotation curves.

The analysis is performed within the context of the hypothesis that velocity fields within spiral discs can be described by generalized power-laws. Rotation curve data is found to impose an extremely strong and detailed correlation between the free parameters of the power-law model, and this correlation accounts for virtually all the variation in the pivotal diagram. In the process, the analysis reveals completely unexpected structure which indicates that galactic dynamics are constrained to discrete states.
Keywords: spiral galaxies, rotation curves

1. Introduction

The following analysis is performed within the context of a prediction arising from a theory of weak-field slow-motion gravitation in material distributions that motions in spiral discs conform to the power-law structure

$$V_{rot} = AR^{\alpha}, \qquad V_{rad} = BR^{\alpha}, \qquad \alpha \geq -1,$$

where V_{rot} and V_{rad} are the rotational velocity and radial velocity respectively, and for constants A and B; since one of these can be absorbed into the scaling of the problem, it can be assumed that there are only two free parameters,

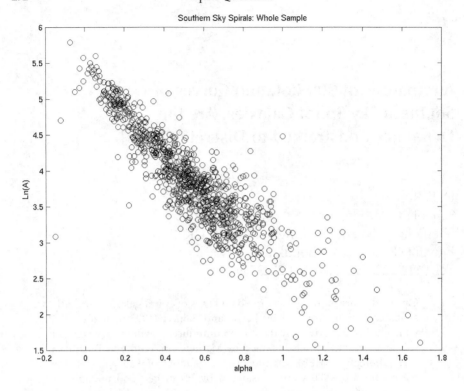

Figure 1. Plot of ln(A) against α for whole sample

(A,α) say. A crucial result, from the point of view of reconciling the with the observations, is the constraint $\alpha \geq -1$, a result which immediately removes any mystery associated with the existence of 'flat' rotation curves.

The foregoing solution was derived purely from an analysis of the dynamics, with mass-conservation being ignored. However, the additional constraint of mass-conservation can do no more than impose an additional constraint on the space of solutions (1). This amounts to a correlation being imposed on the free-parameters, (A,α), of the model, and it can be shown that the existence of a perfect correlation would imply the model is exact for the physics. However, rotation-curve data is extremely noisy, and so we cannot expect perfect correlations; it follows that, since perfect correlations cannot be expected, the whole argument revolves around the *quality* of any correlations uncovered.

From the point of view of the second part of the following analysis §6, it is important to know that it was stimulated as a consequence of a trial investigation using a very small independent sample provided in by Rubin, Ford & Thonnard (1980, RFT hereafter). This trial was sufficient to give a quantitative element to the hypothesis concerning the (A,α) correlation and,

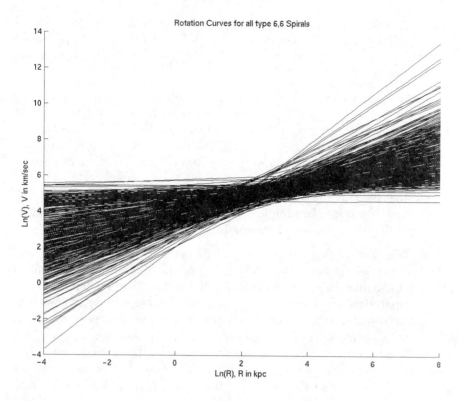

Figure 2. All Type 6 Rotation Curves in (ln(R),ln(V)) Plane

additionally, gives rise to an hypothesis concerning the distribution of ln(A). Referring to these hypotheses as H_1 and H_2 respectively, they can be stated as

- H_1: ln(A) and α are linearly correlated in a negative sense;
- H_2: When linear scales are assigned on the basis of the assumption that $H = 50/km/sec/Mpc$, then ln(A) is constrained to take values which lie within ±0.15 of an integer or half-integer; this is equivalent to the hypothesis that the allowed values of ln(A) have a periodic structure, with period approximately 0.5.

2. The Data

The data given by PS is obtained from the raw Hα data of MFB by deprojection, folding and cosmological redshift correction. For any given galaxy, the data is presented in the form of estimated rotational velocities plotted against angular displacement from the galaxy's centre; estimated linear scales are not given and no data-smoothing is performed.

The analysis proposed here requires the linear scales of the galaxies in the sample to be defined which, in turn, requires distance estimates of the sample galaxies from our own locality. This information is given in the original MFB paper in the form of a Tulley-Fisher (TF hereafter) distance estimate given in km/sec, and assumes $H = 85$km/sec/Mpc for the conversion. We have assumed:

- that the MFB method of presenting TF distances in km/sec, including their use of $H = 85$km/sec/Mpc, gives an accurate estimate to the cosmological component of the redshift in the sample galaxies. This assumption is actually central to MFB's analysis since this analysis was primarily designed to give accurate determinations of peculiar velocities in the sample;
- that the criteria by which RFT selected, observed and processed their very much smaller sample ensured relatively accurate determinations of the corresponding cosmological redshifts.

Given these assumptions, then nominal agreement between the RFT and MFB linear scales can be obtained by converting the MFB distances, as quoted in km/sec, to a linear scale using the RFT value of $H = 50$km/sec/Mpc.

An analysis of the distribution of morphological types in the PS data base shows that the great majority of the selected galaxies are of types 3,4,5 and 6, with only two examples of types 0,1,2 and a tail of 31 examples of types 7,8,9. To maximise the homogeneity of the analysed data, the distribution tails—consisting of the morphological types 0,1,2,7,8 and 9—were omitted, and the remaining 867 galaxies partitioned into the classes {3}, {4,5} and {6}. These contained, respectively, 306, 177 and 384 galaxies. A separate analysis was performed on each of these three partitions.

3. Is There Any Correlation Between A And α?

The basic assumption is simply that rotation velocities behave as $V_{rot} = AR^\alpha$ and the discussion of §1 concluded there should be a correlation between A and α. Since the regression constants arising from a linear regression of $\ln(V_{rot})$ on $\ln(R)$ give estimates of $\ln(A)$ and α, our basic analysis performs a linear regression on each of the 867 rotation curves, and records the pair $(\alpha,\ln(A))$ for each galaxy.

Fig. 1 gives the scatter plot of $\alpha,\ln(A)$ for the full sample and shows that there exists an extremely strong negative $\alpha,\ln(A)$ correlation. The corresponding figures for the individual galaxy type-classes (not shown) are similar in all respects, each occupying similar areas in their respective $\alpha,\ln(A)$ planes and each displaying the same fan-like structure going from a broad spread of points at the bottom right-hand of the figure to a narrow neck at the top left-hand of the figure. For the remainder of this paper, discussion will be restricted to the type-class {6} (that is, late-types) galaxies, since the conclusions arising here are broadly repeated in the two remaining classes.

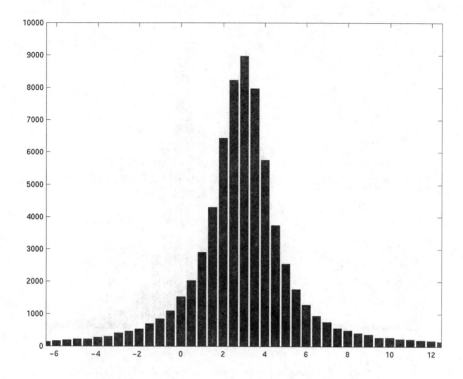

Figure 3. Frequency of Intersections on ln(R) axis

4. A Consequence of Linear Correlation Between α and ln(A)

A tentative interpretation of Figure 1 is that

$$\ln(A) = a_0 + b_0\,\alpha, \tag{1}$$

where a_0 and b_0 are constants which might differ between galaxy type-classes. It is easily shown how this implies that all the rotation curves in any given type-class,

$$\ln(V_{rot}) = \ln(A) + \alpha\ln(R), \tag{2}$$

intersect at the fixed point $(-b_0,a_0)$ in the $(\ln(R),\ln(V))$ plane. If this point is denoted as $(\ln(R_0),\ln(V_0))$, then (2) is more transparently written as

$$\ln(A) = \ln(V_0) - \alpha\ln(R_0) \tag{3}$$

Whilst the idea of a single intersection point for all rotation curves in the class seems rather extreme, it is unambiguously deduced from the most obvious interpretation of Figure 1. The most direct test of the statement is simply to plot the corresponding rotation curves in the $(\ln(R),\ln(V))$ plane, and to observe their actual behaviour.

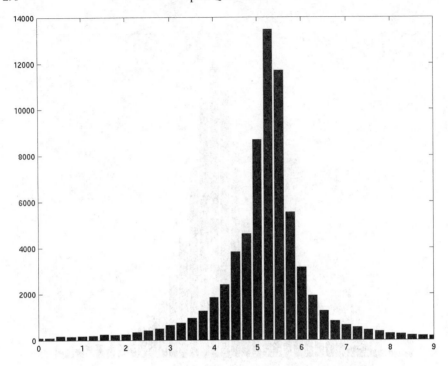

Figure 4. Frequency of Intersections on ln(V) axis

This is done in figure 2 for the type-class {6}—the plots for the type-classes {3} and {4,5} are similar in all details. The figure gives a very clear sense of convergence in a relatively small region of the ln(R),ln(V) plane. As a secondary means of illustrating this apparent convergence, we have calculated the coordinates of intersection between all possible pairs of rotation curves in the plane and have plotted the frequencies of intersection for type-class {6} along the ln(R) and ln(V) axes respectively in Figures 3 and 4. It is clear from these figures that there is a very sharp peak of intersection points at ln(R),ln(V) ≈ (3, 5.2). The results for the type-classes {3} and {4,5} are similar.

A powerful geometric test of the convergence statement can be formed from the realization that, if the rotation curves really do converge on a single point in the ln(R),ln(V) plane, then all of the rotation curves will *transform into each other* under rotations about the convergence point, (ln(R_0),ln(V_0)), in this plane; that is, the individual rotation curves in the set of all rotation curves associated with a given ln(R_0),ln(V_0) are equivalent to within a rotation about this point in the ln(R),ln(V) plane.

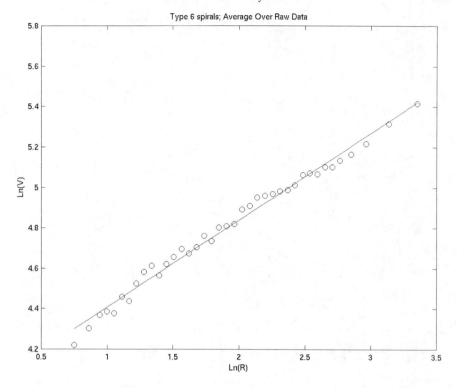

Figure 5. Type 6 spirals; Average over raw data

5. Testing H_1: Is the Linear Assumption Reasonable?

Suppose that, as Figure 1 suggests, all the rotation curves in a given type-class pass through $\ln(R_0), \ln(V_0)$ in the $\ln(R), \ln(V)$ plane. Then an arbitrarily chosen straight line passing through this point can be defined as a standard 'reference line' into which every rotation curve in the class can be transformed by a simple bulk-rotation about $\ln(R_0), \ln(V_0)$. Since, according to this idea, the rotation curves are reduced to equivalence by the rotation, then the process of forming an "average rotation curve" from the set of rotated such curves should greatly reduce the internal noise associated with the individual rotation curves, and we would expect the resulting average curve to be a very close fit to the standard reference line, referred to above.

For the analysis, the galaxies are grouped into type-classes {6}, {4,5} and {3} and, for brevity, we only give the results for type-class {6} here—the results for the other type-classes are similar. The fixed point in the $(\ln(R), \ln(V))$ plane is estimated (using a minimisation procedure not described here) as (3. 615,5. 416).

Figure 5 shows the plot which arises from averaging all rotation curves in each of the type-classes *directly* in their raw state (that is, without rotating

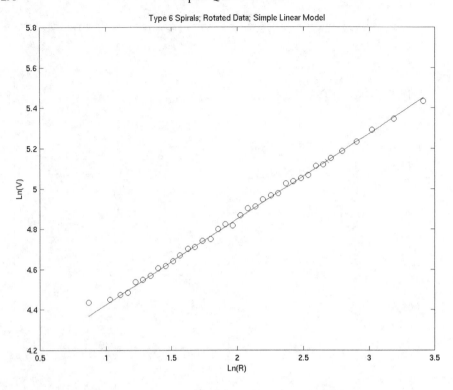

Figure 6. Type 6 spirals; Averaged rotated data testing H_1

about the fixed point $\ln(R_0),\ln(V_0)$. Figures 6 shows the plot which arises when, for type-class $\{6\}$, the rotation curves are rotated about the fixed point so that they coincide *in a least-square sense* with the reference line defined to pass through $(\ln(R_0),\ln(V_0))$, and fixed (arbitrarily) so that $\ln(A) = 4$, and the results averaged over all the rotation curves in the class. The scatter present in Figure 5 is virtually eliminated, providing the strongest possible evidence for the idea of the equivalence of rotation curves with respect to rotations about particular fixed points in the $\ln(R),\ln(V)$ plane.

6.   Testing H_2: An Analysis of the $\ln(A)$ Distribution

A preliminary analysis of the RFT data, mentioned in §1, gave rise to the hypothesis H_2 that the $\ln(A)$ parameter was constrained to lie within ± 0.15 of an integer or half-integer. On the original RFT data, this was observed to occur at a rate which had a by-chance probability of approximately 0.002—which is sufficiently small to make it natural to pose the question 'is this an artifact of the RFT data, or does it reflect some underlying constraint imposed on the physics of rotation curves?' We have analysed the distribution of the $\ln(A)$ values obtained from the PS data base, and present the results in the following.

The specific question posed was *how many of the ln(A) values lie within ±0.15 of either an integer or half-integer value?* Because these ranges cover 60% of the real line, they are referred to as 60% bins in the following. The range of ±0.15 was chosen simply because this is the question raised by the RFT data. The details of the data-reduction used in this analysis are not given here but, briefly, they amount to a *prior-decided* means of minimizing the effect of the nuclear bulge on the rotation-curve calculations. The results of the analysis are condensed into Table 1 and, in this table, a 'Hit' is defined to be when a particular ln(A) lies within one of the 60% bins.

Table 1: Distance Scale = TF3

Galaxy Types	Sample Size	Number Of Hits	Probability of Chance Event
0..9	900	586	0.909×10^{-3}
0..5	485	294	0.410
6..9	415	292	0.731×10^{-5}

From the table, we see that when the analysis is performed over the whole sample, then the number of hits is 586 from a possible 900 trials; the odds of this being a chance event are about 1:1100. As with the RFT data, this is an interestingly small probability, but is not sufficiently conclusive. However, when the data is partitioned into two classes consisting of the type-classes {0,1,2,3,4,5} and {6,7,8,9} a totally different picture emerges: specifically, on the class {0,1,2,3,4,5} there are 294 hits out of 485 trials. The odds of this occurring by chance are about 4:10; this is exactly what is to be expected if there is no effect. By contrast, on the class {6,7,8,9}, there are 292 hits out of 415 trials. The odds of this occurring by chance are about 1:137,000 which is *extremely* small. There remains the possibility that there is a significant effect on the {0,1,2,3,4,5} data which is masked either by being phase-shifted with respect to the effect on the {6,7,8,9} data, or by having a different period. This possibility was tested by running both sets of data through a power spectrum analysis program which searches automatically for periodicities at arbitrary phases. This secondary analysis confirmed the positive results on the late-type spiral data, and found no evidence of any effect of the early-type spiral data.

There are two broad questions which immediately arise:

- Firstly, the analysis concerns the distribution of the ln(A) values. However, A arises originally in the power law $V_{rot} = AR^\alpha$ from which it is clear that its value for any given rotation curve depends on the definition of the linear scale—which, here, has been defined on the basis of the assumption that $H = 50$km/sec/Mpc. This fact makes it highly implausible that ln(A) is constrained to take on exactly integer or half-integer values.
- Secondly, if the effect on the {6,7,8,9} class is real, why does it appear to be absent from the {0,1,2,3,4,5} class?

In answer to the first question, it is sufficient to note that a 'Hit' is defined to occur when a given $\ln(A)$ value is within the 60% bin. The 60% bin, by definition, occupies 60% of the real line so that, although the two parts of the bin are centred on integer and half-integer values respectively, there is plenty of room for $\ln(A)$ to have a preference for discrete values without these necessarily being exactly integer or half-integer values. In answer to the second question, it is sufficient to hypothesise that the effect does exist for early-type spirals, but with a different period. The effect was spotted in the late-type spirals purely because of a fortuitous choice of the Hubble coefficient in the original analysis of Rubin *et al.*

If $\ln(A)$ is constrained to occupy discrete values, as the odds against a chance result of 1:137,000 appear to indicate, then, since $\ln(A)$ and α are known to be extremely strongly correlated, α must also be constrained to occupy discrete values. However, since the range of α is much less than that of $\ln(A)$, it is to be expected that any α-periodicity will be much less than the 0.5 of $\ln(A)$; this means that the 60%-binning method used to test the explicit hypothesis H_2 is not appropriate for testing α. Correspondingly, we used the method employed for the secondary analysis of H_2, the power-spectrum analysis program, as the primary means of searching for α-periodicities. This indicated good evidence for a period in α of 0.245, with a phase-shift of about 0.14 from the origin; noting that the predicted periodicity is about one-half of the $\ln(A)$ period, an alternative test became possible: specifically, if the α-values were multiplied by two (giving a period of 0.49), and shifted by 0.28 (double phase), then a significant effect should be detectable with the 60%-binning analysis. This analysis then recorded 286 hits from 415 trials, which has a by-chance probability of $0.104 \times 10^{-3} \equiv 1{:}9600$. However, it is to be noted that the $\ln(A)$ result was a prior quantitative prediction made on RFT data, and is therefore to be given a much greater weight. The latter results on the α data can only be considered as confirmatory, and used as a basis of formulating a specific hypothesis to be tested against independent data.

The wider implications of the present positive result (1:137,000 for a chance happening) obtained for the type-class {6,7,8,9} are that galactic dynamics are constrained to occupy discrete states.

Conclusions

The most interesting part of the forgoing analysis from the point of view of the ANPA programme is that of §6, which appears to strongly suggest that the dynamics of large scale structures exhibit discrete structure. The purpose of the earlier parts of the analysis is simply to emphasize that the power-law model reveals substantial other structure in the data. One can use the metaphor of spectacles: The wrong spectacles for the short sighted man add nothing to what he sees; but the right spectacles reveal a whole new world. The power-law model has revealed a considerable structure in rotation curve data and has led to the further hypothesis that—maybe—regular discreteness (quantization in some form) exists on very large scales.

References

[1] Mathewson, D. S., Ford, V. L., Buchhorn, M. 1992 A Southern Sky Survey of the Peculiar Velocities of 1355 Spiral Galaxies *Astrophys J. Supp.* **81** 413-659.

[2] Persic, M., Salucci, P., 1995 Rotation Curves of 967 Spiral Galaxies *Astrophys. J. Supp.* **99** 501-541.

[3] Rubin, V. C., Ford, W. K., Thonnard, N. 1980 Rotational Properties of 21 Sc Galaxies with a Large Range of Luminosities and Radii from NGC 4605 (R = 4kpc) to UGC 2885 (R = 122kpc) *Astrophys. J.* **238** 471-487.

Quantum Theory
and Relativity

Schrödinger's "Aether" Unifies Quantum Mechanics and Relativistic Theories

A.P. Bredimas
NCSR "Demokritos," Institute of Materials Science
15310 Aghia Paraskevi, Attiki – Greece
    and
Université Paris 7, UFR de Physique, Tour 33-34,4eme
Case 7008, 2 Place Jussieu, 75251 Paris Cedex 05, France
e-mail: bredimas@ccr.jussieu.fr and bredimas @cicrp.jussieu.fr
Fax: +33144275294 and Tel/Fax: +33160146424

Reminders of "contradictions in special relativity," the definitions of special relativity, of relative energy-rate motion theory, of the relativity of the Schrödinger's aether-material ruled by the transformations-isometries $L(v)$, $B(\varepsilon)$, $A(N)$ are given. The positively right answer to the question 'how to bridge the breach between quantum mechanics and relativistic theories' is given.

Introduction

The aim of the present talk is twofold: (a) to recall some basic aspects, (i) of the "crazy contradictions of special relativity," (ii) of the foundations of quantum mechanics, (iii) of the foundations of relative energy-rate motion (RE-RM) theory resulting from the non linear theory of matter, (iv) of the "relativity theory of the Schrödinger aether-material, and, then, (b) to deduce the unification of all of the above three relativistic theories, in the sense that the admissible respective coordinates transformations are identical for small values of the parameters which caracterize them.

In modern physics, both quantum and classical relativistic, there are two main difficulties: 1) the incredible situation of the *wrong applications of the Lorentz transformations* contradicting the principles of special relativity by the almost globality of physicists, except a few contradictors, and, on the other hand, 2) the very large breach and fracture between quantum mechanics and relativity.

Indeed, as concerns the first one, it is an obvious think that

(A) the commonly accepted, and almost every where reported in books,....
     theoretical explanation by the successive application of *two non-parallel*
     Lorentz-transformations of the old Thomas precession effect contradicts
     the principle of special relativity postulating that "uniquely parallel
     Lorentz-transformations corresponding to parallel 3-vector velocities of
     frames are admissible by the usual standard uniform motion kinematical
     principle": if not, there are non-zero accelerations, and this situation is
     beyond the special relativity area;

(B) The commonly accepted, and also, every where reported in research, as
     well   as   undergraduate,   books....,   application   of   the   Lorentz
     transformations to the circular electron trajectory with the argument
     "one can apply Lorentz transformations because the modulus of the 3-
     vector velocity is constant" again contradicts the above mentionned
     uniform kinematical principle of special relativity;

(C) All of the other theoretical explanations of effects or computations
     involving non-parallel, as well as parallel but non constant, Lorentz
     transformations, as the Sagnac effect,... again contradict the uniform
     motion kinematical principle of special relativity.

One concludes that the relativistic physicists community since 80 years,
and including the founders of special relativity theory, "speak" about
relativity and work outside the area of special relativity and make wrong
applications of Lorentz transformations, and of course, does not accept any
contradictor; is it "a crazy situation," or not?

I will not pursue here this analysis, and I refer the reader to my paper to
be published in *Outopia* [2] and [3 to 12].

The main points I will mention about quantum mechanics [1] are:

(a) Schrödinger's wave functions are solutions of the de Broglie wave-
     equation, $\left\{(N\dfrac{\partial}{\partial t})^2 - (\dfrac{\partial}{\partial x})^2\right\}\Psi(t,x) = 0$, where $N = (2mE_{kin})^{1/2} / E_{tot}$ is
     interpreted as an index of refraction, m, E_{kin}, E_{tot} are the mass, kinematical
     and total energies of the particle-system I consider, and of the
     Schrödinger equation,

     $i\hbar\dfrac{\partial}{\partial t}\Psi(t,x) = H\Psi(t,x)$ where h is the Planck constant and H the
     hamiltonian operator of the particle-system: here, for simplicity, 2-
     dimensional hyperbolic world is postulated, denoted M_2, with signature
     (+,–);

(b) The wave-packet representation will be considered in uniform motion in
     the case of non-localization of the system (free states, E_{kin} = constant ≥ 0),
     while localization of the system implies quantification (bound states,
     $E_{tot} < 0$);

(c)   the quantity $|\Psi(t,x)|^2$ is interpreted as the 'probability of presence of the

particle-system at the point (t,x) of M_2', $\int\limits_{-\infty}^{+\infty} |\Psi(t,x)|^2\,dx = 1$.

Definition 0.1. I will define "Schrödinger's aether-material" to be the medium, the index of refraction N of which characterizes all of its structure.

Remark 0.1. In Schrödinger's QM the characteristic parameter is the energy, which is the best possible physical quantity leading to conservation and Lorentz invariance.

The main points I will mention about the foundations of special relativity are: (a) the M_2 hyperbolic Minkowski manifold will be adopted for describing the physical space-time world; (b) then, the special theory of relativity is defined to be the study of the physics of the given particle-system inside the Lorentz-admissible coordinate-systems determined uniquely and exclusively by the Lorentz transformations representing uniform velocities motions,

$$\begin{bmatrix} t'(v) \\ x'(v) \end{bmatrix} = L(v)\begin{bmatrix} t \\ x \end{bmatrix} = (1-v^2)^{-1/2}\begin{bmatrix} 1 & -v \\ -v & 1 \end{bmatrix}\begin{bmatrix} t \\ x \end{bmatrix} \tag{1}$$

where, $v = Cnt, -1 < v < 1$ and $c = 1$ (choice of units system).

Remark 0.2. Special relativity and its equivalent (by definition) the Lorentz group of transformations of coordinates-systems are characterized by the velocity parameter, which is a 'bad parameter', because it is not conservable and does not lead to Lorentz invariance.

The main points I will mention about the Relative Energy-Rate Motion (RE-RM) theory [2 to 12] are:

(a)   RE-RM theory is originated as a first order dynamical approximation of the non linear theory of matter: the soliton representation of the particle-system is similar to that of the wave-packet and asymptotically tends to the usual point particle, while the "hydrodynamical-like generalized evolution processes" of it lead to the RE-RM theory, where one identifies coordinate-systems with E-energy state-frames of the particle-system: so, the passage from a coordinate-system to another is,here, interpreted as a "transition from the E(1)-energy state-frame to the E(2)-energy state-frame" of this particle-system;

(b)   The "relativity" of RE-RM theory is defined to be ruled by the "local isometries," below, where ε is the "total relative energy-rate function-parameter," the choice of the units system being defined by $c = 1$,

$$\begin{bmatrix} t'(\varepsilon) \\ x'(\varepsilon) \end{bmatrix} = B(\varepsilon)\begin{bmatrix} t \\ x \end{bmatrix} = (1-\varepsilon^2)^{-1/2}\begin{bmatrix} 1 & -\varepsilon \\ -\varepsilon & 1 \end{bmatrix}\begin{bmatrix} t \\ x \end{bmatrix} \tag{2}$$

$$\varepsilon = (2(E(2)-E(1))/E(2))^{1/2}, \ -1 < \varepsilon < 1$$

Remark 0.3. (i) RE-RM theory's version defined by the local isometries $B(\varepsilon)$ given by (3) is essentially characterized by the function-parameter $\mathring{a}$ which is directly connected to the energy parameters: this property is important for physical interpretations and permits to go far and far beyond the area of validity of the special relativity theory. In fact, (2) proves that RE-RM theory is very large extension of the usual special relativity, and in particular RE-RM includes gravitation, quantum effects and explains all the "unclear" until now experimental effects, as well as it solves all the SR's contradictions [2 to 12];

Also, both the non linear theory of matter and RE-RM theories contain a subcase which is equivalent to the Schrödinger's quantum mechanics [7,8,11].

The relativity theory of Schrödinger's aether-material concerns non localized wave-packets in "uniform motion," so, certainly, it is concerned by the conditions, $E_{pot} = 0$, $E_{kin} \geq 0$, $E_{tot} = m + E_{kin}$. In [3,4,7,11], I proved that the structure of the Schrödinger's aether-material can be refered to the "isotropic aether-material," and that the " states-frames " of this aether-material are identified to "coordinate-systems" characterized by the value of the parameter index of refraction N, and that these states-frames are ruled by the following isometries, the system of units always being defined by $c = 1$,

$$\begin{bmatrix} t'(N) \\ x'(N) \end{bmatrix} = (1 - N^2)^{-1/2} \begin{bmatrix} 1 & -N \\ -N & 1 \end{bmatrix} \begin{bmatrix} t \\ x \end{bmatrix} = A(N) \begin{bmatrix} t \\ x \end{bmatrix} \tag{3}$$

$$N = \frac{(2mE_{kin})^{\frac{1}{2}}}{E_{tot}} \text{ and } 0 \leq N < 1$$

All of the above is the minimum necessary to know about QM, SR, RE-RM and the relativity theory of the structure of Schrödinger's aether-material, in order to solve the question "how to bridge the breach between quantum mechanics and relativistic theories?," and I will proceed to survey the solution to this question in the next part.

Bridging the Gap between QM, SR, RE-RM Relativistic Theories

For simplicity of my argument (without any loss of mathematical rigor), I will uniquely consider the three kinds of isometries defined by (1), (2) and (3). Indeed, let us examine the expressions of these isometries, *for low values* of their parameters, v, ε, $N \in [0, 1]$, and *globally* in M_2, because here one is concerned by a *physical particle-system in uniform motion*. Then, one has in the *"weak kinematical energy approximation"* successively from (1), (2), (3) $\varepsilon \cong v$, $N \cong \varepsilon$ and so, one finds:

$$N \cong \varepsilon \cong v \geq 0 \tag{4}$$

Approximate relations (4) prove that the isometries, $L(v)$, $B(\varepsilon)$ and $A(N)$, are approximately identical in the framework of the weak kinematical energy approximation for the same always physical particle-system.

But, now, from Remark 0.2 one deduces that the "relativistic theories SR, RE-RM and the relativity of the Schrödinger's aether-material are approximately equivalent in the weak kinematical energy approximation," certainly, because the corresponding isometries $L(v)$, $B(\varepsilon)$ and $A(N)$, respectively, are approximately equivalent in the sense of (4) in this weak kinematical energy approximation.

This constitutes the establishment of the passage-bridge from QM to relativistic theories, with the help of coherent mathematical computation in group theory, and, consequently, the positively right answer to the above question.

Note that the developments of the non-linear matter and relative energy-rate motion theories are given in [2 to 12] and that the full explanations of the SR contradictions, the unexplained known experimental effects, as well as new effects are analyzed and developed in [2 to 12].

References

[1]   Wigner, E.P., *Group Theory and its Application to Quantum Mechanics of Atomic Spectra*, Academic Press, New York, 1959.

[2]   Bredimas, A.P., De l'équation des ondes aux théories quantiques, à la relativite et la théorie énergétique du mouvement relatif: une synthèse épistémologique de l'utopie à la réalité, to be published in *Outopia*.

[3]   Bredimas, A.P., "From wave equation..." in Greek, Preprint Nov. 1996, 14 pages.

[4]   Bredimas, A.P., "The phenomenology of Schrödinger's aether-material: the passage from quantum mechanics to relativistic theories," Preprint May 1997, 10 pages.

[5]   Bredimas, A.P., *Selected Conferences-Texts in Mathematics and Physics by A.P. Bredimas*, 205 pages,ISBN960-8492-06-8, Athens-Greece,1994.

[6]   Bredimas, A.P., *Preprints 1985-1996 by A.P. Bredimas*, 246 pages, ISBN960-8492-38-6, Athens -Greece, 1997.

[7]   Bredimas, A.P., "Spin-Lorentz-Maxwell transforms and the theory of evolutional classical and exotic space-time manifolds: the evolutional 6-D relativity," to be published.

[8]   Bredimas, A.P., Non linear theory... in Greek, *Proc. Conf. on causality and locality in microphysics*... 27-29 Nov. 1985, Athens-Greece, p.2o7, Edited by the Hellenic Physical Society and Prof. E.I. Bitsakis, Athens-Greece,1988.

[9]   Bredimas, A.P., *Non linear*..... in Greek, ΕΠΙΘΕΩΡΗΣΗΣ ΦΥΣΙΚΗΣ, No 15-16, p.78, Edited by the Hellenic Physical Society, Nov 1989.

[10]   Bredimas, A.P., *Comptes rendus Acad.Sci.Paris*, 272A,p.85,1971; *Let.N.Cimento*, No 24,6,p.569,1973; *Proc.Conf.Journées relativistes d'Angers*, CNRS-SMF, p.107,1979; ZAMM Band 63, p.316 and 318 from the *Proc. J. Bolyai*, Budapest 1982, *Zentral Blatt fur mathematik*, Band 587, p.208-209, 1986.

[11]   Bredimas, A.P., *Non linear theory of matter: mathematical and physical applications*, Series of 10 lectures given at the NCSR "Demokritos,"Dec.1984 the texts of the first 5 ones are published in Vol.1, 489 pages, ISBN (set)960-8492--00-9, ISBN T.1 96o-8492-01-7, Athens-Greece 1988 and Vol.2, p.490-p.771 ISBN(set)96o-8492-00-9,ISBN T.2 960-8492-02-5, Athens-Greece 1990.

[12]   Bredimas, A.P., (A) *Relative Energy-Rate Motion theory: Doppler applications a) to the Pound-Rebka-Snider experiments, b) to low temperature particles plasmas*, lecture given at NCSR "Demokritos," Inst. Nuclear Physics on 05/12/96; (B) *Relative*

Energy-Rate Motion theory: Doppler applications to materials, Lecture given at NCSR "Demokritos," Inst.of Materials Science, on 12/12/1996.

Entangled States and the Compatibilty Between Quantum Mechanics and Relativity

Augusto Garuccio
Dipartimento Interateneo di Fisica
INFN, Sezione di Bari
Via Amendola 173, 70126 Bari, Italy
Augusto.Garuccio@ba.infn.it

It is possible to prove that Quantum Mechanics is incompatible with Special Relativity if the postulate of the reduction of wave packet is included in the axiomes of this theory. In this paper an apparatus based on a Michelson interferometer with a phase-conjugate mirror is described; it is proved that it is able to transmit superluminal signals using the intrinsic non-locality of quantum mechanics. The result is obtained using only the axioms of quantum mechanics and the wave packet reduction postulate.
PACS: 03.65.Bz, 42.50.Wm

1. Introduction

In the last years, there has been increased interest in exploiting the intrinsic nonlocality of quantum mechanics to study new phenomena like teleportation of a quantum state and quantum cryptography.

We can briefly summarize the situation in this way: Most physicists working on the subject of the incompatibility between quantum mechanics and any local realistic theory are convinced that experiments have confirmed, without any reasonable doubt, the validity of quantum mechanical predictions and the intrinsic non-locality of this theory. However, they [1]im that this feature of quantum mechanics cannot give rise to any contradiction with relativity because it is not possible to use the correlated, non-local, quantum mechanical states to transmit signals faster than light.

A small number of physicists criticize these experiments on the basis of the important role played by some supplementary assumptions to [1]im violation of the predictions of Einstein Locality in this [1]ss of experiments, [1]iming that the issue of the compatibility between quantum mechanics and relativity has not yet been resolved.

Open Questions in Relativistic Physics
Edited by Franco Selleri (Apeiron, Montreal, 1998)

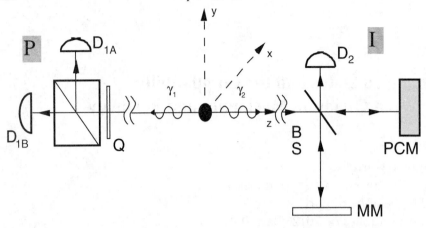

Fig.1. Outline of the superluminal quantum Telegraph (SQT). The first photon impinges on the polarization area P, while the second photon impinges on the interference area I. Q is a quarter-wave plate that can be inserted or removed from beam 1 to detect linear polarization or circular polarization, respectively. In the area I one metallic mirror of a Michelson interferometer has been substituted for a phase-conjugate mirror.

In this paper we will prove that the intrinsic nonlocality of quantum mechanics leads to the possibility of superluminal communication and, therefore, contradicts the postulates of relativity. More precisely, we will describe an apparatus, based on quantum mechanical wave packet collapse, which can permit binary signals to travel nearly instantaneously between distant points.

2.   The apparatus

Let us consider a source S of correlated photon pairs. The two photons are emitted in opposite directions along the z-axis and are correlated in polarization (Fig. 1). The state which describes the polarization of the pair is the entangled singlet state of positive parity

$$|\psi\rangle = \frac{1}{\sqrt{2}}\{|x_1\rangle|x_2\rangle + |y_1\rangle|y_2\rangle\} \tag{2.1}$$

where $|x_1\rangle$, $|x_2\rangle$ and $|y_1\rangle$, $|y_2\rangle$ are the two orthogonal polarization states along an arbitrary $x-y$ frame for the first and second photons, respectively.

The mathematical structure of the state (2.1) remains unchanged if the polarization basis is rotated through an arbitrary angle around the z-axis.

Moreover, the state (2.1) preserves its mathematical structure of a symmetric entangled state even when a circular polarization basis is used to represent it.

Indeed, if we define

$$|R_1\rangle = \frac{1}{\sqrt{2}}\{|x_1\rangle + |y_1\rangle\}, \quad |L_1\rangle = \frac{1}{\sqrt{2}}\{|x_1\rangle - i|y_1\rangle\}, \qquad (2.2)$$

$$|R_2\rangle = \frac{1}{\sqrt{2}}\{|x_2\rangle + |y_2\rangle\}, \quad |L_2\rangle = \frac{1}{\sqrt{2}}\{|x_2\rangle - i|y_2\rangle\}, \qquad (2.3)$$

we can write (2.1) as

$$|\psi\rangle = \frac{1}{\sqrt{2}}\{|R_1\rangle|R_2\rangle + |L_1\rangle|L_2\rangle\}. \qquad (2.4)$$

The equivalence between these two representations has been tested experimentally [1],[2],[3].

Let us suppose now that the "polarization detection region" P is placed in the optical path of the first photon beam at a distance very far from the source S. In this area an experimenter can choose whether to measure linear polarization or circular polarization. In the first case, he will detect the photon along one of the two channels of a linear polarization analyzer oriented along an arbitrary, but fixed, direction x. In the second case, he will insert the quarter-wave plate Q along the path of the beam with the optical axis at a 45° angle with respect the direction x. The quarter-wave plate changes the phase relationships between the linear polarization components of the light, so that, for example, light which enters the waveplate with RHC polarization will emerge linearly polarized along x, and light which enters with LHC polarization will emerge with linear y-polarization. The new x and y components are then separable, as before, by the linear polarization analyzer, and subsequent measurements of these components are equivalent to measuring circular projections of the initial polarization state.

In both cases, the experimenter will perform a polarization measurement on each impinging photon and, therefore, he will induce a quantum wave collapse.

If the experimenter at P measures linear polarization, half he trials will give the result $|x_1\rangle$, and for those trials the second photon can only be found in the state $|x_2\rangle$ due to the entangled form of (2.1). The other half of the trials will find the first photon in state $|y_1\rangle$, forcing the second to be in state $|y_1\rangle$. The action of a linear polarization measurement, therefore, is to collapse the system from the pure state (2.1) into the mixed state represented by the density operator

$$\rho_L = \frac{1}{2}\{|x_1\rangle|x_2\rangle\langle x_1|\langle x_2| + |y_1\rangle|y_2\rangle\langle y_1|\langle y_2|\}. \qquad (2.5)$$

It is commonly pointed out that, since the outcome of each polarization measurement at P is random, there is no way to use this collapse to send a message, because there is no way for the experimenter at P to force a particular outcome for the distant measurement on the other photon.

Vice versa, if the experimenter chooses to measure circular polarizations, the final (mixed) state will be represented by

$$\rho_C = \frac{1}{2}\{|R_1\rangle|R_2\rangle\langle R_1|\langle R_2| + |L_1\rangle|L_2\rangle\langle L_1|\langle L_2|\}.$$

The two mixtures (2.5) and (2.6) are completely indistinguishable from a measurement of polarization performed only on the second photon, since the expectation value of any polarization operator of the second photon is always equal to zero for both the mixtures (2.5) and (2.6). This result is used by many authors to prove the impossibility to transmit signals faster than light *via* the intrinsic nonlocality of quantum mechanics.

In the next section we will show that an interference apparatus based on phase-conjugation is able to distinguish the state (2.5) from (2.6) by performing measurements only on the second photon.

3. The Michelson interferometer with a phase-conjugate mirror

Let us suppose that the photon 2 impinges after a year from its production into the "interference detection region" I. In this area a phase-conjugation interferometer has been constructed; it consists of a Michelson interferometer in which one metallic mirror is replaced by a phase conjugate mirror (PCM). Let us first describe the behavior of this apparatus with an ideal PCM.

In the next section we will describe the theory and the results of an experiment of interferometry with a real PCM.

A PCM is a non linear medium that performs a complex conjugation on the spatial part of the complex amplitude of an impinging electromagnetic field. The effect is to reflect the impinging wave back in the direction of propagation while changing the phase from a to $-a$. Consequently, the PCM maintains the polarization state of the incident wave, more precisely, a linear polarized wave is reflected as a linear polarized wave, while a circular polarized wave is reflected with identical circular polarization [4]. This behavior of a PCM is different from that of a conventional mirror, which inverts the circular polarization state of the light, and has an interesting consequence on the visibility of the interference in the phase-conjugate interferometer.

If the impinging light is linearly polarized, for example in the x-direction, the splitting of the beam by the beamsplitter BS and the reflection of the produced beams on the metallic and phase-conjugate mirrors do not produce any variation of the polarization. So, when the two beams overlap at the beamsplitter, an interference effect will occur with an oscillation of the amplitude, which is a function of the optical path of the two beams. Conversely, if the impinging beam has, for example, right-handed circular polarization $|R_2\rangle$, the beam reflected from the metallic mirror has left-handed circular polarization, while the beam reflected from the PCM maintains its right circular polarization. Hence, when they overlap at the

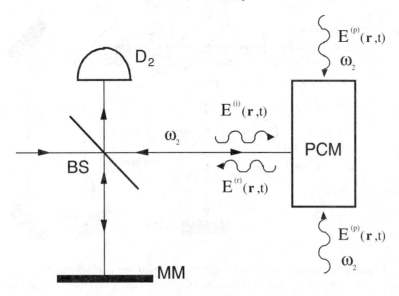

Fig. 2. The modified Michelson interferometer. One of two metallic mirrors is replaced with the phase-conjugate mirror PCM. This mirror is pumped by two optical beams of the same frequency as the impinging wave and with opposite direction.

beamsplitter BS, they will not interfere, since they are in two orthogonal states of polarization.

In conclusion, the phase-conjugate interferometer is able to distinguish states of circular polarization from states of linear polarization.

The interference produced on reflection at phase-conjugate mirror has been studied in detail both theoretically and experimentally [5,6,7]. We will summarize here the main results in order to apply it to our apparatus.

Let us consider (Fig.2) an electromagnetic wave with wave vector $\mathbf{k}$ incident upon the PCM and frequency $\omega/2\pi$

$$\mathbf{E}^{(i)}(\mathbf{r},t) = \mathbf{e}A^{(i)} \exp[i\mathbf{k}\cdot\mathbf{r} - \omega t] \qquad (3.1)$$

where $\mathbf{e}$ is the complex unit polarization vector satisfying the condition $\mathbf{e}\cdot\mathbf{e}^* = 1$.

Let us suppose the PCM is pumped with two beams of identical frequency $\omega/2\pi$. The reflected wave leaving the PCM is given by

$$\mathbf{E}^{(r)}(\mathbf{r},t) = m\mathbf{e}^* A(i)^* \exp[i(-\mathbf{k}\cdot\mathbf{r} - \omega t)], \qquad (3.2)$$

where m is the complex reflectivity of the mirror and depends on the intensity of the pumping beams, the strength of the coupling between pumping beams via the nonlinear susceptibility, and the length of the PCM. In general m is less than 1, but it can be equal to or greater than 1 under well defined conditions [4].

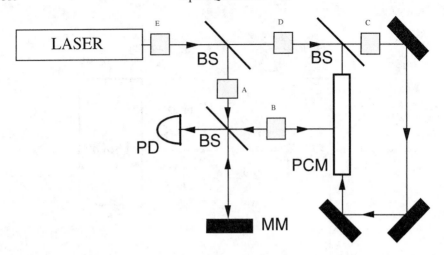

Fig.3. Outline of the experimental setup to verify the existence of interference produced at a reflection at a phase-conjugate mirror. A phase shifter was introduced at one of five different positions A-E, and the position of the interference patter was determined with respect an arbitrary reference point by the photodetector PD. The predictions of formula (10) were completely verified.

The superposition of the incident and reflected fields result in a total field

$$
\begin{aligned}
\mathbf{E}(\mathbf{r},t) &= \mathbf{E}^{(i)}(\mathbf{r},t) + \mathbf{E}^{(r)}(\mathbf{r},t) \\
&= \mathbf{e}A(i)\exp\left[i(\mathbf{k}\cdot\mathbf{r}-\omega t)\right] + m*A(i)*\exp\left[i(-\mathbf{k}\cdot\mathbf{r}-\omega t)\right]
\end{aligned}
\tag{3.3}
$$

and in a total light intensity

$$
I(r,t) = |E(r,t)|^2 = |\mathbf{e}|^2|A(i)|^2|m|^2|\mathbf{e}|^2|A(i)|^2 + m*\mathbf{e}^2A(i)^2\exp\left[-2i\mathbf{k}\cdot\mathbf{r}\right] + c.c. \tag{3.4}
$$

Since $|\mathbf{e}|^2 = 1$, and assuming

$$
\begin{aligned}
m &= |m|\exp[i\phi], \\
A(i) &= |A(i)|\exp[i\alpha], \\
\mathbf{e}^2 &= |\mathbf{e}^2|\exp[i\delta],
\end{aligned}
$$

we have

$$
I(r,t) = |\mathbf{E}(\mathbf{r},t)|^2 = |\mathbf{e}|^2|\mathbf{A(i)}|^2\left(1+|\mathbf{m}|^2\right) + 2|\mathbf{m}||\mathbf{e}^2||\mathbf{A(i)}|^2\cos(2\mathbf{k}-\phi+\delta+2\alpha) \tag{3.5}
$$

and the visibility of the interference

$$
V = \frac{2|m||\mathbf{e}^2|}{1+|m|^2}
$$

exhibits a maximum when $|m| = 1$ and $|\mathbf{e}^2| = 1$.

If $|m| = 1$, and if the impinging wave is x-linearly [y- linearly] polarized, we have $e = i[e = j]$ (i and j are the vectors of the x-axis and y-axis, respectively), and $d = 0$. Then, the scalar product $\mathbf{e}^2$ is equal to 1, and the visibility of the interference V is equal to 1.

Vice versa if the impinging light is right-handed [left-handed] polarized, the polarization vector is $\mathbf{e} = (\mathbf{i} + i\mathbf{j})/\sqrt{2}\left[\mathbf{e} = (\mathbf{i} + i\mathbf{j})/\sqrt{2}\right]$, the scalar product $\mathbf{e}^2$ is equal to zero, and consequently $V = 0$.

Mandel, Wolf, and co-authors [6],[7] confirmed these predictions with an experiment (Fig. 3) in which a PCM is inserted in a Michelson interferometer. In the experiment a phase shifter was introduced in different positions in order to vary the phase of the incident and pumping waves. The displacement of the position of the interference pattern was determined from measurements of the light intensity impinging on the photodetector PD.

TheSuperluminal Quantum Telegraph (SQT)

The apparatus described in the previous section is now able to transmit signal faster than light. Indeed, if the experimenter in P measures linear polarization on the first beam, the state vector (2.1) is collapsed in the mixture (2.5) and the experimenter in the interference region I sees, at the same time, an interference effect on the second beam. *Vice versa*, if the experimenter in P measures circular polarization, the state (2.4) collapses in the mixture (2.6) and the experimenter in I cannot see any interference effect.

If the two experimenters agreed in the past to interpret the non-detection of the interference as a "zero-value" signal and the detection of the interference as a "one-value" signal, the experimenter in P has the possibility of transmitting messages in the binary language with a velocity that is greater than the velocity of the light.

Of course, the experimenter in the interference region I needs to collect a given quantity of photons in order to distinguish interference from non-interference, so the detection of a signal (zero or one) requires a time t. If the photons spend a time T travelling from the source S to the detection areas, then the velocity of transmission of the binary signal is

$$v = \frac{2cT}{t}, \qquad (4.1)$$

where c is the velocity of light. The velocity v can be greater than c, if T/t is greater than 0.5.

In order to make a numerical evaluation of t let us suppose that n is the rate of the emitted pairs per second, and that the mirrors of the interferometer are positioned to obtain constructive interference in the photodetector D_2. When linear polarized photons impinge on the interferometer giving rise to the interference effect, the photons detected in D_2 during the measurement time t will be $N_L = nt$. When circularly polarized photons impinge without producing interference, the detected photons will be $N_C = nt/2$. If we require that the two numbers differ by at least 5 standard deviations and we assume that $\sigma_L = \sqrt{N_L}$, the measuring time t must satisfy the inequality

$$\frac{N_L - N_C}{\sqrt{nt}} > 5 . \qquad (4.2)$$

If, for example, $n = 10,000$ pairs/sec, the measuring time for a single signal will be $t = 0.001$ sec and the minimum distance between the two experimenters at which the quantum telegraph operates as a superluminal apparatus is 300 km.

Moreover, it is worth noting that in our reasoning the assumption of the existence of a physical reality has never been used. In this sense this result is not just a different version of the paradox of Einstein-Podolsky-Rosen (Einstein, 1935), but a new paradox of quantum mechanics.

5. Generalized entangled states

In the experiment of Mandel, Wolf, and co-authors a laser beam, *i.e.* a coherent state, was used in order to prove that the Michelson interferometer with a phase-conjugate mirror is able to verify the interference between a linearly polarized impinging wave and its conjugate wave in the case of real PCM. The interference pattern, as we can see from formula (3.5), is a function of the phase α of the impinging light and it was suggested that in the case of EPR-type entangled states only one photon at time is traveling in the opposite directions. Therefore, the phase of the impinging photons is completely random and the interference pattern vanishes.

This is completely true in the case of EPR-type experiments in which one need to measure correlation functions on pair of single photons, but in our case we do not need to measure coincidence, but single counting rate. Then it is sufficient that the coherence time of our entangled beams τ will be greater than the measuring time t in order to have a stationary interference pattern.

This means, for example, that we can start with a fully entangled coherent state

$$|\psi\rangle = \frac{1}{\sqrt{2}}\left\{|n_{x1}\rangle|n_{x2}\rangle + |n_{y1}\rangle|n_{y2}\rangle\right\} \qquad (5.1)$$

or with a mixed coherent state

$$|\psi\rangle = \frac{1}{\sqrt{2}}\left\{|x_1\rangle|n_{x2}\rangle + |y_1\rangle|n_{y2}\rangle\right\} \qquad (5.2)$$

where $|n_{x1}\rangle$, $|n_{x2}\rangle$ and $|n_{y1}\rangle$, $|n_{y2}\rangle$ are the two orthogonal coherent polarized states along an arbitrary $x-y$ frame for the first and second photons, respectively. With coherent states the phase α of the impinging photons on the Michelson interferometer is a constant and the interference pattern remains stable all the time.

If the existence of superluminal signals were proved experimentally, the complete non-locality of this theory would be confirmed and we would be compelled to reject the relativistic postulate of nonexistence of superluminal signals.

6. Acknowledgments

The author would like to acknowledge the Administration Council of the University of Bari and the Italian Istituto Nazionale di Fisica Nucleare for financial support to this research.

References

[1]   J.F. [1]J.F. Clauuser, *Nuovo Cimento* **338**, 740 (1976).

[2]   A.J. Duncan, H. Kleinpoppen, and Z.A. Sheikh, in: *Bell's Theorem and the Foundations of Modern Physics*, Edited by A. van der Merwe, F. Selleri, and G. Tarozzi p. 161 (Word Scientific, Singapore, 1992).

[3]   J.R. Torgerson, D. Branning, C.H. Monken and L. Mandel. *Phys. Rev* **A 51**, 4400 (1995)

[4]   D. M. Pepper, Nonlinear optical phase coniugation in *Laser Handbook* Vol IV, edited by M.L. Stitch and M. Bass, Nort- Holland , Amsterdam (1985)

[5]   R.W. Boyd, T.M. Habashy, L.Mandel, M. Nieto-Vesperinas, and E.Wolf , *J.Opt. Soc. Am.* **4B**, 1260 (1987)

[6]   A.A. Jacobs, W.R. Tompkin, R.W. Boyd, and E.Wolf , *J.Opt. Soc. Am.* **4B**, 1266 (1987)

[7]   R.W. Boyd, T.M. Habashy, A.A. Jacobs, L.Mandel, M. Nieto- Vesperinas, W.R. Tompkin, and E.Wolf , *Opt. Lett.* **12**, 42, (1987)

[8]   A. Einstein, B. Podolsky and N. Rosen, *Phys. Rev.* **47**, 777 (1935).

Lorentz Symmetry Violation, Vacuum and Superluminal Particles

Luis Gonzalez-Mestres[†††††]
Laboratoire de Physique Corpusculaire, Collége de France
11 pl. Marcellin-Berthelot, 75231 Paris Cedex 05, France
and
Laboratoire d'Annecy-le-Vieux de Physique des Particules
B. P. 110, 74941 Annecy-le-Vieux Cedex, France

If textbook Lorentz invariance is actually a property of the equations describing a sector of the excitations of vacuum above some critical distance scale, several sectors of matter with different critical speeds in vacuum can coexist and an absolute rest frame (the vacuum rest frame) may exist without contradicting the apparent Lorentz invariance felt by "ordinary" particles (particles with critical speed in vacuum equal to c, the speed of light). Sectorial Lorentz invariance, reflected by the fact that all particles of a given dynamical sector have the same critical speed in vacuum, will then be an expression of a fundamental sectorial symmetry (*e.g.* preonic grand unification or extended supersymmetry) protecting a parameter of the equations of motion. Furthermore, the sectorial Lorentz symmetry may be only a low-energy limit, in the same way as the relation ω (frequency) $= c_s$ (speed of sound) k (wave vector) holds for low-energy phonons in a crystal. We show that, in this context, phenomena such as the absence of Greisen-Zatsepin-Kuzmin cutoff and the stability of unstable particles at very high energy are basic properties of a wide class of noncausal models where local Lorentz invariance is broken, introducing a fundamental length. Then, observable phenomena are produced at the wavelength scale of the highest-energy cosmic rays or even below this energy, but Lorentz symmetry violation remains invisible to standard low-energy tests. We discuss possible theoretical, phenomenological, experimental and cosmological implications of this new approach to matter and space-time, as well as prospects for future developments.

[†††††] E–mail: lgonzalz@vxcern. cern. ch
Open Questions in Relativistic Physics
Edited by Franco Selleri (Apeiron, Montreal, 1998)

1.   Introduction

"The impossibility to disclose experimentally the absolute motion of the earth seems to be a general law of Nature"
    H. Poincaré

"Precisely Poincaré proposed investigating what could be done with the equations without altering their form. It was precisely his idea to pay attention to the symmetry properties of the laws of Physics"
    R. P. Feynman

"The interpretation of geometry advocated here cannot be directly applied to submolecular spaces... it might turn out that such an extrapolation is just as incorrect as an extension of the concept of temperature to particles of a solid of molecular dimensions."
    A. Einstein

Is relativity the result of a symmetry of the laws of Nature (Poincaré, 1905), and therefore necessarily broken at some deeper level (Einstein, late period), or does it reflect the existence of an absolute space-time geometry that matter cannot escape (Einstein, early papers on relativity)? Most textbooks teach "absolute" relativity (early Einstein papers) and ignore the possibility of a more flexible formulation (Poincaré, late Einstein thinking) that we may call "relative" relativity (relativity is a symmetry of the laws of Nature expressed by the Lorentz group: whether this symmetry is exact or approximate must be checked experimentally at each new energy scale). In the first case, ether does not exist: light just propagates at the maximum speed allowed by the "absolute" space-time geometry; in the second case, the question of ether remains to be settled experimentally at any new small-distance scale. By introducing important dynamics into the vacuum structure, particle physics has brought about a return to the ether: it would be impossible for the $W^\pm$ and the Z^0 to be gauge bosons with nonzero masses if they did not propagate in a medium where the Higgs fields condense; similarly, modern theories of hadron structure conjecture that free quarks can exist only inside hadrons, due to non-trivial properties (*e.g.* superconducting) of the non-perturbative QCD vacuum. It could still be argued that this new "ether" does not necessarily have a preferred rest frame, and that special relativity is an exact symmetry which prevents us from identifying such a frame. However, this hypothesis does not seem to fit naturally with general physics considerations. Modern dynamical systems provide many examples where Lorentz symmetry (with a critical speed given by the properties of the system) is a scale-dependent property which fails at the fundamental distance scale of the system (*e.g.* a lattice spacing). In practical examples, the critical speed of the apparently relativistic dynamical system is often less than $10^{-5} c$ and "relativity," as felt by the dynamical system, would forbid particle propagation at the speed of light. Light would appear to such a system just like superluminal matter would appear to us.

Furthermore, high-energy physics has definitely found cosmic-ray events with energies above 10^{20} eV (*e.g.* Hayashida *et al.*, 1997). This energy scale is, in orders of magnitude, closer to Planck scale (10^{28} eV) than to the electroweak scale (10^{10} eV). Therefore, if Lorentz symmetry is not an exact symmetry of nature and is instead broken at $\approx 10^{-33}$ cm length scale, the parameters of Lorentz symmetry violation observed (if ever) in the analysis of the highest-energy cosmic-ray events will provide us with direct and unique information on physics at Planck scale. This may be the most fundamental physics outcome of experiments such as AUGER (AUGER Collaboration, 1997) devoted to the study of cosmic rays at $E \approx 10^{20}$ eV: Lorentz symmetry violation at these energies would unravel phenomena originating at higher energy scales, including the possible existence of a fundamental length scale. In the vacuum rest frame, particles of the same type moving at different speeds are different physical objects whose properties cannot be made identical through a Lorentz transformation. This essential property remains true in any other frame, but parameters measured in the vacuum rest frame have an absolute physical meaning. Indeed, assuming that the laboratory frame moves slowly with respect to the vacuum rest frame (which may be close to that suggested by the study of cosmic microwave background radiation), the observed properties of particles at $E \approx 10^{20}$ eV may look more like physics at Planck scale than physics at electroweak or GeV scale (*e.g.* the failure of the parton model and of standard relativistic formulae for Lorentz contraction and time dilation; see Gonzalez-Mestres, 1997h), and basic parameters of Planck-scale physics may become measurable through $E \approx 10^{20}$ eV cosmic-ray events if Lorentz symmetry is violated (though with exact Lorentz symmetry, collisions of very high-energy cosmic rays would be exactly equivalent to collider events at much lower laboratory energies).

Here we review and comment on recent work by the author on Lorentz symmetry violation and possible superluminal sectors of matter. Non-tachyonic superluminal particles (superbradyons) have been discussed in previous papers (Gonzalez-Mestres, 1995, 1996, 1997a and 1997b) and other papers have been devoted to Lorentz symmetry violation (Gonzalez-Mestres, 1997c, 1997d and 1997e) as well as to its astrophysical consequences (Gonzalez-Mestres, 1997f and 1997g), its application to extended objects (Gonzalez-Mestres, 1997h) and its relevance for future accelerator programs (Gonzalez-Mestres, 1997i).

2.   Lorentz symmetry as a low-energy limit

Lamoreaux, Jacobs, Heckel, Raab and Forston (1986) have set an experimental limit from nuclear magnetic resonance measurements which, when suitably analyzed (Gabriel and Haugan, 1990), amounts to the bound $|c_{matter} - c_{light}| < 6 \times 10^{-21} c_{light}$ in the $TH\varepsilon\mu$ model of Lorentz symmetry violation (e.g. Will, 1993). However, the $TH\varepsilon\mu$ model assumes a scale-independent violation of Lorentz invariance (through the non-universality of the critical speed parameter) which does not naturally emerge from dynamics violating Lorentz symmetry at Planck scale or at some other fundamental length scale, where we would naturally expect such an effect to be scale-dependent and possibly vary (like the effective gravitational coupling) according to a E^2 law (E = energy scale). The E^2 law is indeed a trivial and rather general consequence of phenomena such as nonlocality, as can be seen from the generalized one-dimensional Bravais lattice equation (Gonzalez-Mestres, 1997d):

$$\frac{d^2}{dt^2}\big[\phi(n)\big] = -K\big[2\phi(n) - \phi(n-1) - \phi(n+1)\big] - \omega_{rest}^2\phi \tag{1}$$

where n (integer) stands for the site under consideration, $\phi(n)$ is a complex order parameter, K an elastic constant and $(2\pi)^{-1}\omega_{rest}$ the frequency of the chain of oscillators in the zero-momentum limit. In the limit where the lattice spacing a vanishes but Ka^2 remains fixed, (1) becomes a two-dimensional dalembertian equation of the Klein-Gordon type, with two-dimensional Lorentz symmetry and critical speed parameter $c = K^{1/2}a$. In terms of the wave vector k and the frequency $(2\pi)^{-1}\omega$, equation (1) leads to the dispersion relation:

$$\omega^2(k) = 2K\big[10\cos(ka)\big] + \omega_{rest}^2 = 4K\sin^2(ka/2) + \omega_{rest}^2 \tag{2}$$

equivalent to:

$$\begin{aligned}
E^2 &= 2K\big[1 - \cos(2\pi h^{-1}pa) + (2\pi)^{-1}h\omega_{rest}^2\big] \\
&= 4K\big[\sin^2(2\pi h^{-1}pa) + (2\pi)^{-1}h\omega_{rest}^2\big]
\end{aligned} \tag{3}$$

where p stands for momentum and h is the Planck constant. The same procedure can be extended to a wide class of nonlocal models, inclundig those with continuous space, giving:

$$E = (2\pi)^{-1}hca^{-1}e(ka) \tag{4}$$

where $[e(ka)]^2$ is a convex function of $(ka)^2$ obtained from vacuum dynamics. We have checked that this is also a fundamental property of old scenarios breaking local Lorentz invariance (cf. Rédei, 1967), although such a phenomenon seems not to have been noticed by the authors. Expanding equation (4) for $ka << 1$, we can write:

$$e(ka) \cong \big[(ka)^2 - \alpha(ka)^4 + (2\pi a)^2 h^{-1}m^2c^2\big]^{1/2} \tag{5}$$

where α is a model-dependent constant, in the range 0 1 – 0.01 for full-strength violation of Lorentz symmetry at the fundamental length scale (α = 1/12 for the Bravais-lattice model and its isotropic extension to three dimensions), and

$$E \cong pc\left[1 + \alpha\frac{(ka)^2}{2}\right] + m^2c^3(2p)^{-1} \qquad (6)$$

and the new term $\Delta E = -pc\alpha\,(ka)^2/2$ in the right-hand side of (6) implies a Lorentz symmetry violation in the ratio Ep^{-1} varying like $\Gamma(k) \approx \Gamma_0 k^2$ where $\Gamma_0 = -\alpha\,a^2/2$. Such an expression is not incompatible with a possible gravitational origin of Lorentz symmetry violation, where the effective gravitational coupling would rise like E^2 below Planck energy. However, other interpretations are possible (*e.g.* Gonzalez-Mestres, 1997d) where all presently kown "elementary" particles and gauge bosons would actually be composite objects made of superluminal matter at Planck scale.

More generally, a k^2 law for the parameters of Lorentz symmetry violation, as suggested by the above formulae, would lead to substantial changes with respect to conventional models (*e.g.* Will, 1993). In particular, Lorentz symmetry would remain unbroken at $k = 0$. With such a law, an effect of order 1 at $p = 3.\ 10^{20}$ eV c^{-1} (the estimated momentum of the highest-energy observed cosmic-ray event) would become of order $\approx 10^{-25}$ at $p = 100$ MeV c^{-1} (the highest momentum scale involved in nuclear magnetic resonance tests of special relativity). Therefore, very large deviations from special relativity at the highest observed cosmic-ray energies would be compatible with a great accuracy of this theory in the low-momentum region. Thus, the main and most fundamental physics outcome of very high-energy cosmic-ray experiments involving particles and nuclei may eventually be the test of special relativity. If Lorentz symmetry is violated at Planck scale, the highest-energy cosmic ray events may, if analyzed closely and with the expected high statistics from future experiments, provide a detailed check of different models of deformed relativistic kinematics and of the basic physics behind the kinematics.

The same kind of deformed relativistic kinematics arises naturally in soliton models (Gonzalez-Mestres, 1997h). Starting from the equation:

$$c^{-1}\frac{\partial^2\psi}{\partial t^2} - \frac{\partial^2\psi}{\partial x^2} = 2\Delta^{-1}\psi(1-\psi^2) \qquad (7)$$

where Δ is the distance scale characterizing the soliton size (x = space coordinate, t = time coordinate), and writing down the one-soliton solution of this equation:

$$\psi(x,t) = \Phi(y) = \tanh(\lambda_0 y) \qquad (8)$$

where $y = x - vt$, v is the speed of the soliton, $\lambda_0 = \Delta^{-1}\gamma_R$ and γ_R is the standard relativistic Lorentz factor $\gamma_R = (1 - v^2c^{-2})^{-\frac{1}{2}}$, we can introduce a perturbation to the system by adding to the left-hand side of (7) a term $-(a^2/12)\partial^4\psi/\partial x^4$ which corresponds to the lowest-order correction to the continuum limit when the Bravais-lattice version of (7) is expanded in powers of a^2 . This new term in the equation will be compensated at the first order in the perturbation by the replacement:

$$\Phi \to \Phi + \varepsilon \Phi(1 - \Phi^2) \tag{9}$$

where $\varepsilon \propto a^2$. We furthermore replace λ_0 by a new coefficient λ to be determined form the perturbed equation. To first order in the perturbation, we get the solutions:

$$\varepsilon \cong 1 - \lambda^2 \gamma_R^{-2} \Delta^2 \tag{10}$$

$$\lambda^2 \cong [3 \pm (1 - 4a^2 \Delta^{-2} \gamma_R^{4}/3)^{\frac{1}{2}}] (1 + a^2 \Delta^{-2} \gamma_R^{4}/6)^{-1} \Delta^{-2} \gamma_R^{2}/4 \tag{11}$$

leading for $\varepsilon << 1$ to $(\Delta\lambda)^{-2} \cong \gamma_R^{-2} + a^2 \Delta^{-2} \gamma_R^{2}/3$. Thus, relative corrections to standard relativistic Lorentz contraction and time dilation factors are proportional to $a^2 \Delta^{-2} \gamma_R^{4}$ and dominate when this variable becomes ≈ 1 . Above this value of γ_R, we expect departures from special relativity to occur at leading level in many phenomena. Deformed kinematics can be obtained at the lowest order in the perturbation. A simplified calculation, valid for a wide class of soliton models, could be as follows. At the first order in the perturbation with respect to special relativity, we start from an effective lagrangian for soliton kinematics:

$$L = - mc^2 \gamma_R^{-1}(1 - \rho \, \gamma_R^{4}) \tag{12}$$

where ρ is a constant proportional to $a^2 \Delta^{-2}$, according to (10) and (11). From this lagrangian, we derive the expression for the generalized momentum:

$$p = m \gamma_R v(1 + 3\rho \, \gamma_R^{4}) \tag{13}$$

from which we can build the hamiltonian:

$$H = pv - L = mc^2(\gamma_R + 3\rho \, v^2 c^{-2} \gamma_R^{5} - \rho \, \gamma_R^{3}) \tag{14}$$

which leads to a deformed relativistic kinematics defined by the relation:

$$E - pc = mc^2 \gamma_R^{-1}(1 + vc^{-1})^{-1} - \rho \, mc^2 [3vc^{-1}(1 + vc^{-1})^{-1} + 1] \gamma_R^{3} \tag{15}$$

and, when expressed in terms of p at $v \cong c$ and for small values of $\rho \, \gamma_R^{4}$, can be approximated by:

$$E - pc \cong mc^2(2p)^{-1} - 5\rho \, p^3(2m^2c)^{-1} \tag{16}$$

where the deformation term $5\rho \, p^3(2m^2c)^{-1}$ differs from that obtained from phonon mechanics in the Bravais lattice only by a constant factor $\eta \propto 2h^2(2\pi mc\Delta)^{-2}$.

Looking at the low-speed limit of (15), we find a renormalization of the critical speed parameter c, δc, such that $\delta c c^{-1} \approx (\Delta/a)^{-2} \approx 10^{-40}$ for hadrons if $a \approx 10^{-33}$ cm and $\Delta \approx 10^{-13}$ cm . This effect (the only one which survives at $k = 0$) is much smaller than the effects contemplated by other authors (e.g. Coleman and Glashow, 1997 and references therein) and, even assuming that it would be different for different particles, it cannot be excluded by existing data which can only rule out values of $\delta c c^{-1}$ above $\approx 10^{-20}$.

3.   Deformed relativistic kinematics

Assuming that Lorentz symmetry is violated at Planck scale or at some other fundamental length scale, how does the new kinematics apply to different particles, nuclei, atoms and larger objects? Other versions of deformed relativistic kinematics led in the past to controversies (Bacry, 1993; Fernandez, 1996) which can be resolved (Gonzalez-Mestres, 1997h) if the value of α depends on the object under consideration. In the presence of a fundamental (super)symmetry, it may be reasonable to assume that α has the same value for leptons and gauge bosons. From the above example with solitons, we conclude (Gonzalez-Mestres, 1997h) that the value of α for hadrons is naturally of the same order as for "elementary" particles, although not necessarily identical. It can also be different for different hadrons. A crucial question is how to extend deformed relativistic kinematics to nuclei and larger objects. Two different simplified approaches can be considered:

- *Model i* . Due to the very large size of atoms, as compared to nuclei, the transition from nuclear to atomic scale appears as a reasonable point to stop considering systems as "elementary" from the point of view of deformed relativity. α would then have a universal value for nuclei and simpler objects, but not for atoms and larger bodies.
- *Model ii*. The example with Φ^4 solitons suggests that hadrons can have values of α close to that of leptons and gauge bosons, and the transition may happen continuously at fermi scale, when going from nucleons to nuclei. Then, the value of α would be universal (or close to it) for leptons, gauge bosons and hadrons (solitons) but follow a m^{-2} law for nuclei (multi-soliton bound states) and heavier systems, the nucleon mass setting the scale.

Experimental tests should be performed and equivalent dynamical systems should be studied. However, *Model i)* would lack a well-defined criterium to separate systems to which the deformed relativity applies with the same value of α as for leptons and gauge bosons from those to which this kinematics cannot be applied, and to characterize the transition between the two regimes. The above obtained $m^{-2}\Delta^{-2}$ dependence of the coefficient of the deformation term for extended objects, as described in *Model ii)*, seems to provide a continuous transition from nucleons to heavier systems, naturally filling this gap. On the other hand, a closer analysis reveals that there is indeed a discontinuity between nuclei and atoms, as foreseen in *Model i)* . As long as the deformation term in electron kinematics can be neglected as compared to the electron mass term, we can consider that most of the momentum of an atom is carried by the nucleus and *Model ii)* may provide a reasonable description of reality. But, when the electron mass term becomes small as compared to the part of the energy it would carry in a parton model of the atom, such a description becomes misleading. To have the same speed as a nucleon, the electron must then carry nearly the same energy and momentum. We therefore propose (Gonzalez-Mestres, 1997h) a modified version of *Model ii)* with $\Delta \approx 10^{-13}$ cm from hadrons and nuclei where, for atoms and larger neutral systems, the coefficient of the deformation term would be corrected by a factor close to 1 at low momentum and to 4/9 at high momentum if the number of neutrons is equal to that of protons. This model is obviously approximate and should be completed by a detailed dynamical calculation that we shall not attempt here. It assumes that electrically neutral bodies can reach very high energies per unit mass, which is not obvious: spontaneous ionization may occur at speeds (in the vacuum rest frame) for which the deformation term in electron kinematics becomes larger than its mass term.

Then, for bodies heavier than hadrons, the effective value of α would decrease essentially like m^{-2} . Applying a similar mass-dependence to the κ parameter of a different deformed Poincaré algebra considered by previous authors (Bacry, 1993 and references therein), i.e. $\kappa \propto m$ for large bodies, yields the relation:

$$F(M_0,E_0) = F(M_1,E_1) + F(M_2,E_2) \qquad (17)$$

with:

$$F(m,E) = 2\kappa(m)\sinh[2^{-1}\kappa^{-1}(m)E] \qquad (18)$$

where $M = M_1 + M_2$, $\kappa(m)$ is our above mass-dependent version of the κ parameter of the deformed Poincaré algebra used by these authors, and E_0 is the energy of a system with mass M made of two non-interacting subsystems of energies E_1 and E_2 and with masses M_1 and M_2 . Defining mass as an additive parameter, the rest energy $E_{i,rest}$ (in the vacuum rest frame) of particle i ($i = 0,1,2$) is given by the equation:

$$M_i c^2 = 2\kappa(M_i)\sinh[2^{-1}\kappa^{-1}(M_i)E_{i},rest] \qquad (19)$$

and tends to $M_i c^2$ as $\kappa(M_i) \to \infty$. Equations (17) and (18) lead to additive relations for the energy of macroscopic objects if the proportionality rule $\kappa(m) \propto m$ is applied. From our previous discussion with a different deformation scheme, such a choice seems to naturally agree with physical reality. Then, contrary to previous claims (Bacry, 1993; Fernandez, 1996), the rest energies of large systems would be additive and no macroscopic effect on the total mass of the Universe would be expected.

4.   Phenomenological implications

As initially stressed, very high-energy cosmic rays can open a unique window to Planck scale if Lorentz symmetry is violated. Contrary to standard prejudice which would suggest that energy-dependent effects of Lorentz symmetry violation at Planck scale can be detected only at energies close to this scale, it turns out that such effects are detectable at the highest observed cosmic-ray energies. As discussed in Section 2, we expect standard relativistic formulae for Lorentz contraction and time dilation for a proton to fail at energies such that $a^2\Delta^{-2}\gamma_R^4 \approx 1$, i.e. $E \approx 10^{19}$ eV for $a \approx 10^{-33}$ cm and $\Delta \approx 10^{-13}$ cm (Gonzalez-Mestres, 1997h). Similarly, with the same figures and taking $\alpha \approx 0.1$, the proton mass term $m^2c^3(2p)^{-1}$ in the expression for the proton energy becomes smaller than the deformation term $\Delta E = -pc\alpha(ka)^2/2$ for E above

$\approx 8.10^{18}$ eV and, even if both terms are very small as compared to the total energy, kinematical balances (which depend crucially on these nonleading terms) are drastically modified (Gonzalez-Mestres, 1997d). The standard parton picture of hadrons is equally disabled by the new kinematics at very high energy (Gonzalez-Mestres, 1997h), due to the impossibility for "almost-free" constituents carrying arbitrary fractions of the total energy and momentum to travel at the same speed. Apart from the failure of the standard parton model for hadrons at wave vectors above $\approx (8\pi^2\alpha^{-1})^{1/4}(mch^{-1}a^{-1})^{1/2}$ (*i.e.* at energies above $\approx 10^{19}$ eV if $a \approx 10^{-33}$ cm, $\approx 10^{20}$ eV for $a \approx 10^{-35}$ cm and $\approx 3.10^{17}$ eV for $a \approx 10^{-30}$ cm), the following new effects at leading level would occur assuming a universal value of α for leptons, gauge bosons and hadrons:

a)   The Greisen-Zatsepin-Kuzmin (GZK) cutoff on very high-energy cosmic nucleons (Greisen, 1966; Zatsepin and Kuzmin, 1966) does no longer apply (Gonzalez-Mestres, 1997d and 1997f). Very high-energy cosmic rays originating from most of the presently observable Universe can reach the earth and generate the highest-energy detected events. Indeed, fits to data below $E = 10^{20}$ eV using standard relativistic kinematics (*e.g.* Dova, Epele and Hojvat, 1997) predict a sharp fall of the event rate at this energy, in contradiction with data (Bird *et al.*, 1993 and 1996; Hayashida *et al.*, 1994 and 1997; Yoshida *et al.*, 1995) which suggest that events above 10^{20} eV are produced at a significant rate. Lorentz symmetry violation from physics at Planck scale provides a natural way out. The existence of the cutoff for cosmic nuclei will then depend crucially on the details of deformed relativistic kinematics, beyond the accuracy of the present discussion.

b)   Unstable particles with at least two massive particles in the final state of all their decay channels (neutron, Δ^{++}, possibly muons, charged pions and τ's, perhaps some nuclei...) become stable at very high energy (Gonzalez-Mestres, 1997d and 1997f). In any case, many unstable particles live longer than naively expected with exact Lorentz invariance and, at high enough energy, the effect becomes much stronger than previously estimated for nonlocal models (Anchordoqui, Dova, Gómez Dumm and Lacentre, 1997) ignoring the small violation of relativistic kinematics. Not only particles previously discarded because of their lifetimes can be candidates for the highest-energy cosmic-ray events, but very high-energy cascade development can be modified (for instance, if the π^0 lives longer at energies above $\approx 10^{18}$ eV, thus favoring hadronic interactions and muon pairs and producing less electromagnetic showers).

c)   The allowed final-state phase space of two-body collisions is modified at very high energy when, in the vacuum rest frame where expressions (4)-(6) apply, a very high-energy particle collides with a low-energy target (Gonzalez-Mestres, 1997d). Energy conservation reduces the final-state phase space and can lead to a sharp fall of cross sections starting at incoming-particle wave vectors well below the inverse of the fundamental length, essentially above $E \approx (E_T a^{-2} \hbar^2 c^2)^{1/3}$ where E_T is the energy of the target. For $a \approx 10^{-33}$ cm, this scale corresponds to: $\approx 10^{22}$ eV if the target is a rest proton; $\approx 10^{21}$ eV if it is a rest electron; $\approx 10^{20}$ eV for a ≈ 1 keV photon, and $\approx 10^{19}$ eV if the target is a visible photon. For a proton impinging on a $\approx 10^{-3}$ eV photon from cosmic microwave background radiation, and taking $\alpha \approx 1/12$ as in the Bravais-lattice model, we expect the fall of cross sections to occur above $E \approx 5 \times 10^{18}$ eV, the critical energy where the derivatives of the mass term $m^2 c^3 p^{-1}/2$ and of the deformation term $\alpha\, p(ka)^2/2$ become equal in the expression relating the proton energy E to its momentum p. With $a \approx 10^{-30}$ cm, still allowed by cosmic-ray data (Gonzalez-Mestres, 1997e), the critical energy scale can be as low as $E \approx 10^{17}$ eV; with $a \approx 10^{-35}$ cm (still compatible with data in the region of the predicted GZK cutoff), it would be at $E \approx 5 \times 10^{19}$ eV. Similar considerations lead to a fall of radiation under external forces (*e.g.* synchrotron radiation) above this energy scale. In the case of a very high-energy γ ray, taking $a \approx 10^{-33}$ cm, the deformed relativistic kinematics inhibits collisions with $\approx 10^{-3}$ eV photons from cosmic background radiation above $E \approx 10^{18}$ eV, with $\approx 10^{-6}$ eV photons above $E \approx 10^{17}$ eV and with $\approx 10^{-9}$ eV photons above $E \approx 10^{16}$ eV. Taking $a \approx 10^{-30}$ cm would lower these critical energies by a factor 100 according to the previous formulae, whereas the choice $a \approx 10^{-35}$ cm would raise them by a factor of 20.

d)   In astrophysical processes, the new kinematics may inhibit phenomena such as GZK-like cutoffs, decays, radiation emission under external forces (similar to a collision with a very low-energy target), momentum loss (which at very high energy does not imply deceleration) through collisions, production of lower-energy secondaries, photodisintegration of some nuclei... *potentially solving all the basic problems raised by the highest-energy cosmic rays* (Gonzalez-Mestres, 1997e and 1997g). Due to the fall of cross sections, energy losses become much weaker than expected with relativistic kinematics and astrophysical particles can be pushed to much higher energies (once energies above 10^{17} eV have been reached through conventional mechanisms, synchrotron radiation and collisions with ambient radiation may start to be inhibited by the new kinematics); similarly, astrophysical particles will be able to propagate to much longer astrophysical distances, and many more sources (in practically all the presently observable Universe) can produce very high-energy cosmic rays reaching the earth; as particle lifetimes are much longer, new possibilities arise for the nature of these cosmic rays. Models of very high-energy astrophysical processes cannot ignore a possible Lorentz symmetry violation at Planck scale, in which case observable effects are predicted for the highest-energy detected particles.

e)   If the new kinematics can explain the existence of $\approx 10^{20}$ eV events, it also predicts that, above some higher energy scale (around $\approx 10^{22}$ eV for

$a \approx 10^{-33}$ cm), the fall of cross sections will prevent many cosmic rays (leptons, hadrons, gauge bosons) from depositing most of its energy in the atmosphere (Gonzalez-Mestres, 1997e). Such extremely high-energy particles will produce atypical events of apparently much lower energy. New analysis of data and experimental designs are required to explore this possibility. Again, the interaction properties of nuclei will depend on the details of deformed kinematics.

Velocity reaches its maximum at $k \approx (8\pi^2{}^{-1})^{1/4}(mch^{-1}a^{-1})^{1/2}$. Observable effects of local Lorentz invariance breaking arise, at leading level, well below the critical wavelength scale a^{-1} due to the fact that, contrary to previous models (*cf.* Rédei, 1967), we directly apply non-locality to particle propagators and not only to the interaction hamiltonian. In contrast with previous patterns (*cf.* Blokhintsev, 1966), s–t–u kinematics ceases to make sense and the motion of the global system with respect to the vacuum rest frame plays a crucial role. The physics of elastic two-body scattering will depend on five kinematical variables. Noncausal dispersion relations (Blokhintsev and Kolerov, 1964) should be reconsidered, taking into account the departure from relativistic kinematics. As previously stressed (Gonzalez-Mestres, 1997d), this apparent nonlocality may actually reflect the existence of superluminal sectors of matter (Gonzalez-Mestres, 1996) where causality would hold at the superluminal level (Gonzalez-Mestres, 1997a). Indeed, electromagnetism appears as a nonlocal interaction in the Bravais model of phonon dynamics, due to the fact that electromagnetic signals propagate much faster than lattice vibrations.

Very high-energy accelerator experiments (especially with protons and nuclei) can play a crucial role in the test of a possible Lorentz symmetry violation. To fit with cosmic-ray events, they should be performed in the very-forward region. At LHC, FELIX (*e.g.* Eggert, Jones and Taylor, 1997) could provide a crucial check of special relativity by comparing its data with cosmic-ray data in the $\approx 10^{16} - 10^{17}$ eV region. VLHC experiments would be expected to lead to fundamental studies in the kinematical region which, according to special relativity, would be equivalent to the collisions of $\approx 10^{19}$ eV cosmic protons. With a 700 TeV per beam $p - p$ machine, it would be possible to compare the very-forward region of collisions with those of cosmic protons at energies up to $\approx 10^{21}$ eV. Thus, it seems necessary that all very high-energy collider programs allow for an experiment able to cover secondary particles in the far-forward and far-backward regions. A model-independent way to test Lorentz symmetry between collider and cosmic-ray data sould be carefully elaborated, but the basic phenomena involved in the case of Lorentz symmetry violation can be (Gonzalez-Mestres, 1997d and 1997h):

i)   failure of the standard parton model (in any version, even incorporating radiative corrections and phase transitions);

ii) failure of the relativistic formulae for Lorentz contraction and time dilation;

iii) longer than predicted lifetimes for some of the produced particles (*e.g.* the π^0).

The role of high-precision data from accelerators would then be crucial to establish the existence of such phenomena in the equivalent cosmic-ray events. To reach the best possible performance, cosmic-ray experiments should (if ever feasible) install in coincidence very large-surface detectors (providing also the largest-volume target) with very large-volume underground or undewater detectors and with balloon or satellite devices able to study early cascade development. It would then be possible to perform unique tests of special relativity involving violations due to phenomena at some fundamental scale close to Planck scale, and even to determine the basic parameters of Lorentz symmetry violation (*e.g.* of deformed kinematics) and of physics at the fundamental length scale.

5.   Superluminal particles

Lorentz invariance can be viewed as a symmetry of the motion equations, in which case no reference to absolute properties of space and time is required and the properties of matter play the main role (Gonzalez-Mestres, 1996). In a two-dimensional galilean space-time, the equation:

$$\alpha\frac{\partial^2\phi}{\partial t^2}-\frac{\partial^2\phi}{\partial x^2}=F(\phi) \tag{20}$$

with $\alpha = 1/c_o^2$ and c_o = critical speed, remains unchanged under "Lorentz" transformations leaving invariant the squared interval:

$$ds^2 = dx^2 - c_o^2\,dt^2 \tag{21}$$

so that matter made with solutions of equation (20) would feel a relativistic space-time even if the real space-time is actually galilean and if an absolute rest frame exists in the underlying dynamics beyond the wave equation. A well-known example is provided by the solitons of the sine-Gordon equation, obtained taking in (20):

$$F(\phi) = -(\omega_0/c_o)^2\sin\phi \tag{22}$$

where ω_0 is a characteristic frequency of the dynamical system. A two-dimensional universe made of sine-Gordon solitons plunged in a galilean world would behave like a two-dimensional minkowskian world with the laws of special relativity. Information on any absolute rest frame would be lost by the solitons, as if the Poincaré relativity principle (Poincaré, 1905) were indeed a law of Nature, even if actually the basic equation derives from a galilean world with an absolute rest frame. The actual structure of space and time can only be found by going beyond the wave equation to deeper levels of resolution. At this stage, a crucial question arises (Gonzalez-Mestres, 1995): is c (the speed of light) the only critical speed in vacuum, are there particles with a critical speed different from that of light? The question clearly makes sense, as in a perfectly transparent crystal it is possible to identify at least two critical speeds: the speed of light and the speed of sound. It has been shown (Gonzalez-Mestres, 1995 and 1996) that superluminal sectors of matter can be consistently generated replacing in the Klein-Gordon equation the speed of light by a new critical speed $c_i >> c$ (the subscript i stands for the i-th superluminal sector). All standard kinematical concepts and formulas (Schweber, 1961) remain correct, leading to particles with positive mass and energy which are not tachyons. We call them *superbradyons* as, according to standard vocabulary (Recami, 1978), they are bradyons with superluminal critical speed in vacuum. The rest energy of a superluminal particle of mass m and critical speed c_i will be given by the generalized Einstein equation:

$$E_{rest} = mc_i^2 \qquad (23)$$

Energy and momentum conservation will in principle not be spoiled by the existence of several critical speeds in vacuum: conservation laws will as usual hold for phenomena leaving the vacuum unchanged. Each superluminal sector will have its own Lorentz invariance with c_i defining the metric. Interactions between two different sectors will break both Lorentz invariances. Lorentz invariance for all sectors simultaneously will at best be explicit (*i.e.* exhibiting the diagonal sectorial Lorentz metric) in a single inertial frame (*the vacuum rest frame, i.e.* the "absolute" rest frame). If superluminal particles couple weakly to ordinary matter, their effect on the ordinary sector will occur at very high energy and very short distance (Gonzalez-Mestres, 1997c), far from the domain of successful conventional tests of Lorentz invariance (Lamoreaux, Jacobs, Heckel, Raab and Forston, 1986; Hills and Hall, 1990). In particular, superbradyons naturally escape the constraints on the critical speed derived in some specific models (Coleman and Glashow, 1997; Glashow, Halprin, Krastev, Leung and Pantaleone, 1997). High-energy experiments can therefore open new windows in this field. Finding some track of a superluminal sector (*e.g.* through violations of Lorentz invariance in the ordinary sector) may be a unique way to experimentally discover the vacuum rest frame. Furthermore, superbradyons can be the fundamental matter from which Planck-scale strings would actually be built. Superluminal particles lead to consistent cosmological models (Gonzalez-Mestres, 1997d), where they may well provide most of the cosmic (dark) matter. Although recent criticism to this suggestion has been emitted in a specific model on the grounds of gravitation theory (Konstantinov, 1997), the framework used is crucially different from the multi-graviton approach suggested in our papers, where each dynamical sector would generate its own graviton.

Conventional tests of special relativity are performed using low-energy phenomena. The highest momentum scale involved in nuclear magnetic resonance tests of special relativity is related to the energy of virtual photons exchanged, which does not exceede the electromagnetic energy scale $E_{em} \approx \alpha_{em} r^{-1} \approx 1$ MeV, where α_{em} is the electromagnetic constant and r the distance scale between two protons in a nucleus. However, the extrapolation between the 1 MeV scale and the 1 – 100TeV scale (energies to be covered by LHC and VLHC) may involve a very large number, making compatible low-energy results with the possible existence of superluminal particles above TeV scale. Assume, for instance, that between $E \approx 1$ MeV and $E \approx 100$ TeV the mixing between an "ordinary" particle (*i.e.* with critical speed in vacuum equal to the speed of light c in the relativistic limit) of energy E_0 and a superluminal particle with mass m_i, critical speed $c_i >> c$ and energy E_i is described in the vacuum rest frame by a non-diagonal term in the energy matrix of the form (Gonzalez-Mestres, 1997c):

$$\varepsilon \approx \varepsilon_0 pc_i \rho(p^2) \tag{24}$$

where p stands for momentum, ε_0 is a constant describing the strength of the mixing and $\rho(p^2) = p^2(p^2 + M^2c^2)^{-1}$ accounts for a threshold effect with $Mc^2 \approx 100$ TeV due to dynamics. Then, the correction to the energy of the "ordinary" particle will be $\approx \varepsilon^2(E_0 - E_i)^{-1}$ whereas the mixing angle will be $\approx \varepsilon(E_0 - E_i)^{-1}$. Taking the rest energy of the superluminal particle to be $E_{i,rest} = m_i c_i^2 \approx 1$ TeV, we get a mixing $\approx 0.5 \, \varepsilon_0$ at $pc = 100$ TeV, $\approx 10^{-2}\varepsilon_0$ at $pc = 10$ TeV and $\approx 10^{-4} \, \varepsilon_0$ at $pc = 1$ TeV. Such figures would clearly justify the search for superbradyons at LHC and VLHC ($E \approx 100$ TeV per beam) machines provided low-energy bounds do not force ε_0 to be too small. With the above figures, at $pc = 1$ MeV one would have a correction to the photon energy less than $\approx 10^{-32} \, \varepsilon_0^2 pc_i$ which, requiring the correction to the photon energy not to be larger than $\approx 10^{-20}$, would allow for large values of ε_0 if c_i is less than $\approx 10^{12}c$. In any case, a wide range of values of c_i and ε_0 can be explored. More stringent bounds may come from corrections to the quark propagator at momenta ≈ 100 MeV. There, the correction to the quark energy would be bounded only by $\approx 10^{-24} \, \varepsilon_0^2 pc_i$ and requiring it to be less than $\approx 10^{-20}$ pc would be equivalent to $\varepsilon_0 < 0.1$ for $c_i = 10^6 c$. Obviously, these estimates are rough and a detailed calculation of nuclear parameters using the deformed relativistic kinematics obtained from the mixing would be required. It must be noticed that the situation is fundamentally different from that contemplated in the $TH\varepsilon\mu$ formalism and, in the present case, Lorentz invariance can remain unbroken in the low-momentum limit, as the deformation of relativistic kinematics for "ordinary" particles is momentum-dependent. Therefore, it may be a safe policy to explore all possible values of c_i and ε_0 at accelerators (including other possible parametrizations of ε) without trying to extrapolate bounds from nuclear magnetic resonance experiments.

   The production of one or two (stable or unstable) superluminal particles in a high-energy accelerator experiment is potentially able to yield very well-defined signatures through the shape of decay products or "Cherenkov" radiation in vacuum events (spontaneous emission of "ordinary" particles). In the vacuum rest frame, a relativistic superluminal particle would have energy $E \cong pc_i$, where $c_i >> c$ is the critical speed of the particle. When decaying into "ordinary" particles with energies $E_\alpha \cong p_\alpha c$ ($\alpha = 1,...,N$) for a N-particle decay product), the initial energy and momentum must split in such a way that very large momenta $p_\alpha >> p$ are produced (in order to recover the total energy with "ordinary" particles but compensate to give the total momentum p. This requires the shape of the event to be exceptionally isotropic, or with two jets back to back, or yielding several jets with the required small total momentum. Similar trends will arise in "Cherenkov-like" events, and remain observable in the laboratory frame. It must be noticed that, if the velocity of the laboratory with respect to the vacuum rest frame is $\approx 10^{-3} c$, the laboratory velocity of superluminal particles as measured by detectors (if ever feasible) would be $\approx 10^3 c$ in most cases (Gonzalez-Mestres, 1997a).

The possibility that superluminal matter exists, and that it plays nowadays an important role in our Universe, should be kept in mind when addressing the two basic questions raised by the analysis of any cosmic ray event: a) the nature and properties of the cosmic ray primary; b) the identification (nature and position) of the source of the cosmic ray. If the primary is a superluminal particle, it will escape conventional criteria for particle identification and most likely produce a specific signature (e.g. in inelastic collisions) different from those of ordinary primaries. Like neutrino events, in the absence of ionization (which will in any case be very weak) we may expect the event to start anywhere inside the detector. Unlike very high-energy neutrino events, events created by superluminal primaries can originate from a particle having crossed the earth. As in accelerator experiments (see the above discussion), an incoming, relativistic superluminal particle with momentum p and energy $E_{in} \cong pc_i$ in the vacuum rest frame, hitting an ordinary particle at rest, can release most of its energy into two ordinary particles or jets with momenta (in the vacuum rest frame) close to $p_{max} = \frac{1}{2} pc_i c^{-1}$ and oriented back to back in such a way that the two momenta almost cancel, or into several jets with a very small total momentum, or into a more or less isotropic event with an equally small total momentum. Then, an energy $E_R \cong E_{in}$ would be transferred to ordinary secondaries. Corrections due to the earth motion must be applied (Gonzalez-Mestres, 1997a) before defining the expected event configuration in laboratory experiments (AUGER, AMANDA...). At very high energy, such events would be easy to identify in large-volume detectors, even at very small rate. If the source is superluminal, it can be located anywhere (and even be a free particle) and will not necessarily be at the same place as conventional sources of ordinary cosmic rays. High-energy cosmic ray events originating form superluminal sources will provide hints on the location of such sources and be possibly the only way to observe them. At very high energies, the GZK cutoff does not in principle hold for cosmic ray events originating from superluminal matter: this is obvious if the primaries are superluminal particles that we expect to interact very weakly with the cosmic microwave background, but is also true for ordinary primaries as we do not expect them to be produced at the locations of ordinary sources and there is no upper bound to their energy around 100EeV. Besides "Cherenkov" deceleration, a superluminal cosmic background radiation may exist and generate its own GZK cutoffs. However, if there are large amounts of superluminal matter around us, they can be the main superluminal source of cosmic rays reaching the earth.

References

Anchordoqui, L., Dova, M. T., Gómez Dumm, D. and Lacentre, P., *Zeitschrift für Physik C* 73, 465 (1997).

AUGER Collaboration, *The Pierre Auger Observatory Design Report* (1997).

Bacry, H., Marseille preprint CPT-93/P. 2911, available from KEK database.

Bird, D. J. *et al.*, *Phys. Rev. Lett.* 71, 3401 (1993).

Bird, D. J. *et al.*, *Ap. J.* 424, 491 (1994).

Blokhintsev, D. I. and Kolerov, G. I., *Nuovo Cimento* 34, 163 (1964).

Blokhintsev, D. I., *Sov. Phys. Usp.* 9, 405 (1966).

Coleman, S. and Glashow, S. L., *Phys. Lett. B* 405, 249 (1997).

Dova, M. T., Epele, L. N. and Hojvat, C., *Proceedings of the 25th International Cosmic Ray Conference* (ICRC97), Durban, South Africa, Vol. 7, p. 381 (1997).

Fernandez, J., *Phys. Lett. B* 368, 53 (1996).

Eggert, K., Jones, L. W. and Taylor, C. C., *Proceedings of ICRC97*, Vol. 6, p. 25 (1997).

Gabriel, M. D. and Haugan, M. P., *Phys. Rev. D* 41, 2943 (1990).

Glashow, S. L., Halprin, A., Krastev, P. I., Leung, C. N. and Pantaleone, J., "Comments on Neutrino Tests of Special Relativity," *Phys. Rev. D* 56, 2433 (1997).

Gonzalez-Mestres, L., "Physical and Cosmological Implications of a Possible Class of Particles Able to Travel Faster than Light," contribution to the *28th International Conference on High-Energy Physics*, Warsaw July 1996 . Paper hep-ph/9610474 of LANL (Los Alamos) electronic archive (1996).

Gonzalez-Mestres, L., "Space, Time and Superluminal Particles," paper physics/9702026 of LANL electronic archive (1997a).

Gonzalez-Mestres, L., "Superluminal Particles and High-Energy Cosmic Rays," *Proceedings of the 25th International Cosmic Ray Conference* (ICRC97), Vol. 6, p. 109 . Paper physics/9705032 of LANL electronic archive (1997b).

Gonzalez-Mestres, L., "Lorentz Invariance and Superluminal Particles," March 1997, paper physics/9703020 of LANL electronic archive (1997c).

Gonzalez-Mestres, L., "Vacuum Structure, Lorentz Symmetry and Superluminal Particles," paper physics/9704017 of LANL electronic archive (1997d).

Gonzalez-Mestres, L., "Lorentz Symmetry Violation and Very High-Energy Cross Sections," paper physics/9706022 of LANL electronic archive (1997e).

Gonzalez-Mestres, L., "Absence of Greisen-Zatsepin-Kuzmin Cutoff and Stability of Unstable Particles at Very High Energy, as a Consequence of Lorentz Symmetry Violation," *Proceedings of ICRC97*, Vol. 6, p. 113 . Paper physics/9705031 of LANL electronic archive (1997f).

Gonzalez-Mestres, L., "Possible Effects of Lorentz Symmetry Violation on the Interaction Properties of Very High-Energy Cosmic Rays," paper physics/9706032 of LANL electronic archive (1997g).

Gonzalez-Mestres, L., "High-Energy Nuclear Physics with Lorentz Symmetry Violation," paper nucl-th/9708028 of LANL electronic archive (1997h).

Gonzalez-Mestres, L., "Lorentz Symmetry Violation and Superluminal Particles at Future Colliders," paper physics/9708028 of LANL electronic archive (1997i).

Greisen, K., *Phys. Rev. Lett.* 16, 748 (1966).

Hills, D. and Hall, J.L., *Phys. Rev. Lett.* 64, 1697 (1990).

Hayashida, N. *et al.*, *Phys. Rev. Lett.* 73, 3491 (1994).

Hayashida, N. *et al.*, *Proceedings of ICRC97*, Vol. 4, p. 145 (1997).

Konstantinov, M. Yu., "Comments on the Hypothesis about Possible Class of Particles Able to Travel faster than Light: Some Geometrical Models," paper physics/9705019 of LANL electronic archive (1997).

Lamoreaux, S.K., Jacobs, J. P., Heckel, B.R., Raab, F. J. and Forston, E.N., *Phys. Rev. Lett.* 57, 3125 (1986).

Poincaré, H., Speech at the St. Louis International Exposition of 1904, *The Monist* 15, 1 (1905).

Recami, E., in *Tachyons, Monopoles and Related Topics*, Ed. E. Recami, North-Holland, Amsterdam (1978).

Rédei, L.B., *Phys. Rev.* 162, 1299 (1967).

Schweber, S.S., *An Introduction to Relativistic Quantum Field Theory*, Row, Peterson and Co., Evanston and Elmsford, USA (1961).

Will, C.M. *Theory and Experiment in Gravitational Physics*, Cambridge University Press (1993).

Yoshida, S. *et al.*, *Proc. of the 24th International Cosmic Ray Conference*, Rome, Italy, Vol. 1, p. 793 (1995).

Zatsepin, G.T. and Kuzmin, V.A., *Pisma Zh. Eksp. Teor. Fiz.* 4, 114 (1966)

On Superluminal Velocities

A. Jannussis
Department of Physics
University of Patras,
26110 Patras, Greece

S. Baskoutas
University of Patras
School of Engineering,
Engineering Science Department
26110 Patras, Greece

In the present work we investigate the Dirac equation in two dimensions for free particles and we obtain subluminal and superluminal velocities. Studying also the above equation with zero initial mass, for Caldirola-Montaldi (C.M.) and small-distance derivative (S.D.D.) model, we prove the existence of superluminal velocities.

1. Introduction

According to Diener [1] it is a cornerstone of our present physical theory that particles, energy or information can never be transported with velocities exceeding that of the vacuum of light. All our fundamental equations are such that causal connections can occur only within or on the light cone. On the other hand, it has been shown both theoretically and experimentally that superluminal group velocities exist. Ten years ago Jannussis [2] pointed out the existence of superluminal velocities by using the Caldirola-Montaldi (C.M.) model [2], and the small distance derivative (S.D.D.) model [3,4]. In (C.M) model the deformed four momentum operator takes the form

$$p_j \Rightarrow \frac{\hbar}{L_j}\sin\frac{L_j p_j}{\hbar}, j = 1,2,3 \tag{1.1}$$

$$p_4 \Rightarrow \frac{i\hbar}{c\tau}\sinh\tau\frac{\overset{o}{\partial}}{\partial t} \tag{1.2a}$$

or

$$\frac{\overset{o}{\partial}}{\partial t} \Rightarrow \frac{1}{\tau}\sinh\tau\frac{\overset{o}{\partial}}{\partial t} \tag{1.2b}$$

where L_j are the Caldirola lengths in 3-dimensions, τ is the Caldirola "chronon" and p_j the usual momentum operator. Since the Caldirola "length represents the critical distance which can cause the space interaction to begin, the Caldirola "chronon" represents the time of the interaction between two physical systems. The Caldirola length and chronon can be connected with the corresponding Max-Planck constants. Also in the (S.D.D.) model the deformed four momentum operators are given by

$$p_j \Rightarrow p_j \cos\frac{L_j p_j}{\hbar} \quad j = 1,2,3 \tag{1.3}$$

$$p_4 \Rightarrow \frac{i\hbar}{c}\frac{\dot{o}}{\partial t}\cosh\tau\frac{\dot{o}}{\partial t} \tag{1.4a}$$

or

$$\frac{\dot{o}}{\partial t} \Rightarrow \frac{\dot{o}}{\partial t}\cosh\tau\frac{\dot{o}}{\partial t} \tag{1.4b}$$

It should be noticed that the above models e.g. (C.M.) and (S.D.D.) are by their nature dissipative, due to the existence of the constants L_j and τ.

Both of the above models have recently used by Jannussis et al. [5], for the 'Deformed gauge problem of electrons in a uniform magnetic field', in the context of nonrelativistic and relativistic quantum mechanics.

Recently, the literature in the topic concerning the existence of superluminal velocities has been increased significantly, as theoretically as experimentally. We mention the more important works by Diener [1], Chiao [6] and coworkers, which discuss in detail the possibility of superluminal group velocities in amplifying media with inverted atomic populations [6-8]. Esposito [9] proposed recently a very simple but general method to construct solutions of Maxwell's equations with group velocity $v_{gr} \neq c$ and discussed an application concerning wavepackets in a waveguide. Superluminal tunneling of wavepackets has been observed experimentally in the case of microwaves [10], as well as for single photons [11], and coherent laser pulser [12]. Also a significant number of observations in Astronomy [13], have been interpreted as the appearance of objects moving with superluminal velocities (velocities greater than the velocity of light). Finally, we mention the existence of tachyons, according to Recami's theory, as well the recent published paper by M. J. Park and Y.J. Park [15], regarding a relativistic dynamical study with tachyons.

The present paper is arranged as follows. In section 2 we investigate the Dirac equation in two dimensions for free particles, and we obtain subluminal and superluminal velocities. Also in section 3 we investigate the same equation without initial mass by using (C.M.) and (S.D.D.) models and we prove the existence of superluminal velocities.

2.   Subluminal and Superluminal velocities.

For simplicity we shall consider the two-dimentional case for particles with initial mass $m_o = 0$. This case corresponds to the well known equation

$$\left(\frac{1}{c^2} \frac{\partial^2}{\partial t^2} - \frac{\partial^2}{\partial q^2} \right) \psi(q,t) = 0 \tag{2.1}$$

In the following we take into account the equations

$$\left(\frac{1}{c^2} \frac{\partial^2}{\partial t^2} - \frac{\partial^2}{\partial q^2} \pm k_o{}^2 \right) \psi(q,t) = 0 \tag{2.2}$$

which are partial case of the Klein-Gordon equation

$$\left(\frac{1}{c^2} \frac{\partial^2}{\partial t^2} - \frac{\partial^2}{\partial q^2} + V(q) \right) \psi(q,t) = 0 \tag{2.3}$$

often used to describe wave propagation in curved spaces, with $V(q)$ describing the scattering of waves by the geometry. According to Ching *et al.* [16], the above equation describes the quasinormal models (QNM'S) in gravitational systems.

For $V(q) = \pm k_0^2$ we obtain the equations (2.2) which are of Dirac type. The case $-k_0^2$ makes some troubles in our establishment, and lead us to some restrictions.

From Dirac energy of free particles

$$E = \pm c \left(p^2 + m_o{}^2 c^2 \right)^{\frac{1}{2}} \tag{2.4}$$

and $p = \hbar k_o$ we obtain

$$\frac{E^2}{c^2} = \hbar^2 k_0^2 + m_o{}^2 c^2 \tag{2.5}$$

or

$$k_0^2 = \frac{1}{\hbar^2 c^2} \left(E^2 - m_o{}^2 c^4 \right) > 0 \tag{2.6}$$

The case $k_0^2 > 0$ corresponds to the usual Dirac theory, since the other case $E^2 < m_0^2 c^4$ leads to the appearance of imaginary values for k_0.

Due to the fact that the operators of eq. (2.1) and (2.2) commute with each other, they have a common set of eigenfunctions of the form

$$\psi(q,t) \sim e^{i(kq - \omega t)} \tag{2.7}$$

and the corresponding dispersion relations are the following

$$\omega^2 = k^2 c^2 \tag{2.8}$$

$$\omega^2 = c^2 \left(k^2 \pm k_0^2 \right) \tag{2.9}$$

From the above relations we obtain the group velocities

$$v_{gr} = \frac{d\omega}{dk} = \pm c \tag{2.10}$$

$$v_{gr} = \frac{d\omega}{dk} = \pm c\left(1 \pm \frac{k_0^2}{k^2}\right)^{-\frac{1}{2}} \tag{2.11}$$

The case

$$v_{gr} = \pm c\left(1 - \frac{k_0^2}{k^2}\right)^{-\frac{1}{2}} \tag{2.12}$$

for $k^2 > k_0^2$ describes the superluminal group velocities, since the case of subluminal group velocity is given by

$$v_{gr} = \pm c\left(1 + \frac{k_0^2}{k^2}\right)^{-\frac{1}{2}} \tag{2.13}$$

for all the real values of k.

As it is clearly seen, formula (2.13) arises from formula (2.12) with the transformation

$$k_o \Rightarrow ik_o \tag{2.14}$$

Furthermore from the relation (2.9), we can caltulate the phase velocities, *i.e.*

$$v_{ph} = \frac{\omega}{k} = \pm c\left(1 \pm \frac{k_0^2}{k^2}\right)^{\frac{1}{2}} \tag{2.15}$$

which satisfy the well known relation

$$v_{ph}v_{gr} = c^2 \tag{2.16}$$

In our opinion the transformation (2.14) can be explained as follows: In the region between $\pm m_0 c^2$ which is not allowed by Dirac's theory, according to the conventional relativistic quantum mechanics, can be assumed that takes place a tunnel effect, where the particles moving with velocities greater than c. An analogous effect has been pointed out by Diener [1], who gives a proof that superluminal group velocities result from the analytic continuation of wave modulations transmitted within the light cone.

3.   Group superluminal velocities in C.M. and S.D.D. models.

In this section we will study the eq. (2.1), considering the C.M and S.D.D. models. Let us begin with the S.D.D. model:

The corresponding eq. (2.1) has the form

$$\left(\frac{1}{c^2}\frac{\partial^2}{\partial t^2}\cosh^2\tau\frac{\partial}{\partial t} - \frac{\partial^2}{\partial q^2}\cosh^2 L\frac{\partial}{\partial q}\right)\Psi(q,t) = 0 \tag{3.1}$$

The dispersion relation which arises from the above equation, with solution $\psi(q,t) \sim e^{i(kq - \omega t)}$, is the following

$$\omega^2 \cos^2 \tau\omega = c^2 k^2 \cos^2 Lk \tag{3.2a}$$

or

$$\omega \cos \tau\omega = \pm ck \cos Lk \tag{3.2b}$$

For $\tau = 0$ we obtain

$$E = \hbar\omega = \pm c\hbar k \cos Lk = \pm cp \cos\frac{Lp}{\hbar} \tag{3.3}$$

which is exactly the new radiation energy. For $Lp/\hbar = Lk = \pi\left(n+\tfrac{1}{2}\right)$ we get $E = 0$, since for $Lp/\hbar = Lk = n\pi$, where n is an integer, we obtain $E = \pm n\pi c\hbar/L$, which is the energy of the Planck Lattice.

Taking into account the above results, we are in the position to conclude that the energy of particles with $m_0 = 0$ on the light cone is not continuous, as in the Dirac's theory, but is quantized. Also from the relation

$$\omega = \pm ck \cos Lk \tag{3.4}$$

we obtain the group velocity, *i.e.*

$$\frac{d\omega}{dk} = \pm c(\cos Lk - Lk \sin Lk) \tag{3.5}$$

which for $Lk = 0$ takes the value $d\omega/dk = \pm c$ since for $Lk = \pi/2$ takes the value $d\omega/dk = \mp(\pi/2)c$ which is greater than the light velocity c.

We are reaching to the same result, considering the more general expression (3.2). In fact, using the Lagrange method [17] for the equation

$$\tau\omega \cos \tau\omega = \pm c\tau k \cos Lk \tag{3.6}$$

we obtain

$$\tau\omega = \sum_{n=1}^{\infty} \frac{B_n}{n} (\pm c\tau k \cos Lk)^n \tag{3.7}$$

where

$$B_n = \frac{d^{n-1}}{dx^{n-1}}\left(\frac{1}{\cos x}\right)^n \Bigg|_{x=0} \tag{3.8}$$

From the above formula it is evident that

$$B_{2n} = 0, \quad B_1 = 1, \quad B_3 = 3, \quad B_5 = 5 \dots \tag{3.9}$$

Finally eq. (3.7) takes the form

$$\omega = \pm ck \cos Lk \left[1 + \frac{1}{2}(c\tau k \cos Lk)^2 + \frac{17}{4}(c\tau k \cos Lk)^4 + \dots\right] \tag{3.10}$$

or

$$\omega = \pm ck \cos Lk \; f(k,\tau,L) \tag{3.11}$$

where

$$f(k,\tau,L) = \left[1 + \frac{1}{2}(c\tau k \cos Lk)^2 + \frac{17}{4}(c\tau k \cos Lk)^4 + \dots\right] \tag{3.12}$$

The corresponding group velocity is

$$\frac{d\omega}{dk} = \pm c(\cos Lk - Lk \sin Lk) f(k) \pm ck \cos Lk \frac{df}{dk} \tag{3.13}$$

For $Lk = \pi/2$ we obtain exactly the formula (3.6), *i.e.*

$$\frac{d\omega}{dk} = \pm \frac{\pi}{2}c \tag{3.14}$$

More details about the S.D.D. model and its applications in physics can be found in ref. [2,5]. Another interesting example, in which the group velocity is also greater than the light velocity, is obtained by (C.M.) model.

The characteristic equation corresponding to eq. (2.1) in (C.M.) model is the following

$$\left(\frac{1}{c^2 \tau^2} \sinh^2 \tau \frac{\partial}{\partial \tau} - \frac{1}{L^2} \sinh^2 L \frac{\partial}{\partial q} \right) \Psi(q,t) = 0 \tag{3.15}$$

The solution of this equation has the form

$$\psi(q,t) \sim e^{i(kq - \omega t)} \tag{3.16}$$

Inserting (3.16) in (3.15) we obtain

$$\sin \tau\omega = \pm \frac{c\tau}{L} \sin Lk \tag{3.17}$$

and

$$\omega = \pm \frac{1}{\tau} \arcsin\left(\frac{c\tau}{L} \sin Lk \right) \tag{3.18}$$

The group velocity takes the form

$$\frac{d\omega}{dk} = \pm c \frac{\cos Lk}{\left(1 - \frac{c^2 \tau^2}{L^2} \sin^2 Lk \right)^{\frac{1}{2}}} \quad \text{for} \quad \frac{c^2 \tau^2}{L^2} \sin^2 Lk < 1 \tag{3.19}$$

For $L = 0$ we obtain

$$\frac{d\omega}{dk} = \pm c\left(1 - c^2 \tau^2 k^2 \right)^{-\frac{1}{2}} \quad \text{for} \quad c^2 \tau^2 k^2 < 1 \tag{3.20}$$

which is greater than the light velocity c.

The above results for the two models (C.M and S.D.D.) appear to confirm Santilli's hypothesis [18,19] that the maximal speed of propagation of causal signals is not an absolute constant, but depends on the local physical conditions. It takes the value c in vacuum and larger (or smaller) than c in the interior of hadronic (or nuclear) matter. This hypothesis was subsequently investigated by De Sabata and Gasperini [20], who found the value $75c$ for the speed of propagation of causal signals, in the interior of a hadron.

4. Conclusion

In the present paper we investigated the Dirac equation in two dimensions for free particles, obtaining superluminal velocities, after the use of the analytic continuation transformation $k_0 \Rightarrow ik_0$. In our opinion this transformation can be explained as follows: In the region between $\pm m_0 c^2$, which is not allowed according to Dirac theory (conventional relativistic quantum mechanics), can be assumed that takes place a tunnel effect, where the particles are moving with velocities greater than c. An analogous effect has been pointed out by Diener [1], who gives a proof that superluminal group velocities result from the analytic continuation of wave modulations transmitted within the light cone. Studying also the Dirac equation with zero initial mass by using (C.M) and (S.D.D.) models (which represent open dissipative conditions), we prove the existence of superluminal velocities.

References

[1] G. Diener: *Phys. Lett.* A223, 327 (1996) and references therein.

[2] A. Jannussis: *Hadr. J.* 10, 79 (1986); A. Jannussis, V. Papatheou, N.Cim. B85, 17 (1985); P. Caldirola, E. Montaldi, *N. Cim.* B53, 241 (1979); A. Jannussis, A. Sotiropoulou, *Frontiers of Fundamental Physics*, edited by M. Barone and F. Selleri (Plenum New York, London 1994) pp. 347-357 (Procc. Int. Conf. Olympia, Greece, Sept. 27-30 1993, invited paper).

[3] P. Gonzalez-Diaz: *Lett. N. Cim.* 41, 481 (1984); *Hadr. J.* 9, 199 (19860; *Hadr. J. Suppl.* 2, 437 (1986).

[4] A. Jannussis: *N. Cim.* B90, 58 (1985), *Hadr. J. Suppl.* 2, 258 (1986).

[5] A. Jannussis, G. Kliros, A. Sotiropoulou, *Comm. Theor. Physics* 5, 1 (1996).

[6] R. Chiao: *Phys. Rev.* A48,34 (1993).

[7] A. Steinberg, R. Chiao: *Phys. Rev.* A49, 2071 (1994).

[8] E. Bolda, J. Garrison, R. Chiao: *Phys. Rev.* A49, 2983 (1994).

[9] S. Esposito: *Phys. Lett.* A225, 203 (1997).

[10] A. Enders, G. Nimtz: *J. Phys.* (Paris) 12, 1963 (1992); *Phys. Rev.* E48, 632 (1993); *J. Phys.* (Paris) 13, 1089 (1993); *Phys. Rev.* B47, 9605 (1993); V. Olkhovsky, E. Recami, *Comm. Theor. Phys.* 5, 71 (1996) and references therein; S. Baskoutas, A. Jannussis, R. Mignani, *J. Phys.* A27, 2189 (1994).

[11] A. Steinberg, P. Kwiat, R. Chiao: *Phys. Rev. Lett.* 71, 708 (1993); A. Steinberg, R. Giao, *Phys. Rev.* A51, 3525 (1995).

[12] C. Spielmann, R. Szipocs, A. Stingl, F. Krausz, *Phys. Rev. Lett.* 73, 2308 (1994).

[13] A. Harpaz: *The Physics Teacher*, 34, 496 (1996).

[14] E. Recami: *Riv. N. Cim.* 9, 6 (1986) and references therein.

[15] Mu-In Park, Young-Jai Park: *N. Cim.* B111, 1333 (1996).

[16] E. Ching, P. Leung, W. Suen, K. Young, *Phys. Rev.* D54, 3778 (1996).

[17] G. Valiron, *Equations Fonctionelless Applications* p. 98, (Paris, Masson et Cie 1945).

[18] R. Santilli: *Lett. N. Cim.* 33, 145 (1982); 37, 545 (1983).

[19] R. Santilli: *Elements of Hadronic Mech.* Vol. I *Mathmatical Foundations* (1993), Vol. II *Theoretical Foundations* (1994), Vol. III *Experimental Verifications* (to appear), Academy of Science of the Ukraine, Kiev, Ukraine, Second Edition (1995).

[20]   V. De Sabata, M. Gasperini: *Lett. N. Cim.* 37, 545 (1983).

Are Quantum Mechanics and Relativity Theory really Compatible?

José L. Sánchez-Gómez
Departamento de Física Teórica
Universidad Autónoma de Madrid

In this paper we discuss some aspects of quantum nonseparability related to the theory of relativity. It is argued that Quantum Mechanics (QM) and Special Relativity (SR) are incompatible unless the so-called Projection Postulate is given up. We also discuss the philosophy underlying decoherence models based on gravitation, in particular the nature of fluctuations of the space-time metric.

1. The projection postulate and relativistic covariance

As is known, according to the "orthodox" (*i.e.*, Copenhagen) interpretation of QM, the reduction of the quantum state takes place instantaneously (von Neumann's projection postulate); but this cannot be so in a relativistic theory because of the constraints imposed by relativistic covariance and causality. For measurements involving local observables this problem was solved by Hellwig and Kraus long time ago [1]; they put forward the so-called covariant reduction postulate (CRP), which states that the measurement of physical properties in a given space-time region changes ("reduces") the corresponding state in the future and side cones of such a region (the side cone being constituted of all the points separated from the said region by space-like intervals). They then showed that CRP preserves covariance and causality and leads to results which are the natural extension of those of non-relativistic quantum mechanics. However, as already said, this is true for local observables only, and then the question still remained as to what happens in the case of non-local observables.

The question of non-local observables was raised by Landau and Peierls [2] as early as 1931. They concluded that all non-local properties of the quantum state at a definite time should not be regarded as observables for relativistic quantum systems, since, they said, their measurement would violate causality. Nevertheless, Aharonov and Albert [3] have explicitly shown that in some cases one can measure certain non-local properties of the quantum state at a definite time without any violation of the causality. In fact, such properties are those characterizing the non-separable aspects of QM, which in some way is the main feature that distinguishes the latter from classical theories. Now, taking into account the apparent impossibility of finding a reduction procedure that simultaneously preserves covariance and causality (*i.e.*, a genuine "covariant collapse" of the quantum state), one is naturally driven to the conclusion that the reduction (projection) postulate should be abandoned if a "peaceful coexistence" of QM and SR is to be preserved, which in turn leads us to the so-called decoherence models.

2.   Quantum decoherence and classical gravitation

The concept of decoherence has turned out to be most helpful in analyzing several profound questions in the field of the foundations of QM and quantum cosmology (QC)—see for instance [4] for an interesting review of quantum decoherence and its relevance, when used in the formalism of the sum over quantum histories, to QC.

Now the question remains unresolved as to the real physical origin of decoherence in the case of ideally closed systems (as we think the Universe must be).

In recent years, the feeling is growing that gravitation might be deeply involved in very basic questions about the foundations of QM (see, for instance, the comments in Penrose's popular book *The Emperor's New Mind* [5]), not only in what concerns the "natural" realm of the Planck length, where a quantum theory of the gravitational field is certainly necessary, but also in the province of macroscopic and mesoscopic systems where the classical—albeit stochastic—aspects of the gravitational field might play a prominent role. This stochasticity is usually accounted for by putting some amount of randomness in the space-time metric, that is by introducing stochastic metric fluctuations. It should be noted that the fact that the fluctuations must be stochastic is crucial, since otherwise the classical gravitational field would be absolutely unable to produce decoherence or, as first pointed out by Károlyházy *et al.*, to play any significant role in the reduction of the wave-packet [6].

In this communication I shall present a model of stochastic metric fluctuations which gives rise to decoherence in the case of macroscopic systems, while quantum coherence is preserved in microscopic ones—this fact being quite relevant to understanding the classical limit of of QM. The technical aspects of the model have been discussed elsewhere (see [7] and references to previous work therein), so that here I will comment only on the main features from a physical point of view.

The main idea of the model is the existence of universal fluctuations of the gravitational field (vacuum fluctuations) of a stochastic nature which preserve the vacuum symmetries; in other words, they are *conformal*. Then one has

$$g_{\mu\nu}(x) = e^{\phi(x)}\langle g_{\mu\nu}(x)\rangle \tag{1}$$

where $\phi(x)$ is a classical stochastic field. In (1), $\langle\ \rangle$ stands for the stochastic mean; the background metric will be taken as Minkowskian by simplicity. From the above equation one easily derives

$$\langle e^{\phi(x)}\rangle = 1, \quad \langle e^{\phi(x)}\phi_\mu(x)\rangle = 0 \tag{2}$$

where $\phi_\mu(x) = \partial_\mu\phi(x)$. Also, the fluctuations being conformal implies

$$\langle\phi_\mu(x)\rangle = 0, \quad \langle\phi(x)\phi_\mu(x)\rangle = 0 \tag{3}$$

From this equation one gets $\langle\phi^2(x)\rangle = \text{const}$. Then it seems natural to assume $\phi(x)$ to be a gaussian process whose mean value can, without any loss of generality, be taken as zero, and with square deviation σ^2, *i.e.*

$$\langle\phi^2(x)\rangle = \sigma^2 \tag{4}$$

For physical reasons, it is also possible to assume the stochastic noise to be white, and thus write the correlation function in the following way

$$\langle\phi(\vec{x},t)\phi(\vec{x}',t)\rangle = \sigma^2 e^{-(t-t')^2/\rho^2}\, e^{-(\vec{x}-\vec{x}')^2/L^2} \tag{5}$$

where the correlation time τ is supposed to be much shorter that the characteristic time of the corresponding quantum system. Also, as it seems natural, one sets $L = c\tau$.

With all the above machinery, and after some calculations whose technical details have been already presented elsewhere (see [7]) one is able to show that this model of conformal metric fluctuations produces decoherence in the case of macroscopic systems whereas the coherence properties of microscopic systems are not affected. Also, a transition region (for masses about 10^{-13} g) is predicted wherein some peculiar behaviour—neither purely quantum nor strictly classical—should appear.

Concerning the free parameters of the model, one can show that σ and L are related to each other in the following way

$$\sigma^2 \sim \frac{\hbar G}{L^2 c^3} \tag{6}$$

where G is the gravitational constant and "$\sim$" means equal up to a numerical factor of the order of unity. Hence one has just one free parameter left, say L. In order to get the results referred to above, one has to take $L \sim 10^{-3}$ cm, which somehow defines the correlation length of the metric fluctuations.

3.   Related quantum-cosmological aspects

Let us now discuss some related cosmological points which are relevant to the consistency of models, in general, of metric fluctuations.

It has been pointed out in ref. [8] that the first model of wave-packet reduction through metric fluctuations, the so-called K-model, already referred to above [6] has the unpleasant feature of predicting a very large mean value of the space-time curvature—such a curvature would imply a mean energy density of the universe much larger than that of a neutron star! This unpleasant fact does not occur in the present model, due mainly to the assumed conformal nature of metric fluctuations. One can see that the mean value of the space-time curvature is given by ($c = 1$)

$$\widetilde{R} = -\frac{1}{2}\left\langle \exp(-2\phi)\left\{ 6\Box\phi + 3\left[\left(\frac{\partial\phi}{\partial t}\right)^2 - (\nabla\phi)^2\right]\right\}\right\rangle \tag{7}$$

Now, conformality—plus the correlation function (5)—implies

$$\left\langle\left(\frac{\partial\phi}{\partial t}\right)^2\right\rangle = \left\langle(\nabla\phi)^2\right\rangle = \frac{4\sigma^2}{L^2} \tag{8}$$

Then $R = 0$ up to order ϕ^4 (recall that the background metric is Minkowskian), so that no problem appears in the present model regarding the space-time curvature. One can also compute the 00-component of the stress (energy-momentum) tensor, which of course is not a Minkowski scalar. Note, however, that the relative dynamic velocity (that is, cosmic expansion not included) between any pair of physically relevant systems seems to be, as a matter of fact, much smaller than the velocity of light; hence one is allowed to think that $<T_{00}>$, computed in the reference frame (Ether?) where (5) holds (note that (5) is not a covariant equation) should give a reliable estimate of the mean energy density of the universe stored in the fluctuating vacuum. Upon using standard techniques one obtains

$$\left\langle\frac{T_{00}}{c^2}\right\rangle = \frac{c^2\sigma^2}{32\pi GL^2} \tag{9}$$

Taking $L = 10^{-3}$ cm one gets $\left\langle T_{00}/c^2\right\rangle \cong 10^{-30}$ g/cm^3 which is a reasonable value.

One can also analyze some other physical and cosmological aspects of the metric fluctuations—see [9]—with the conclusion that to get sensible results one is practically restricted to considering only conformal fluctuations, which, as already said in Sect. 2, have, moreover, the attractive feature of preserving—maximally—the symmetries of the classical gravitational vacuum.

One question, central to quantum cosmology, which remains unsolved in spite of the intense work devoted to it is how to get a classical space-time out of a "purely quantum world," which is only supposed to exist in the Planck age. This is indeed quite a formidable problem; to help solve it, one may formulate another related problem, in principle more tractable: how the space-time becomes classically defined from the quantum superposition of all possible "space-times" present in the corresponding semiclassical solutions of the Wheeler-DeWitt equation.

It has been sugested that the space-time becomes classical by a process similar to the emergence of classical properties of macroscopic systems in decoherence models (such as the one discussed here)—see [10] and references therein. Decoherence would then arise from the entanglement of the states of the system with its environment. (Now the system is the Universe, so that it is not in principle easy to imagine what the environment could possibly be; see however [10].) In the model introduced in this communication, the environment is provided just by the classical gravitational vacuum (being then universal). By following a similar philosophy we have shown [11] that it is possible to understand the "classicity" of space-time if one assumes the existence of *quantum* conformal metric fluctuations, in the sense of the existence of some allowed degrees of freedom in the evolution of the manifold representing the properties of empty gravitational solutions. This seem to be the only required environment in the problem of the "self-measurement" of space-time in quantum gravity.

4. Final comments

At the begining of this communication it was stated that QM and SR are incompatible unless the projection postulate (wave packet reduction) is taken out of the quantum formalism. Then, in order to recover the classical limit, the concept of decoherence is introduced. Now, to lay a physical foundation for decoherence, we have resorted in some way to general relativity (GR) or, actually, to some modification of GR, since the space-time metric is regarded as having stochastic fluctuations. Nevertheless, the model presented here is not in complete agreement with the theory of relativity, and not even with SR, because the correlation function shown in eq. (5) is not covariant. (As a matter of fact, (5) is supposed to hold in some "privileged" system; the Ether say.) Such a discrepancy arises just as a consequence of the impossibility (for the time being, at least) of formulating a fully relativistic theory of stochastic processes. Perhaps this fact is telling us that QM and RT are only "approximate" theories, and that a more general (stochastic?) theory is waiting to be discovered. But this is mere speculation at present.

References

[1]   K. E. Hellwig and K. Kraus, *Phys. Rev.* **D1**, 566 (1970).

[2]   L. D. Landau and R. Peierls, *Z. Phys.* **69**, 56 (1931).

[3]   Y. Aharonov and D. Z. Albert, *Phys. Rev.* **D21**, 3316 (1980); **D24** 359 (1981).

[4]   J. B. Hartle, in *Quantum Cosmology and Baby Universes*, edited by S. Coleman *et al.*, World Scientific (Singapore, 1991).

[5]   R. Penrose, *The Emperor's New Mind*, Oxford University Press (Oxford, 1989), in particular, Chapter 6.

[6]   F. Károlyházy, A. Frenkel and B. Lukács, in *Physics as Natural Philosophy*, eds. A. Shimony and H. Feshbach, MIT Press (Cambridge, MA, 1982)

[7]   J. L. Sánchez-Gómez and J. L. Rosales, *Ann. N. Y. Acad. Sc.* **755**, 500 (1995).

[8]   L. Diosi and B. Lukács, Budapest report KFKI-1993-05/AB, and *Phys. Lett.* A **181**, 366 (1993).

[9]   J. L. Rosales and J. L. Sánchez-Gómez, *Phys. Lett.* A **199**, 320 (1995).

[10] J. J. Halliwell, in *Complexity, Entropy and the Physics of Information*, ed. W. H. Zurek, Addison Wesley (Redwood City, CA, 1990).

[11] J. L. Rosales and J. L. Sánchez-Gómez, *Ann. Fond. L. de Broglie*, to be published.

Behind the Scenes at the EPR Magic Show

Caroline H Thompson
Department of Computer Science, University of Wales
Aberystwyth, UK
Email: cat@aber.ac.uk
Web Pages: http://www.aber.ac.uk/~cat

If we drop the assumption that the atoms in Aspect's "atomic cascade" source act independently, then his subtraction of "accidental coincidences" cannot be justified. Newly discovered data from an unsummarised table in his PhD thesis shows that the "correction" is much more important than hitherto thought. If we drop the assumption that light is corpuscular, we may easily have "variable detection probabilities." Dropping both allows realist interpretations of all three of Aspect's EPR experiments - no magic is needed. Similar considerations apply to current experiments using parametric down conversion sources. This "accidental subtraction" or "emission loophole" is quite different from the much better known "detection loophole" and is numerically of great importance. It can and should be investigated, instead of continuing in the obsessional pursuit of the "loophole-free" experiment - the end of the rainbow. If we *must* have such experiments, let us insist that all assumptions, all data adjustments, all limitations and "anomalies" be clearly stated—in other words, insist on a return to valid scientific standards.

Introduction

In this paper I shall discuss some points from Aspect's experiments from two viewpoints - a "photon" version, as near as I can to the way he saw things, and my own realist version, similar to Stochastic Optics [1] or CWN [2]. The experiment I shall concentrate on is his first EPR one [3], this being the one that (*a*) uses a Bell test that is basically valid and (*b*) is the *only* experiment for which I have managed to access the relevant raw data. His last one [4], with time-switching, is from the point of view of my model identical and so susceptible to the same explanation. Aspect's is the work I know best, having read his thesis, but I have every reason to think that similar criticisms can be levelled at *all* EPR experiments. They all share basic problems, in that the experimenter is given a free hand to adjust many parameters to get the

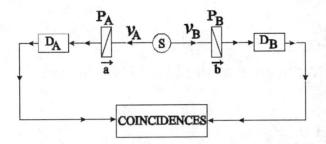

Fig. 1: Scheme for 1981 Experiment (from one of Aspect's published papers). S is source, P's polarisers and D's detectors. Polariser axes are at angles a and b respectively.

strongest violation of a Bell inequality he can manage, in the faith that the theory he is defending must be correct because so many other people have agreed with its predictions. This approach does not lead to valid science except when, by happy chance, the theory *is* correct!

Experimenters seem unaware of the full range of realist possibilities, and hence of the critical role played by their assumptions. The fact that many of these are unproven has become lost in the mists of time. They have become accepted simply on the grounds that they have caused no conflict with quantum theory. An experiment in Geneva, involving a parametric down conversion source and EPR correlations over 10 km, has recently received some publicity. It is that by Tittel *et al.* [5], now reported in the quant-ph archive. If the subtraction of accidentals is disallowed the results are compatible with local realism, yet the possibility is not even mentioned.

Aspect's 1981 Paired-Photon Experiment

The QT story is that pairs of "photons" are produced by the source in an atomic cascade, pass through the polarisers if their polarisation is suitable, and a fixed proportion of those that pass through are detected.

My information comes largely from Aspect's PhD thesis [6]—a document that, had it been in English, might by now have become a best-seller. I have made translations of two sections available through my Web pages.

Subtraction of Accidentals

I shall start at the end, with the data that I find so dramatic. As you will see, the adjustment subtracting "accidentals" is large. The raw data follow a nice sine curve; the adjustment shifts it down, which increases the visibility from 0.55 to 0.88. If you calculate the Bell test you will find that the raw data does *not* infringe it. The visibility is in fact only slightly greater than the prediction (0.5) of the simplest realist model - the model that assumes the counts obey Malus' Law *exactly* and that you get the expected coincidence

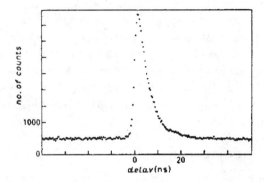

Fig. 2: A time spectrum (histogram) of differences between detection times.

probabilities by multiplying the two singles probabilities, for fixed polarisation angle. The basic QT prediction for visibility is 1.0, but it is not difficult to find excuses to reduce this to the figure observed!

Now one might be forgiven for thinking that such an evidently important adjustment would have been the subject of vigorous debate and fully investigated. It *has* been discussed - by Marshall, Santos and Selleri, and Aspect and Grangier responded with a paper in 1985 [7]. But Aspect used theoretical arguments that few, if any, could be expected to follow. Moreover, he quoted figures not for the experiment I am concerned with but for his second experiment, which had two outputs for each polariser and used a different Bell test, one in which each term is estimated using a ratio in which the denominator is the sum of four coincidence rates. The use of this denominator invalidates the test - it can produce a bias that allows it to be infringed relatively easily, whenever there are "variable detection probabilities" (of which more later). I have published a paper [8] explaining by means of an intuitive analogy (a "Chaotic Ball") this well-known but frequently forgotten fact.

To return to our story: Aspect was able to find a case in which a Bell test was violated even when he did not do the subtraction, but this was irrelevant to the real problem, which concerns violation of the more stringent tests used in single-channel experiments. These tests have the weakness that they involve comparisons between results with and without polarisers in place, but there is no *a priori* reason to expect bias. To my knowledge, no *experimental*

Raw and adjusted coincidence rates

Angle between polarisers	0	22.5	45	67.5	90	One absent	Both absent
Raw	96	87	63	38	28	126	248
Accidental	23	23	23	23	23	46	90
Adjusted	73	64	40	16	5	81	158

investigation has been done on the subtraction. There is no theoretical difficulty: we just have to adjust our source so as to have negligible accidentals (something that Freedman was able to do back in 1972, though his experiments had a different fault - coincidence windows that were far too small [9,10]). The subject, it would appear, has attracted little attention because the data I used above has not been made public. I summarised it from a table (table VII-A-1) in Aspect's thesis that was presented in confused order, that in which it was collected, which was impossible to assess by eye.

What *are* these "Coincidences"?

Let us consider what coincidences really are, in order to assess whether or not subtraction of accidentals is reasonable.

The experiment outputs a time-spectrum, of which one is shown in Fig. 2. Those displayed on a VDU during the running of each subexperiment would have had rather greater scatter as the accumulation time was shorter.

Aspect's picture is quite simple, though his remarks in his thesis show that he knew full well that it amounted to a pure assumption that, if untrue, invalidated his analysis.

The QT Story

The source produces pairs of "photons" as a result of individual atoms being stimulated into an excited state then relaxing by two stages, as indicated below:

The diagram implies that A is emitted then B, after an interval governed by the "lifetime of the intermediate state of the cascade." Thus it is natural for Aspect to assume that the falling part of the spectrum simply mirrors the time of emission of B, with the peak at 0, corresponding to zero delay. Time is measured from the time of *emission* of A. He followed established practice in assuming the rising part represented just error, which he confirmed to be of a suitable magnitude in various subsidiary experiments. (Or so he thought.)

The basic idea in defining "coincidences" is to chose a start and end time relative to the peak and count the number of events in this "window." Note, incidentally, that the QT model is inadequate in practice, as it only ever mentions one parameter, the window size.

But what of the "shoulders"? For a valid EPR experiment (see Clauser and Horne's 1974 paper [11]), we should have organised the source so that coincidences

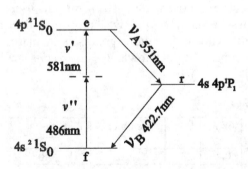

Fig. 3: Atomic cascade energy levels and frequencies.

are easy to identify, with the shoulders negligible. Aspect's clearly were not, so he had to fall back on a model (Fig. 4 (a)), valid only if emissions are stochastically independent (and, incidentally, detectors are able to detect events arbitrarily close together, which they are not!).

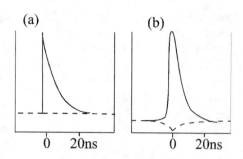

Fig. 4: (a) Aspect's model of time spectrum (from his thesis);
(b) A realist speculation.
The region below the dashed line is taken to represent "accidental coincidences."

In this model, the shoulders, together with the whole base of the spectrum, correspond to emissions by "other" atoms. To estimate these "accidentals," we need to know the probability of events happening this close together just by chance, when there is no synchronisation, so we artificially delay one channel (the B one) by 100 ns, sufficient to destroy this synchronisation. The "coincidences" as measured using the delayed channel give the required estimate.

My Picture

My picture might be more like Fig. 4 (b), or even more extreme. It comes from questioning whether the atomic cascade idea might not be completely wrong. After all, how much direct evidence is there for it? Perhaps large atoms in a solid may act individually, but might not these atoms - heated to 800 degrees centigrade and illuminated by two lasers, with polarisations parallel to the beam direction - be behaving more like a plasma? Or more like sound? If it is the whole EM field that oscillates, from time to time attaining a resonant state involving primarily the two frequencies of interest, then there can only be one emission at a time. Possibly there is a natural minimum interval between emissions. Certainly there is a minimum interval for detections, as the electronics ensure this. The pair of signals approaching the two detectors might be visualised as in Fig. 5:

The detection process is modelled by assuming EM noise to be added to the signals, mainly at the photocathode of the detectors, with detections occurring when a threshold is crossed. There is, in a sense, plenty of experimental evidence for the importance of this noise: we know that temperature is critical; screening is important (and it might not have been possible [2] to screen out EM noise from local electronic equipment); physical proximity of detectors can increase correlations. The assumption that it is only intensity that matters seems adequate for our purpose. It lends itself to computer simulation, which can easily confirm that the output will be at least qualitatively as observed. There is more than one way in which the shoulders might arise, and further experimentation is needed. They are very likely for

Aspect's experiments to include a large contribution from signals that were emitted at only the one frequency, unpaired, as the system produced three times as many B detections as A ones. Aspect may well have been wrong in assuming this was just because detection efficiencies were lower for A.

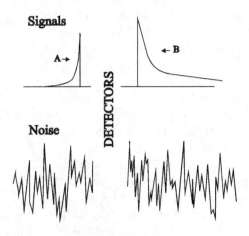

Fig. 5: Intensity profiles of two signals approaching their detectors, and their accompanying noise.

This model is, incidentally, entirely compatible with results that purport to demonstrate the particle bevaviour of light [12].

There will be a large element of random error in the time of detection, but there will also be systematic effects, with higher intensities (per signal) being detected sooner (only the *first* detection counts, as a result of dead times). This feature gives a testable difference from QT: if we insert an extra polariser, say, to decrease the intensity, in the QT model we decrease the number of photons. The *shape* of the time spectrum stays the same. In my model, the shape will change. I must emphasise that this effect is only expected with detection systems similar to that used in this particular experiment, in which dead times would have been large enough to suppress later detections and only the first is registered.

Thus the assumption of independence of emissions seems unlikely. Under realist reasoning, the infringement of Bell inequalities is always evidence of false assumptions, and this seems a prime suspect. The subtraction of accidentals is numerically important whenever it is applied, but it is not sufficient on its own to account for violation of the Bell inequality in Aspect's second, two-channel, experiment [13]. As mentioned earlier, the Bell test appropriate here is easily violated if we have "variable detection probabilities," which we can have if Malus' Law does not apply exactly to *counts* [8].

Does Malus' Law apply to *counts*?

Under classical theory, Malus' Law applies to EM amplitudes and hence gives a rule relating *intensities* to relative polarisation angle. Experiments with "single-photon" light operate with detectors set in "Geiger Mode." The detectors have various parameters (temperature, voltage of photomultiplier, threshold voltage of discriminator) chosen so that they approximate as closely as possible to a linear response of counts to intensity. Output counts therefore show the same pattern as input intensities, and it does not matter whether we think of counts or intensities (hence the success of the probabilistic interpretation of QT!). But in reality this linearity *can* only be approximate. We can organise things so that when we pass light through two polarisers at various angles we get counts following very closely to Malus' Law. If we leave the setting untouched, though, we are likely to get deviations if we alter the intensity of our source. The curve we see is always, in any case, the weighted mean of curves representing different intput intensities. They *cannot*, under my model of detection, all be the same shape. For weak signals, we might have the dashed curve of Fig. 6.

The observed curve is obtained in the course of calibrating the experimental polariser by inserting an extra one near the source and varying the relative angle. The population emerging from the first polariser will not all have the same intensity, especially if the source produced signals of random polarisation direction. It could well happen (I think I have found slight evidence) that the population of interest - those entering into coincidences - is not the whole one. A pattern such as the dashed curve would increase the visibility of the coincidence curve, though not to the extent that it would cause violation of a valid Bell test. The one used in two-channel experiments I dismiss as invalid.

There is, I believe, some experimental evidence that Malus' Law does not hold perfectly for counts. Aspect reports in his thesis several "anomalies," such as reversal of the roles of A and B producing changes in coincidence rates, and total of the four coincidence rates in his two-channel experiment not being quite constant. (In relation to the former, he makes the highly questionable decision that it does not matter that the separate values do not conform to QT since it is only the total that is needed for his Bell test.) Considering that he has made settings so that all the singles rates appear to be behaving correctly and he gets, as expected, a doubling of coincidence rate when he removes a polariser, the most likely explanation of the anomalies is slight deviations from the rule. They are all small, but they must have been reproducible or he would not have felt the need to report them at all.

Conclusion

Thus we have two straightforward factors - deviation from linearity in detector response and failure of the independence assumption - accounting in large measure for the observations. Published information may not be sufficient to *prove* this, but it is hard to see why experimenters have not investigated these factors (and *published the results!*). The first is surely well known to the experts, though it has been played down to the point of extinction in the story put out to the public. The second appears to have been totally discounted, though it is relatively easy to test, just by seeing if the Bell statistic decreases towards the classical value as we decrease count rates and hence "accidentals." The subtraction has become completely routine, which is perfectly acceptable if all you want is a measure of the contrast of your time spectrum, but not if you want a figure that will be valid in a Bell test.

In EPR experiments, the experimenters do what they are asked to do: find conditions in which Bell inequalities were infringed! Nobody, it seems, puts any restraints on the methods they used, or asks them to publish full data, including the runs that do not quite work. The magicians know how to *produce* their illusions (albeit not quite perfectly - witness those "anomalies"), but why do they still not *understand* them?

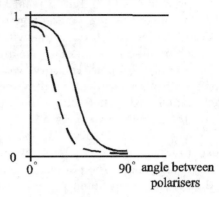

singles count   (normalised)

Fig. 6: Dashed curve: relationship between counts and angle suggested by my model for low intensity signals. Solid curve: observed relationship, for whole ensemble in actual calibration run.

References

[1]   De la Peña, L and Cetto, A M: *The Quantum Dice: an Introduction to Stochastic Electrodynamics*, Kluwer (1996).

[2]   Gilbert, B and Sulcs, S: "The measurement problem resolved and local realism preserved *via* a collapse-free photon detection," *Foundations of Physics* 26, 1401 (1996).

[3]   Aspect, A, Grangier, P and Roger, G: *Physical Review Letters* 47, 460 (1981).

[4]   Aspect, A: "Experimental Test of Bell's Inequalities Using Time-Varying Analyzers," *Physical Review Letters* 49, 1804-1807 (1982).

[5]   Tittel, W *et al.*: "Experimental demonstration of quantum-correlations over more than 10 kilometers," submitted to *Physical Review Letters* 1997. Available at http://xxx.lanl.gov, ref quant-ph/9707042.

[6]   Aspect, A: *Trois Tests Expérimentaux des Inégalités de Bell par mesure de corrélation de polarisation de photons*, PhD thesis No. 2674, Université de Paris-Sud, Centre D'Orsay (1983).

[7]   Aspect, A and Grangier, P: *Lettere al Nuovo Cimento* 43, 345 (1985).

[8]   Thompson, C H: "The Chaotic Ball: An Intuitive Analogy for EPR Experiments," *Foundations of Physics Letters* 9, 357 (1996). Available from my Web page or from http://xxx.lanl.gov, where it is ref 9611037 of the quant-ph archive.

[9]   Freedman, S J: *Experimental Test of Local Hidden-Variable Theories*, PhD thesis (available on microfiche), University of California, Berkeley (1974).

[10] Thompson, C H: "Timing Accidentals and Other Artifacts in EPR Experiments," submitted April 1997 to *Physical Review Letters*. Revised version available from my Web page or as ref 9711044 of http://xxx.lanl.gov quant-ph archive.

[11] Clauser, J F, and Horne, M A: *Physical Review D* 10, 526 (1974).

[12] Grangier, P, Roger, G and Aspect, A: "Experimental Evidence for a Photon Anticorrelation Effect on a Beam Splitter: a New Light on Single-Photon Interference," *Europhysics Letters* 1, 173 (1986).

[13] Aspect, A: "Experimental Realization of Einstein-Podolsky-Rosen-Bohm *Gedankenexperiment*: A New Violation of Bell's Inequalities," *Physical Review Letters* 49, 91-94 (1982).

Relativistic Physics
and Quantum Measurement Theory

M.A.B. Whitaker
Department of Physics
Queen's University, Belfast

Shimony has talked of a "peaceful coexistence" between relativity and
quantum (measurement) theory. Various aspects of this relationship are
studied: the collapse postulate, transfer of information in EPR pairs,
and the quantum Zeno effect.

I. Introduction

Shimony [1] has suggested that there may be a "peaceful coexistence"
between quantum theory and relativity. He is referring to quantum
measurement theory, and, in particular, to the issues surrounding Bell's
theorem and its experimental study [2,3,4].

To sketch his reasons, it may be said that Bell's theorem establishes a
contradiction between quantum theory and "locality" [5]. "Locality," in fact,
entails two conditions. The first is *parameter independence*: the action of an
experimenter at one point in space cannot influence the outcome of a
measurement made simultaneously at another point. (In practice, we must
usually consider the expectation value of the result for two different
experimental settings, though see section III below.) Conversely, the
experimental result can give no information about the experimental setting.

The second condition is *outcome independence*: the result of a
measurement at one point cannot influence the outcome of a measurement
made simultaneously at another point.

Breaking the first condition implies the possibility of sending a signal at
infinite speed, and so disobeys special relativity, but breaking the second
condition does not. It is found that EPR-Bell type experiments may be
regarded as disobeying outcome independence, but not parameter
independence.

For example, let us consider "straight" EPR. If we measure S_z in each
wing of the experiment, then, since the results are correlated, if we assume no

hidden variables, we have violated outcome independence, but clearly cannot use this to send signals.

Now imagine, though, that we may measure *either* S_{z1} or S_{x1} in the first wing. We also measure S_{x2}, and, for either choice of measurement in the first wing, $\langle S_{x2} \rangle = 0$. We cannot use our choice of experiment to send a signal from one wing to the other, and so we do not violate parameter independence or special relativity.

Bell showed by generalising EPR that, even if hidden variables *are* allowed, quantum theory conflicts with locality, but Jarrett [6] and Shimony [1] showed that it is *outcome dependence* not *parameter independence* that is violated, hence the "peaceful coexistence." The remainder of this paper explores various aspects of this coexistence.

II. Collapse

Collapse is thought by most (though certainly not all) to be an essential ingredient of quantum measurement theory. To discuss measurement according to the collapse procedure, we assume a measurement of observable O, with associated operator $\hat{O}$, which has eigenvalues and eigenfunctions O_n and α_n. Then if the initial wave-function of the system is one of the eigenfunctions, α_m, the measurement process may be represented by

$$\alpha_m ; \phi_0 \rightarrow \psi_f \phi_m \tag{1}$$

where ϕ_0 and ϕ_m are initial and final wave-functions of the measuring device. ψ_f is the final wave-function of the measured system. This may or may not be related to α_m and ϕ_m (which are directly related). (Indeed the measured system often does not have independent existence after the measurement.) In this paper, we assume the measurement is what was called by Pauli a "measurement of the first kind," for which $\psi_f = \alpha$. In this case an immediate repeated measurement will yield the same result.

This is all very satisfying, but things are very different if the initial wave-function of the measured system is a linear combination of eigenfunctions, *i.e.* $\sum_n c_n \alpha_n$. The linearity of the Schrödinger equation then results in the interaction being represented by

$$\sum_n c_n \alpha_n ; \phi_0 \rightarrow \sum_n c_n \alpha_n \phi_n \tag{2}$$

and it appears that the coupled system is left in an unphysical combination of states.

To cope with this difficulty, the collapse hypothesis assumes that, at the time of the measurement, the right-hand-side of (2) collapses to a single term;

$$\sum_n c_n \alpha_n \phi_n \rightarrow \alpha_p \phi_p \tag{3}$$

the probability of collapse to $\alpha_p \phi_p$ being given by $\left|c_p\right|^2$. This ensures again that immediate repetition of the measurement yields the same result.

Despite, though, the apparent centrality of the idea of collapse, it is interesting that Peres [7] has written that "The notion of collapse.....appears to have no meaning whatever in a relativistic context." (omitted matter later). It must be said that much work has actually been done on putting collapse into a relativistic framework—collapse along past light cone, future light cone, or more complicated hypersurface [7].

Yet Peres' argument is simple and quite convincing. It follows EPR, and recognises that relativistically we cannot say that observer 1, say, necessarily performs the measurement before observer 2 in *all* frames. To be specific, let us imagine that observer 1 measures S_{1z} and observer 2 measures S_{2x}, and, for convenience, consider only cases where *both* observers obtain the answer $+\hbar/2$. The initial state may be written as $(1/\sqrt{2})\{|+_z -_z\rangle - |-_z +_z\rangle\}$. In frames in which observer 1 measures first, he will be described as collapsing the state-vector to $|+_z -_z\rangle|+_{1z}\rangle$ (the second ket being a measuring device), but in frames where observer 2 measures first, she will be described as collapsing it to $(1/\sqrt{2})\{|+_z +_x\rangle - |-_z +_x\rangle\}|+_{2x}\rangle$. In both cases, of course, the final collapse is to $|+_z +_x\rangle|+_{1z}\rangle|+_{2x}\rangle$, but, if one thinks of collapse as a real physical process, we have reached a contradiction/paradox. There is no disagreement about what is actually observed, but different collapses "take place," and for the different observers, the state-vector passes through different states.

To make sense of this, it seems we must (a) get rid of "collapse," at least as a "real" happening, as distinct possibly from a mathematical convenience. Though we are presenting the argument from relativity here, collapse has been much criticised from other points of view as well. It assumes that quantum systems develop in two distinct ways: (1) *via* the Schrödinger equation, but (2) (at measurement) by collapse. Yet measurement is an ill-defined concept. To complete Peres' quote [7]: "The notion of collapse, which is of dubious value in non-relativistic quantum mechanics...."

How can we avoid having to talk of collapse? Popular answers would be ensembles [8] or many worlds [9], but I would suggest that these do not really solve the problems [10,11]. Other possibilities—environment, decoherent histories—are not discussed here. We may not need to accept such methods if we are prepared to accept (b) The wave-function of a system is not necessarily directly related to the (physical) "state" of the system, but should only be used to predict experimental results. This is a form of words used by quite a few people in different contexts—Ballentine, Stapp (advocating Copenhagen), Peres. Here we follow Ref.12. Suppose a state-vector $|\psi\rangle$ is given by $|\psi\rangle = a_1|\psi_1\rangle + a_2|\psi_2\rangle$, where $|\psi_1\rangle$ and $|\psi_2\rangle$ are eigenvectors of $\hat{O}$, the operator corresponding to observable O, which is being measured, the appropriate eigenvalues being O_1 and O_2.

After a measurement of O, and assuming collapse, $|\psi\rangle$ is given by $|\phi_1\rangle|A_1\rangle$ or $|\phi_2\rangle|A_2\rangle$ with probability $|a_1|^2$ and $|a_2|^2$ respectively, where $|A_1\rangle$ and $|A_2\rangle$ are apparatus states. This assumes that the measurement is of the first kind, and so an immediate second measurement will give the same result as the first.

If no collapse is assumed, the state-vector after the first measurement is

$$|\psi\rangle = a_1|\phi_1\rangle|A_1\rangle + a_2|\phi_2\rangle|A_2\rangle \tag{4}$$

and, after the second measurement,

$$|\psi\rangle = a_1|\phi_1\rangle|A_1\rangle|B_1\rangle + a_2|\phi_2\rangle|A_2\rangle|B_2\rangle \tag{5}$$

where $|B_1\rangle$ and $|B_2\rangle$ are apparatus states for the second measurement.

Thus the probability of obtaining O_1 (O_2) in each measurement is $|a_1|^2$ ($|a_2|^2$), and there is zero probability of obtaining different results. One thus obtains the same results as without the use of collapse as with its use, *but* without it, it is not possible to write down a state-vector for any period between the measurements which acknowledges that, at the first measurement, a particular result was obtained. *i.e.* the state-vector is a good *predictor* but a bad *representor*.

Is this good or bad for realism? At first sight it looks bad; the nearest thing to something "real" has become a lot less real. At second sight, it may be good; the wave-function was always an uninspiring candidate for "reality"; if we lose that hope, must we have something more concrete—a hidden variable—to give some indication of where the particle is and what it is doing?

Two last comments: (1) The above may not be really new—it is actually similar to the ideas of Everett [13], but it is interesting that we are pushed in a direction in which we may already want to go by relativistic considerations. (2) From (a) above, we should definitely avoid talk of "collapse" in conceptual analysis. In practice, though, if it is *predictions* that are important, we may perhaps use the word pragmatically. The main problem in (a) was that it led to non-uniqueness of wave-functions. If the same experimental results occur, (b) may take some of the sting out of (a).

A Possible Example of Parameter Dependence [14]

Consider an EPR-Bell experiment with N pairs of spins, where N is initially 2; later it may vary but is always even. In the first wing, σ_θ is measured for each spin, where θ is the angle made with the z-axis. Pairs are only selected for which 1 particle (A) gives $\sigma_\theta = 1$, and the other (B) gives $\sigma_\theta = -1$..

In the second wing, σ_z is measured. For the pair of spin A, we have $P(+) = \sin^2(\theta/2)$, $P(-) = \cos^2(\theta/2)$. For the pair of spin B, we have $P(+) = \cos^2(\theta/2)$, $P(-) = \sin^2(\theta/2)$. Thus with S the total spin along the z-direction in this wing in units of the particle spin, we have

$$P(S=2) = P(S=-2) = (1/4)\sin^2\theta; \; P(S=0) = 1-(1/2)\sin^2\theta \qquad (6)$$

so over a number of experiments: $\langle S \rangle = 0$ (of course); $\langle S^2 \rangle = 2\sin^2\theta$; s.dev $= \{\langle S^2 \rangle - \langle S \rangle^2\}^{1/2} = \sqrt{2}\sin\theta$

Clearly the results in the second wing depend on the value of θ chosen in the first wing. As N varies, we find that the standard deviation becomes $\sqrt{N}\sin\theta$, increasing with increasing N.

We must not forget that we are selecting runs from information in the first wing. *i.e.* we require $N_+ = N_- = N/2$, so we must be careful drawing conclusions. The experiments themselves may be performed as quickly as required, but there must be a subsequent signal from the first to the second wing to confirm that $N_+ = N_-$ in the first wing, so we cannot claim faster-than-light signalling.

However, we do suggest that the experiment does demonstrate parameter dependence. The above signal does not provide information on θ, so that information must have passed at the first stage of the experiment, and this may be faster than light.

IV The Quantum Zeno Effect

The quantum Zeno effect (QZE) is well known; an important early paper was by Chiu *et al.* [15] while a recent substantial account is in Ref.16. In this paper we restrict the term quantum Zeno effect to experiments analogous to that of Ref.15, where a macroscopic measuring device has an effect on a macroscopically separated microscopic system. The usual quantum Zeno set-up will always give such an effect (except if the decay is perfectly exponential—which it never is!). We do not require to work in a t^2-region of decay, nor do we look for total freezing, or even necessarily a dramatic reduction in decay. Indeed a dramatic reduction in decay is not sufficient to justify the term QZE. For example the well-known and excellent experiments of Dehmelt [17], Itano *et al.* [18] and Kwiat *et al.* [19] do not qualify. These experiments show clearly reduction in decay, *but* there is no macroscopic distance between "decaying" system and "measuring" or "perturbing" system. These experiments are brilliantly conceived and executed. They demonstrate clearly the existence of the t^2-region. But they are in no way difficult to understand, and do not relate directly to the conceptual problem raised by Ref. 15. The suggestion of Inagaki *et al.* [20], on the other hand, does qualify as a possible example of a QZE.

We discuss two variants of the QZE. The first is the simple idea of Chiu *et al.* [15] with the detector completely surrounding the system. Naively we might say this represents continuous measurement, and should lead to frozen decay. But the concept of continuous measurement is difficult to define, and so we may use a variant with discrete measurement.

In variant 2, the detector is on the inside of a sphere of balloon-type material; which may be expanded and then collapsed extremely quickly. The decaying system is at the centre of the sphere. We wish to test for survival/decay at discrete times: T/n, $2T/n$........T. For convenience we assume that all decay particles have the same speed v, and we assume that $v<<c$.

From $t=0$ to $t=t_1$, the radius of the balloon is held at R_2, where $R_2>>vt_1$. Then at t_1, the radius is decreased, at speed u such that $v<<u<<c$, to R_1, a macroscopic distance. Here $t_1=T/n + R_1/v$. Decay particles detected in the sweep-in of the balloon will be ascribed to decays in period up to $t=T/n$. At this point the radius of the balloon will be immediately returned to R_2.

Similar contractions and expansions are performed at $t=2T/n + R_1/v$, $3T/n + R_1/v$ and so on. Thus decay statistics are built up as a function of n, or one may use unequally spaced measurements, or measurements at random times. The quantum Zeno prediction is that the decay profile depends on the times of the various measurements.

We now consider the relationship between the QZE and local realism. Realism implies that the state of each decaying atom has associated with it an additional (hidden) variable which determines (either deterministically or stochastically) the instant at which the atom will decay. The survival probability is thus pre-programmed, again either deterministically or stochastically. The distribution of decay-times should agree, within statistical uncertainties, with the predicted quantum distribution.

We now discuss the effect of the measurement on the hidden variables. The quantum Zeno prediction will be obeyed if the following rule applies. Assume that the experiment begins at $t=0$, and in the absence of measurement, a particular nucleus is scheduled to decay at t_0. If there is a (negative) measurement at t_m ($t_m<t_0$), then at t_m, the hidden variable must be adjusted as follows. If $p(t)$ is the usual survival probability, and t_{00} is the scheduled time of decay as modified by the measurement, then $p(t_{00}-t_m) = p(t_0)/p(t_m)$. In this way, the decay profile following the measurement will be of the same shape as the initial profile; the atoms decay in the same sequence as if there had been no measurement, but the decay profile is altered appropriately.

Local realism implies that there can be no instantaneous connection between measurement and re-setting of hidden variable. There are two approaches to local realism broadly analogous to Aspect experiments without and with switching of polarisers. Without switching, there is a possibility of "conspiratorial" exchanges of information between polarisers and photons at $>c$. With switching this possibility is removed.

Let us consider variant 1 of quantum Zeno, ruling out, for the moment, "conspiracy." Inhibition of decay will be caused by a two-way passage between decaying particle and detector: (i) (non-)passage of decay particle from decaying atom to detector, followed by (ii) return signal. The time taken must be greater than $2R_1/c$, assuming locality. Variant 1, however, requires an instantaneous connection between (non-)decay and suppression of further decay. Thus it must be non-local.

However, allowing "conspiracy," we note that the detector has a permanent presence. It is conceivable that there may be a connection between its presence at $(t - R/c)$, and re-setting of the hidden variable at t. In this sense, variant 1 does not rule out local realism. Variant 2, though, has the potential to cover the point of Aspect's switching. The movement of the detector balloon may render any "conspiracy" between detector and hidden variables inoperative. The times of movement of the balloon must be random, and, to establish lack of local realism, it is necessary to establish agreement, to statistical accuracy, with precise predictions of the QZE for the particular sequence of measurement times.

The above, though, cannot easily be put in terms of "parameter dependence" or "outcome independence." To study this point, we may consider the following experiment. On one side of a decaying system, we have a relatively small "primary detector" separated from the decaying system by a macroscopic distance. On the other side, we have a "secondary detector" in the form of part of a sphere centred on the decaying system, with radius a comparable macroscopic distance. The secondary detector is of substantial size, subtending a solid angle of perhaps 2π at the decaying system.

The secondary detector may be present or "absent"; by the latter it is meant that various sections of the detector may be rotated to lie along the line of sight from the decaying system. It is assumed that the presence of the secondary detector will slow down decay, *via* the QZE, thus increasing the number of surviving atoms. It is also assumed that the rate of signals at the primary detector depends only on the number of surviving atoms, not *directly* on the presence or absence of the secondary detector. However, *via* the decaying system, information *does* pass from secondary to primary detector, and indeed, in principle, one could cope with switching by making periods of the presence or absence of the secondary detector random.

It must take a substantial period of time for information to be assimilated at the primary detector, and therefore one must ask: can one manage to do this faster than light would travel from secondary to primary detector? The answer, in principle, must be yes, because one may make the distance between primary and secondary detectors as large as we wish. We may claim for this thought-experiment then both parameter dependence—the results we obtain at the primary detector depend on what we do with the secondary detector, and faster-than-light signalling.

V. Conclusions

We have presented suggestions against parameter independence. They may perhaps be questioned as being crafty (section 3), or reliant on rather unlikely experiments, with some additional assumptions (section 4). A fair assessment might be that the question is still open. Even in this area, the relationship between quantum theory and relativity may not be quite as cosy as sometimes thought.

References

[1]   A. Shimony in *Quantum Concepts in Space and Time*, R. Penrose and C. Isham (eds.) (Oxford University Press, Oxford, 1986).

[2]   J.S. Bell, *Physics* **1**, 195 (1964).

[3]   A. Aspect, P. Grangier and G. Roger, *Phys. Rev. Lett.* **49**, 91 (1982).

[4]   F. Selleri, *Quantum Paradoxes and Physical Reality* (Kluwer, Dordrecht, 1990).

[5]   A. Whitaker, *Einstein, Bohr and the Quantum Dilemma* (Cambridge University Press, Cambridge, 1996).

[6]   J. Jarrett, *Nous* **18**, 569 (1984)

[7]   A. Peres, *Annals of the New York Academy of Sciences* **755**, 445 (1995).

[8]   L.E. Ballentine, *Rev. Mod. Phys.* **42**, 358 (1970).

[9]   B.S. de Witt, in *Battelle Rencontres 1967 – Lectures in Mathematics and Physics* C. de Witt and J.A. Wheeler (eds.) (Benjamin, New York, 1968)

[10] D. Home and M.A.B. Whitaker, *Phys. Reports* **210**, 223 (1992).

[11] M.A.B. Whitaker, *J. Phys.* A **18**, 253 (1985).

[12] D. Home and M.A.B. Whitaker, *Phys. Lett.* A **210**, 5 (1996).

[13] H. Everett, *Rev. Mod. Phys.* **29**, 454 (1957).

[14] D. Home and M.A.B. Whitaker, *Phys. Lett.* A **187**, 227 (1994).

[15] C.B. Chiu, E.C.G. Sudarshan and B. Misra, *Phys. Rev.* D **16**, 520 (1977).

[16] D. Home and M.A.B. Whitaker, *Ann. Phys.* (N.Y.) **258**, 237 (1997).

[17] H. Dehmelt, *Proc. Natl. Acad. Sci.* USA **83**, 2291 (1986).

[18] W.M. Itano, D.J. Heinzen, J.J. Bollinger and D.J. Wineland, *Phys. Rev. A* **41**, 2295 (1990)

[19] P. Kwiat, H. Weinfurter, T. Herzog, A. Zeilinger and M.A. Kasevich., *Phys. Rev. Lett.* **74**, 4763 (1995).

[20] S. Inagaki, M. Namiki and T. Tajiri, *Phys. Lett.* A **166**, 5 (1992)

Index of Proper Names

A

Aharonov, Y., 347, 351
Albert, D.Z., 54, 101, 125, 347, 351
Allan, D.W., 32, 34, 38
Alvarez, E., 290
Ambartsumian, V., 272
Anandan, J., 3, 5
Anchordoqui, L., 329, 336
Anderson, R., 9, 18
Arp, H., 214, 223, 267, 268, 271, 274, 281
Ashby, N., 102
Aspect, A., 239, 240, 360, 370
Assis, A.K.T., 177, 185
Avenarius, R., 133, 134, 139

B

Bacry, H., 327, 328, 336
Bailey, J., 5, 6, 18, 80
Ballentine, L.E., 365, 370
Barashenkov, V.S., 24, 195, 200
Bargmann-Michel-Telegdi equation, 191
Barone, M., 18, 24, 93, 101, 102, 114, 345
Barrow, J.D., 4, 6
Bartlett, D.F., 184, 185
Barut, A.O., 202, 203, 207
Baskoutas, S., 339, 345
Bastos Filho, J.B., 103, 113, 114
Bates, H.E., 18
Bell inequalities, 360
Bell, J., 68, 101, 110, 114, 149, 159, 226, 234, 239, 240, 353, 354, 355, 358, 359, 360, 363, 364, 366, 370
Bergia, S., 101
Bernasconi, C., 9, 18, 101
Bilger, H.R., 5, 6, 29, 38
Bird, D.J., 329, 336
Bitsakis, E.I., 111, 113, 114, 115, 125, 309

Bjorken, J.D., 207
Blokhintsev, D.I., 331, 336
Bohm, David, 3, 4, 5, 110, 114, 225, 226, 239, 240
Bohr, Nils, 109, 110, 111, 114, 129, 150, 151, 153, 155, 156, 159, 207, 239
Bolda, E., 345
Bollinger, J.J., 370
Boltzmann, L., 127, 157
Bondi, H., 21, 24
Bonnet, G., 184
Bopp, V.P., 202, 203, 205, 207
Borel, E., 115, 125
Born, Max, 127, 149, 165, 239, 240
Bozic, M., 202, 207
Bradley, J., 261, 263
Brandes, J., 275, 281
Bravais-lattice model, 324, 330
Bredimas, A.P., 305, 309
Broberg, H., 209, 223
Brodsky, S.J., 202, 207
Brown, R.H., 60, 63
Buchhorn, M., 291, 301
Burt, E.G.C., 33, 38

C

Caldirola-Montaldi model, 339
Carnap, L., 153, 156, 157, 159
Carnap, R., 146
Casimir, H., 202, 207
Cavalleri, T., 9, 18, 99, 101
Cetto, A.M., 360
Chase and Tomashek experiment, 40
Cherenkov radiation, 334
Chiao, R., 340, 345
Ching, E., 345
Chiu, C.B., 367, 368, 370
Christoffel symbols, 16
Chu, Y., 270, 274
Clauser, J.F., 356, 360
Cohen-Tannoudji, C., 240
Coleman, S., 326, 333, 336, 351

Québec, Canada
1999